# 2nd National Conference on Advanced Materials and Applications (NCAMA 2020)

AA007030

IOP Conference Series: Materials Science and Engineering
Volume 1120

Raipur, India
28 - 29 December 2020

**Editors:**

**Ayush Khare**
**Bijay K Sahoo**
**Sapan M Saini**

ISBN: 978-1-7138-3406-9
ISSN: 1757-8981

**Printed from e-media with permission by:**

Curran Associates, Inc.
57 Morehouse Lane
Red Hook, NY 12571

**Some format issues inherent in the e-media version may also appear in this print version.**



Institute of Physics
Dirac House, Temple Back
Bristol BS1 6BE UK

Phone: 44 1 17 929 7481
Fax: 44 1 17 920 0979

techtracking@iop.org

**Additional copies of this publication are available from:**

Curran Associates, Inc.
57 Morehouse Lane
Red Hook, NY 12571 USA
Phone: 845-758-0400
Fax: 845-758-2633
Email: curran@proceedings.com
Web: www.proceedings.com

# TABLE OF CONTENTS

## LUMINESCENT MATERIALS



## NANO MATERIALS

**SOLAR CELL MATERIALS**



**SOLAR MATERIALS**



**SEMICONDUCTOR MATERIALS**

## DIELECTRIC MATERIALS



## MATERIAL CHARACTERIZATION



## THERMOELECTRIC MATERIALS

**MEDICAL PHYSICS**



**DENSITY FUNCTIONAL THEORY**

IOP Conf. Series: Materials Science and Engineering 1120 (2021) 011001 doi:10.1088/1757-899X/1120/1/011001

## PREFACE

The $2^{nd}$ National Conference on Advanced Materials and Applications (NCAMA-2020) has successfully been organized by Department of Physics, National Institute of Technology (NIT), Raipur, India during December 28-29, 2020. NCAMA-2020 continues to be a forum mainly for the Physics, Chemistry, Material Science professional and academic communities engaged in research and development of smart luminescence materials, dielectric materials, photovoltaic cells, nano materials, thermoelectric device engineering, and related areas of engineering.

The major goal of NCAMA-2020 was to bring together scientists and practitioners, engineers and industry researchers to share their experience and innovations. The two-day online conference provided a great opportunity to initiate discussions on topical issues of research & development on advanced materials testing and applications, and discussed the challenges encountered and the solutions adopted.

Scientists and practitioners, mostly from Chhattisgarh, Maharashtra, Gujarat, Madhya Pradesh, Karnataka, Rajasthan and other nearby states of India submitted over 45 papers. All the papers were reviewed by the editors (Dr. Ayush Khare, Dr. Bijaya Kumar Sahoo and Dr. Sapan Mohan Saini) and invited experts. The most up-to-date and comprehensive manuscripts were selected for this volume. We sincerely hope that the NCAMA-2020 proceedings will show the readers a broad overview of the latest research results from the field of photovoltaic, luminescence, thermoelectrics and nanotechnology.

We thankfully acknowledge and express our sincere gratitude to Director and Dean (Research and Development) of NIT Raipur, India for their financial support. At last, we would like to express our deep appreciation and gratitude to all the members of the NCAMA-2020 scientific committees for their strenuous efforts. Without their assistance and dedication, it would have not been possible to hold the discussion at such a high level and to collect quality papers for compiling the conference proceedings. Finally, we want to show appreciation of all the authors and participants for their critical contributions to the work, holding the conference and to the preparation of the papers.

*With kindest regards,*

***Editors***
*Ayush Khare,*
*Bijay K Sahoo,*
*Sapan M Saini*

IOP Conf. Series: Materials Science and Engineering 1120 (2021) 011001 doi:10.1088/1757-899X/1120/1/011001

# List of Committee

**Chief Patron**

Dr. A. M. Rawani
Director, NIT Raipur

**Patron**

Dr. Shubhrata Gupta
Dean, Research and Consultancy

**Chief Advisors**

Dr. Sadhana Agrawal
Dr. A. K. Shrivastava
*Professors of Physics*

Chairman

Dr. Ayush Khare

Organizing Secretaries

Dr. B. K. Sahoo
Dr. S. M. Saini

| Advisory Committee | Organizing Committee |
|---|---|
| Dr. S. Auluck, IIT Roorkee | Dr. K. S. Ojha |
| Dr. Mukesh Jewariya, NPL Delhi | Dr. Anju Pansari |
| Dr. P. D. Sahare, Delhi University | Dr. Sudha Kumari |
| Dr. P. K. Dwivedi, IIT Kanpur | Dr. Abhishek Johri |
| Dr. B. R. Sankapal, VNIT Nagpur | Dr. Sanjay Kumar Sahu |
| Dr. V.R.K Murthy, IIT Madras | Dr. Sheetal Soni |
| Dr. Somnath Nag, IIT BHU | |
| Dr. S. K. Sharma, IIT Dhanbad | |
| Dr. D. S. Mehta, IIT Delhi | |
| Dr. A. K. Meikap, NIT Durgapur | |
| Dr. D. Behera, NIT Raorkela | |
| Dean (R & C), NIT Raipur | |
| Registrar, NIT Raipur | |
| Dean (AA), NIT Raipur | |

IOP Conf. Series: Materials Science and Engineering 1120 (2021) 011002 doi:10.1088/1757-899X/1120/1/011002

# Peer review declaration

All papers published in this volume of *IOP Conference Series: Materials Science and Engineering* have been peer reviewed through processes administered by the Editors. Reviews were conducted by expert referees to the professional and scientific standards expected of a proceedings journal published by IOP Publishing.

- **Type of peer review: Single-blind**
- **Conference submission management system: Not Applicable**
- **Number of submissions received: 45**
- **Number of submissions sent for review: 45**
- **Number of submissions accepted: 39**
- **Acceptance Rate (Number of Submissions Accepted / Number of Submissions Received X 100): 87%**
- **Average number of reviews per paper: 1**
- **Total number of reviewers involved: 17**
- **Any additional info on review process:**

  1. First we have checked the plagiarism of each paper by turnit software subscribed by NIT Raipur. Above 20% similarity papers, have been return to authors to reduce to less than 20%.
  2. After plagiarism papers were sent to various reviewers.
  3. Reviewer's comments and editorials suggestions were sent to authors for revision.
  4. Revised manuscripts were checked by editor. On the basis of response to the reviewers comments papers were accepted/rejected.
  5. Please note that 39 papers were accepted for publication out of 45 received.

- **Contact person for queries: Dr. Ayush Khare** akhare.phy@nitrr.ac.in

IOP Conf. Series: Materials Science and Engineering 1120 (2021) 012001 doi:10.1088/1757-899X/1120/1/012001

# Evaluation of trapping parameters of europium doped tristrontium dialuminium oxide phosphor by different thermoluminescent analysis methods

**Piyush Jha* and Ayush Khare**

Department of Physics, National Institute of Technology, GE Road, Raipur – 492 010, India

*Corresponding author's e-mail address: piyushjha22@rediffmail.com

**Abstract:** In the present work, we have synthesized Europium (Eu) doped Tristrontium Dialuminium Oxide ($Sr_3Al_2O_6$:Eu) phosphor by solid state reaction method in air atmosphere. The thermoluminescence (TL) properties of the synthesized phosphor are investigated and presented. Theoretical analysis of glow curves of $Sr_3Al_2O_6$:Eu was done by computerized glow curve deconvolution (GCD) method, and the trapping parameters of the synthesized phosphor, such as activation energy, frequency factor, order of kinetics, etc. are discussed and explained. These trapping parameters are determined by different methods, such as Chen's peak shape method, initial rise method, and Ilich method. The synthesized phosphor is found to be a good candidate for dosimetric applications.

## 1. Introduction

$Sr_3Al_2O_6$:Eu is one of the well-known materials, which is used in WLEDs [1], transparent polycrystalline ceramic [2], anti-counterfeiting devices [3], stretching/strain sensors [3], etc. It shows red light emission owing to the presence of rare-earth doped $Eu^{3+}$ ions in host $Sr_3Al_2O_6$ phosphor [4,5]. Akiyama and coworkers [6, 7] reported that $Eu^{2+}$ ions activated $Sr_3Al_2O_6$emitted green light as it was prepared in the reducing atmosphere. Sharma et al. [8] found that up to 1mol% the photoluminescence (PL) intensity is found to be maximum. Its red emission made this phosphor useful in luminous paints for highways, airport, and warning signs [8]. For the production of a large amount of phosphors solid state reaction is used [9]. The emission colour in Eu doped $Sr_3Al_2O_6$ phosphor is decided by the $Eu^{2+}$ and $Eu^{3+}$ is depends on the preparation of this phosphor in normal air or reducing atmosphere [3]. The shallow and deep traps level decides the time of afterglow of the phosphor [10]. To know the trap level (activation energy) of the phosphor, the thermoluminescence (TL) analysis method is used. TL is also known as thermally stimulated luminescence, is a relatively complex process and widely used in radiation dosimetry since it involves a trap, a luminescent centre, and the transfer (during warm - up) of an electron or a hole from the trap to the luminescent centre [11]. TL is a type of delayed phosphorescence, where the photon energy is released when a material is heated after subjecting it to ionizing radiations. For a better understanding of transition of charge carriers from trap levels in $Sr_3Al_2O_6$:Eu phosphor, it is essential to study the TL parameters by which a suitable model for TL may be predicted. The present paper reports the TL of $Sr_3Al_2O_6$:Eu phosphor and calculate the trapping parameters such as order of kinetics, activation energy and frequency factor by using peak shape method, initial rise method and Ilich method.

## 2.Experimental

For the synthesis of $Sr_3Al_2O_6:Eu^{3+}$ (8mol%) phosphor, the solid-state reaction method was employed [12, 13]. $SrCO_3$, $Al_2O_3$ and $Eu_2O_3$ were used as the raw materials in the present investigation. All precursor materials were weighed as per their stoichiometric ratios. The materials are mixed by using mortar and pestle and add appropriate amount of ethanol as a dispersing agent. Then, the mixture was carefully transferred to an alumina crucible and sintered at 1400 °C for 4 h in air. At room temperature the sample was ground for later use. The phase structure of $Sr_3Al_2O_6$:Eu phosphor is identified using PANalytical 3 kW X'pert Powder - Multifunctional X-ray Diffractrometer using Cu Kα radiations ($\lambda$ =0.15406 nm) in the 2θ range from 10° to 80°. The thermoluminescence glow curve was recorded on a Nucleonics TLD reader (Model: 1009I). The heating rate in TL experiment was kept at 5°C/s.

## 3. Results and Discussions

Figure 1 shows the photograph of $Sr_3Al_2O_6$:Eu phosphor showing red glow under UV exposure.

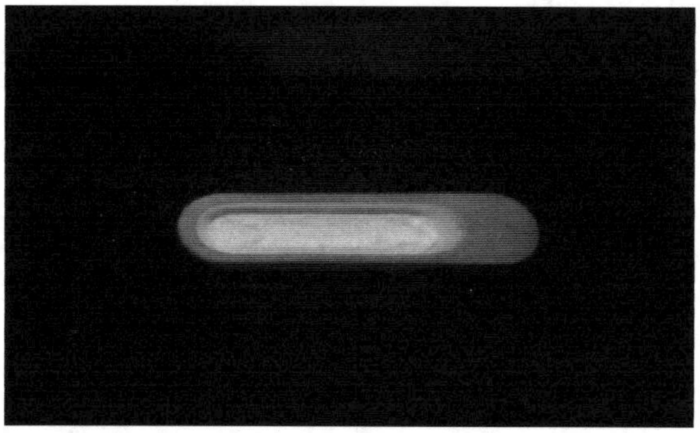

**Figure 1.** Photograph of red glow from $Sr_3Al_2O_6$:Eu phosphor under UV exposure.

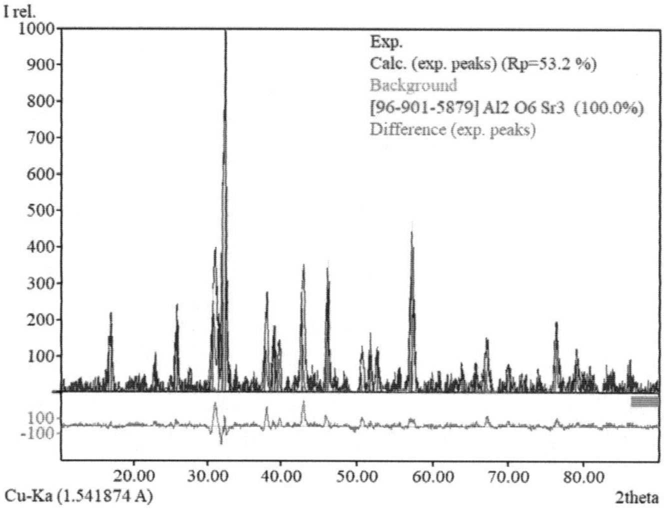


**Figure 2.** XRD patterns of $Sr_3Al_2O_6$:Eu phosphor.

The cubic phase of $Sr_3Al_2O_6$:Eu phosphor is confirmed by the XRD patterns as shown in figure 2 [12,13]. These diffractograms match well with COD card no. 96-901-5879. The average size of the crystallites in $Sr_3Al_2O_6$:Eu phosphor is calculated using Debye-Scherrer formula D= Kλ/βcosθ, where K is a constant, λ the wavelength of X-rays, β-the full width at half maxima (FWHM) and θ is the Bragg's angle and it is found to be 66.5 nm.

The TL glow curve (under 5 min UV exposure) exhibits a single emission peak of $Sr_3Al_2O_6$:Eu phosphor as shown in figure 3. The TL peak is found to be broad and the area covered under it suggests that it contains multiple numbers of TL trap depths.

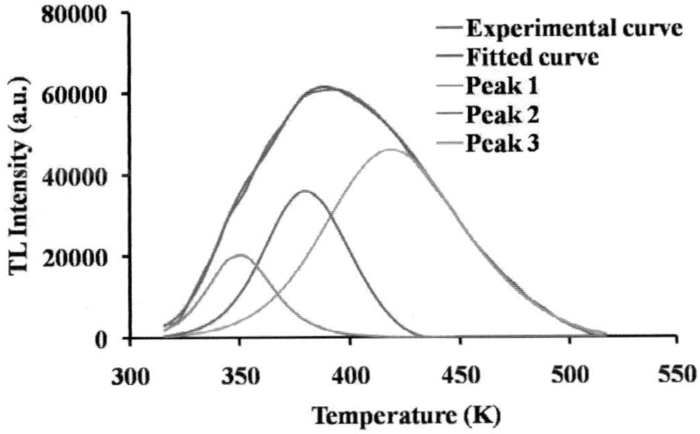


**Figure 3.** TL glow curve deconvolution peaks of $Sr_3Al_2O_6$:Eu phosphor.

But, sometimes the TL glow curves are complex and consists of a number of main peaks and some satellite peaks. The process used to separate complex glow curves and to get information about the traps from this glow-curve is known as deconvolution.

*3.1 Chen's peak shape method*

Chen's peak shape method is a very popular method for analyzing TL parameters. This depends on the shape and size of the TL glow curve and is useful to detect a number of energy levels formed due to defects and their positions in the forbidden energy-gap of the host materials. This method is based on the equations developed by Chen to determine trapping parameters given below [14]:

$$E_\alpha = c_\alpha\left(\frac{kT_M^2}{\alpha}\right) - b_\alpha(2kT_M) \quad (1)$$

where $\alpha$ is $\tau$, $\delta$, or $\omega$. The values of $c_\alpha$ and $b_\alpha$ are summarized as below:

$$c_\tau = 1.510 + 3.0(\mu - 0.42)\, b_\tau = 1.58 + 4.2(\mu - 0.42) \quad (2)$$

$$c_\delta = 0.976 + 7.3(\mu - 0.42)\, b_\delta = 0 \quad (3)$$

$$c_\omega = 2.52 + 10.2(\mu - 0.42)\, b_\omega = 1 \quad (4)$$

The value of geometrical factor μ can be calculated using formula given below:

$$\mu = \frac{T_2 - T_M}{T_2 - T_1} \quad (5)$$

where $T_m$ is the temperature corresponding to peak TL intensity, $T_1$ and $T_2$ are temperatures at half the maximum TL intensity. For first order kinetics, μ = 0.42 and for second order kinetics, μ = 0.52 [14].

According to order of kinetics, the values of activation energy and frequency factor using Chen's $\tau$ -equation, Chen's $\delta$-equation and Chen's $\omega$-equation can be calculated [14] where $\tau = T_m - T_1$, $\delta = T_2 - T_m$, and $\omega = T_2 - T_1$.

Figure 4 (a-c) shows the TL glow curve of $Sr_3Al_2O_6:Eu^{3+}$ phosphor used for peak shape method. The order of kinetics, activation energy and frequency factor were calculated by using Chen's peak shape method and the corresponding values are enlisted Table 1.

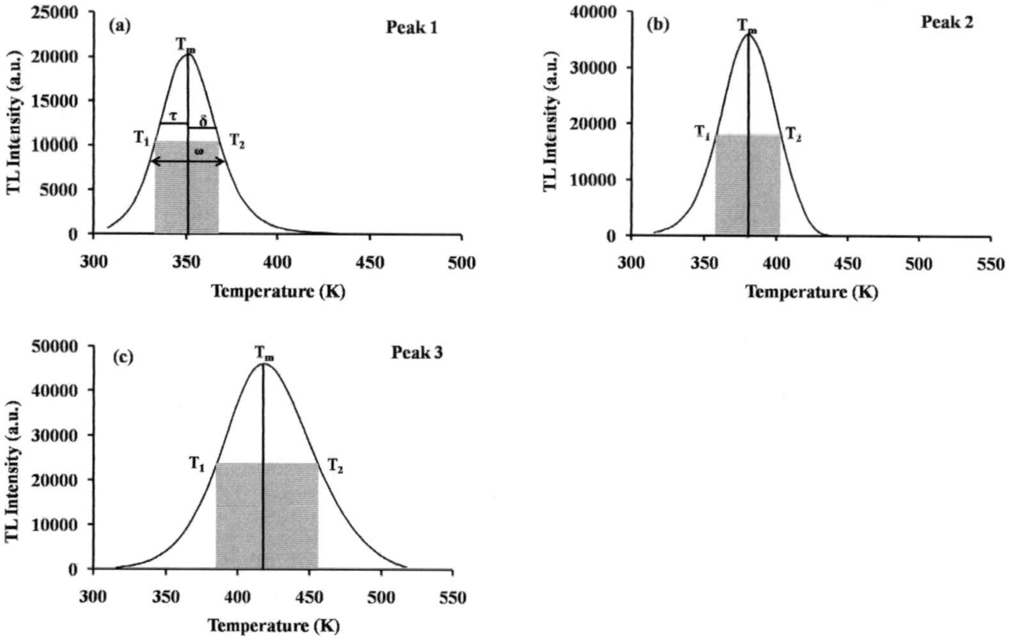


**Figure 4.** (a-c) TL glow curves of $Sr_3Al_2O_6$:Eu phosphor used for peak shape method.

**Table 1.** Trapping parameters of $Sr_3Al_2O_6$:Eu phosphor calculated by Chen's peak shape method.

| Sample Name | Peak Number | Geometrical factor (μ) | Order of kinetics (b) | Method of calculation: Chen's τ equation Activation energy (eV) | Chen's τ equation Frequency factor (s) | Chen's δ equation Activation energy (eV) | Chen's δ equation Frequency factor (s) | Chen's ω equation Activation energy (eV) | Chen's ω equation Frequency factor (s) |
|---|---|---|---|---|---|---|---|---|---|
| $Sr_3Al_2O_6$:Eu | Peak 1 | 0.45 | 1.67 | 0.75 | $8.4 \times 10^{10}$ | 0.79 | $1.99 \times 10^{11}$ | 0.77 | $1.11 \times 10^{11}$ |
| | Peak 2 | 0.50 | 1.92 | 0.82 | $7.93 \times 10^{10}$ | 0.85 | $1.70 \times 10^{11}$ | 0.84 | $1.36 \times 10^{11}$ |
| | Peak 3 | 0.54 | 2.08 | 0.70 | $3.3 \times 10^{8}$ | 0.71 | $3.56 \times 10^{8}$ | 0.71 | $3.96 \times 10^{8}$ |

*3.2 Initial rise method*

The evaluation of trapping parameters through an initial rise method was first reported by Garlick and Gibson in 1948 [15]. In this method, the initial rise part of the TL glow curve was used to determine the trapping parameters. The graph is plotted between log of TL intensity versus 1/kT, where k is the Boltzmann constant and T is the temperature and the slope of the graph gives the activation energy [16]. The initial rise method is applied only when the TL glow curve is well separated by another one. The advantage of this method is, it is applied on any order of kinetics and does not require any knowledge of frequency factor. TL glow curve of $Sr_3Al_2O_6:Eu^{3+}$ phosphor used for initial rise method shown in figures 5 (a), (b) and (c) and their analysis by initial rise method shown in figures 5 (d), (e) and (f).

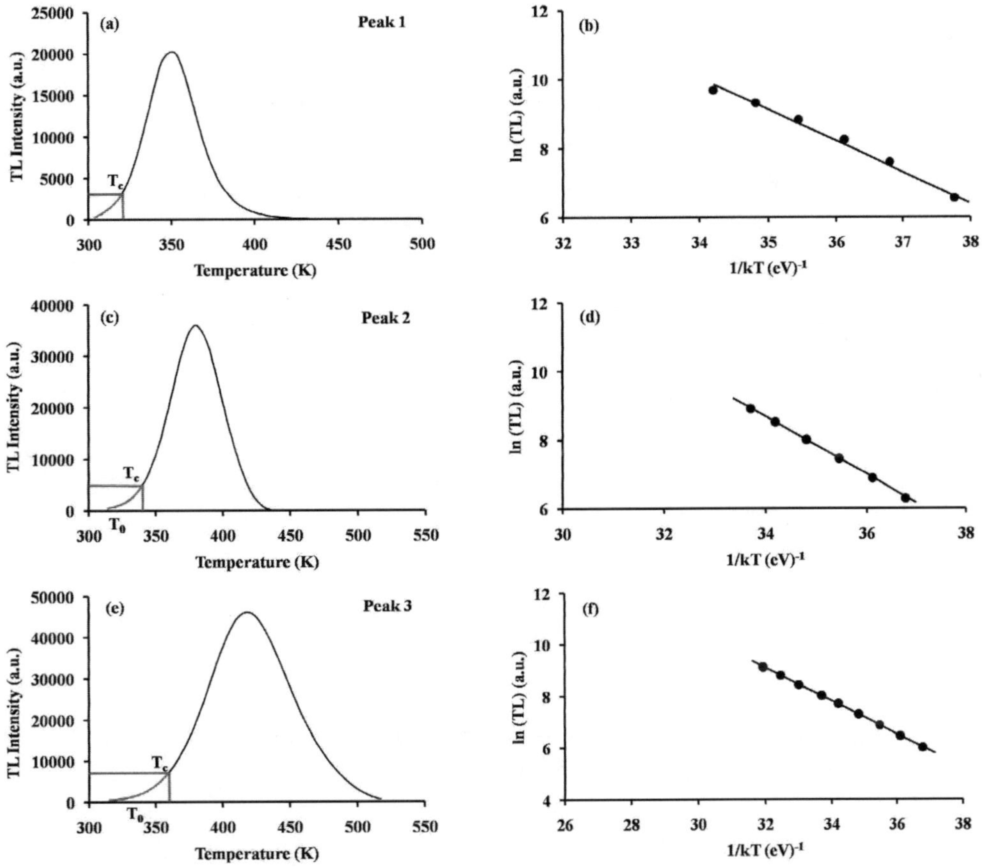


**Figure 5.** (a-c) TL glow curves of $Sr_3Al_2O_6$:Eu phosphor by initial rise method, (d-f) TL glow curve analysis of $Sr_3Al_2O_6$:Eu phosphor by initial rise method.

### *3.3 Ilich method*

For the calculation of activation energy from TL glow curve, Ilich method is employed, which is independent of the order of kinetics. In this method, the activation energy is calculated by using critical temperature $T_c$ and the starting temperature $T_0$ [11]. Thus, the activation energy is expressed as:

$$E = k\frac{T_c^2}{T_c - T_0} \quad (6)$$

The trapping parameters of $Sr_3Al_2O_6$:Eu phosphor by using Ilich method are shown in figure 6.

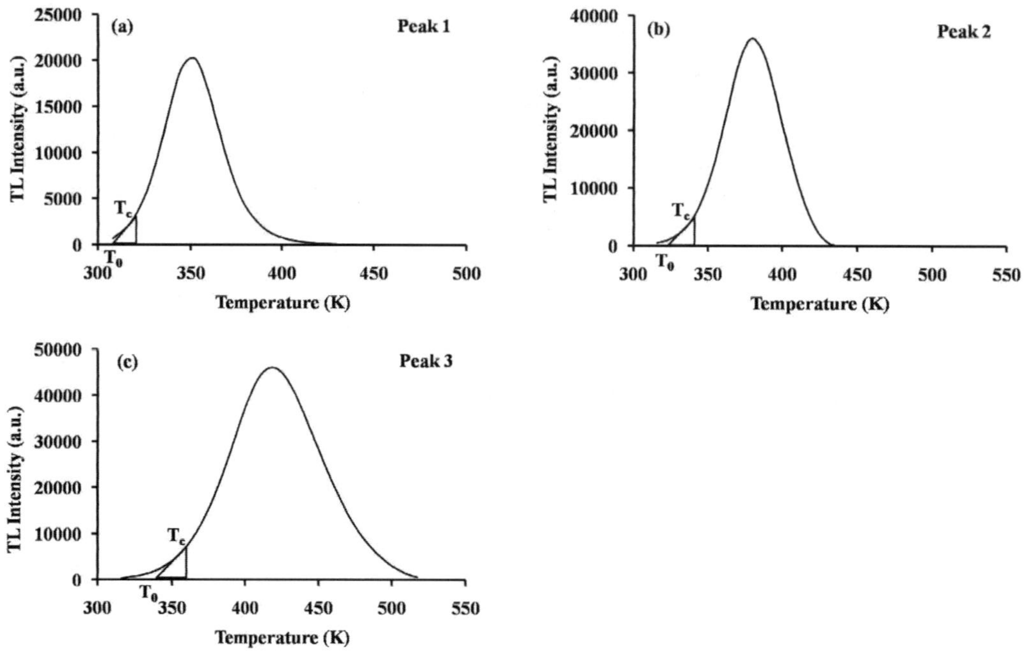


**Figure 6.** Peaks obtained after the deconvolution of TL glow curves for $Sr_3Al_2O_6$:Eu phosphor (analysis of trapping parameters done using Ilich method).

**Table 2.** Trapping parameters of $Sr_3Al_2O_6$:Eu phosphor calculated by Initial rise and Ilich methods.

| S. No. | Sample Name | Peak Number | Peak Position (K) | Activation energy (eV) | |
|---|---|---|---|---|---|
| | | | | Initial rise method | Ilich Method |
| 1 | $Sr_3Al_2O_6$:Eu | Peak 1 | 352 | 0.88 | 0.74 |
| | | Peak 2 | 381 | 0.85 | 0.75 |
| | | Peak 3 | 417 | 0.75 | 0.73 |

## 4. Conclusions

In the present investigation $Sr_3Al_2O_6$:Eu phosphor was prepared by solid state reaction method. The TL glow curve is found single and broad. The activation energy calculated by peak shape method is found in the range from 0.70 to 0.85 eV, through initial rise method it is found to range from 0.75 to 0.88 eV and by Ilich method the values lie between 0.73 and 0.75 eV. All the results suggest that the synthesized phosphor is a good candidate for dosimetric applications.

## References

[1] Li X, Pan H, Tang A, Zhang J, Guan L, Su H, Dong G, Yang Z, Wang H and Teng F 2016 *J. Nanoscience Nanotechnology* **16** 3474

[2] Alahraché S, Al Saghir K, Chenu S, Veron E, de Sousa Meneses D, Becerro A I, Ocana M, Moretti F, Patton G, Dujardin C and Cusso F 2013 *Chem. Mater.* **25** 4017

[3] Wu C, Zeng S, Wang Z, Wang F, Zhou H, Zhang J, Ci Z and Sun L 2018 *Adv. Funct. Mater.* **28** 1803168

[4] Page P, Ghildiyal R and Murthy K V R 2006 *Mater. Res. Bull.* **41** 1854

[5] Pan Y, Sung H H Y, Wu H, Wang J, Yang X, Wu M and Su Q 2006 *Mater. Res. Bull.* **41** 225
[6] Akiyama M, Xu C N, Nonuka K and Watanabe T 1998 *Appl. Phys. Lett.* **73** 3046
[7] Akiyama M, Xu C N, Taira M, Nonaka K and Watanabe T 1999 *Philos. Mag. Lett.* **79** 735
[8] Sharma S K, Pitale S S, Malik M M, Qureshi M S and Dubey R N 2009 *J. Alloys Compd.* **482** 468
[9] Kim D, Kim H E and Kim C H 2020 *ACS Omega* **5** 10909
[10] Van der Heggen D, Vandenberghe D, Moayed N K, De Grave J, Smet P F and Joos J J 2020 *J. Lumin.* **226** 117496
[11] Pagonis V, Kitis G and Furetta C 2006 Numerical and practical exercises in thermoluminescence. Springer Science & Business Media.
[12] Zhang W, Liu F, Feng S, Hao S, Xu Y and Guan J 2014 *J. Mater. Sci.: Mater. Electron.* **25** 2355
[13] Zhang W, Zhang P, Li J and Qiu K 2014 *Optoelectron. Adv. Mat.* **8** 37
[14] Chen R 1969 *J. Electrochem. Soc.* **116** 1254
[15] Garlick G F J and Gibson A F 1948 *Proc. Phys. Soc.* **60** 574
[16] Parauha Y R and Dhoble S J 2021 *J. Molecular Struct.* **1211** 127993

IOP Conf. Series: Materials Science and Engineering 1120 (2021) 012002 doi:10.1088/1757-899X/1120/1/012002

# Optical properties of rare earth (Ce) and transition metal (Ti) doped $ZrO_2$ phosphors

**Ugendra Kurrey[1]*, Nameeta Brahme[2] and D.P. Bisen[2]**

[1] Govt. C.L.C. Arts & Science College., Patan, Durg, Chhattisgarh, India
[2] School of Studies in Physics and Astrophysics, Pt. Ravishankar Shukla University Raipur, Chhattisgarh, India

* Corresponding author E-mail: kurrey7947@gmail.com, namitabrahme@gmail.com

**Abstract.** Thermoluminescence and Photoluminescence of un-doped as well as Titanium ($Ti^{3+}$) & Cerium ($Ce^{3+}$) doped $ZrO_2$ phosphors, prepared via solid-state reaction method. For the characterization of the sample FTIR, XRD, EDAX and FESEM study were done. FTIR confirms the presence of conventional impurities (such as NO3, OH-) present in the prepared sample. Structural characterization technique (XRD) shows the monoclinic formation and reveals the average grain size in the nano region. EDAX study confirms the chemical composition and through FESEM morphological status is studied. For kinetic parameters study themoluminescence studies is carried out and try to correlate the XRD results with thermoluminescence behavior of prepared phosphors. Photoluminescence emission spectra show the blue and green emission band for Ti and Ce doped $ZrO_2$ phosphors respectively.

## 1. Introduction

Phosphors are most commonly used materials in photonics as well as in optics fields due to its superior hardness, high refractive index, optical transparency, chemical stability, high thermal expansion coefficient, low thermal conductivity and high thermo-mechanical resistance. Zirconium Oxide ($ZrO_2$) phosphors are most promising host material in the field of electro-optical materials, due to its electrical, mechanical, chemical and optical characteristics. Zirconia has a wide band gap (= 5.4 ev) semiconductor material, has good optical transparency and low phonon energy of 470 $cm^{-1}$ [1]. This will increase the number and probability of radiative transition and reduce non radiative multiphonon relaxation [2,3]. And hence, it has vast application areas in oxygen sensor, Solid State electrolytes, thermal barrier coatings, and in the field of photonics [4,5,6]. Zirconia is a IV group member, in periodic database table and the main source of zirconia is Baddeleyite ($ZrO_2$) and zircon (zirconium ortho-silicate $ZrSiO_4$). Hafnium, Hamatite ($Fe_2O_3$) Fe and Ti are the major impurities found in a commercial zirconia due to closeness in their crystal radii ($Zr^{4+}$ 0.79$A^0$, $Hf^{4+}$ 0.78 $A^0$) [7].

$ZrO_2$ exhibits three polymorphs with increasing temperature at a normal atmospheric pressure: the monoclinic phase (m-$ZrO_2$) that is stable from room temperature (RT) to 1175$^0$C; the tetragonal phase (t-$ZrO_2$) stability ranges between 1175$^0$C to 2370$^0$C and cubic phase (c-$ZrO_2$), exceptional stable at 2370$^0$C to 2750$^0$C (melting point). However, the substitution of dissimilar elements into the $ZrO_2$ host develops oxygen ion vacancies and as a consequence the phase of the phosphor material stabilized through charge compensation mechanism [8,9]. Phosphors or luminescent materials are mostly inorganic materials consisting of a host lattice intentionally doped with impurities or activators.

Impurities are undesirable elements or substances that lower the things quality or value but depending on its amount and nature of impurity it enhances the mechanical, electrical or optical properties of the host lattice. Oxides, halides, sulfides, nitrides, nitride-carbides or oxy-nitrides are used as host materials for the preparation of optically enhanced luminescent materials. Oxides lattices have proved to be an excellent host material due to its good chemical, thermal and mechanical stability. $ZrO_2$ is a well-known wide band gap material and hence it consider as a semiconductor material. $ZrO_2$ host works as a good sensitizer because it absorbs the incident light and transfers this energy to $RE^{3+}$ ions. Basically $RE^{3+}$ ions are weakly fluorescent due to the parity forbidden f-f transition [10].

Doping of RE ions into the host lattice does not change the structure of the prepared phosphors, however due to the ionic radii of the different RE ions the average crystalline size varies, and hence the luminescence property also varies with their size. For the selection of doping elements we must have clear knowledge about its optical activity and the stability of its valence state within the crystalline host. Doping of RE ions have been considered the most important optical activators for luminescence devices [11].

In present work we report the optical properties of Solid State synthesized undoped as well as Ti and Ce doped $ZrO_2$ phosphors.

## 2. Experimental

Synthesis of desired products is a basic step for the research output, whatever the method we adopt, the product must be reproduce with the same parameters i.e its grain size and morphology of the final product should be same every time. Solid –State reaction synthesis is very useful for the production of undoped as well as RE doped $ZrO_2$ nanophosphors. In nanometer-scale structures it is possible to control the fundamental properties of the material, such as their melting temperature, magnetic properties, charge capacity and even their color without changing the chemical composition [12,13].

For the preparation of undoped as well as doped phosphors through Solid State reaction high purity oxide material (from sigma Aldrich) is used. Zirconium dioxide $ZrO_2$ is used as a host material and, Cerium (IV) oxide $CeO_2$ and Titanium (IV) dioxide $TiO_2$ were used as starting material for doping, all the samples are weighted and mixed thoroughly with agate mortal for 3 to 4 hours and heated at $1200^0$C for 3 hour in an air atmosphere. The basic mechanism of solid state reaction is diffuse control reaction. The controlled atmosphere is necessary to master the valance of the activators.[14]

After the synthesis characterization is needed, to confirm the morphology and grain size of the phosphor product. In present study FTIR and XRD is carried out for every prepared sample to confirm the reproduction. The FTIR spectra were taken through Shimadzu IR-Affinity between IR region i.e 400 to 4000 $cm^{-1}$ with a resolution of 4 $cm^{-1}$, before taking FTIR spectra samples are mixed and ground with IR grade KBr powder with a ratio of 1: 100 mg. The crystal structure and phase purity is conformed through X-ray powder diffraction (XPD). The patterns were collected through Bruker D-2 Phaser (Table Top Model) with CuKα radiation 1.5405 $A^0$. The 2θ values are $20^0$ to $60^{\,0}$ and the increment is 0.02, For the morphological and elemental composition study FEI Quanta 200 FEG-SEM instrument is used.

Optical studies is carried out by taking thermoluminescence glow curve and photoluminescence spectra of the prepared sample. Nucleonix TL10091 TL analyzer is used to record TL glow curve, before taking TL all the sample is heated at $300^0$C to clear the previous stored radiation. PL properties of the samples are taken through Shimadzu spectrophotometer RF-5301PC equipped with 150 W xenon light source. The system is operated between wavelength scale 220-900 nm with ± 1.5 nm accuracy and S/N ratio is 150 which make it highly sensitive over the range. Excitation spectra were recorded within 250-350 nm range whereas the emission spectra were recorded in the 400- 700 nm range. For taking emission spectra four different excitation wavelength is used i.e 220 nm, 250 nm, 295 nm and 320 nm.

## 3. Results and Discussion

*3.1 X-ray Diffraction Studies*
XRD is the basic tools to analyze the formation of crystal structure and its grain size. The synthesis of solids possessing desired structure, composition and properties continues to be a challenge to chemist, scientist and engineers. Synthesized phosphors materials exhibit the luminescence properties only if it has crystalline nature, the amorphous nature doesn't show any kind of luminescence. Doping of $RE^{3+}$ ions because rare earth ions are the most important optical activators for luminescence devices [11]. into the host $ZrO_2$ doesn't influence the structure of prepared phosphors, however due to the ionic radii of the different $RE^{3+}$ ions the average crystalline size varies, M. Hartmanova et al reported that "The growth of Zirconia crystals stabilized with rare earth oxides of large ionic radii (Pr,Nd and Ce ) leads to a significant decrease of the crystal size" [15]. With the variation in the size of synthesized phosphors we observed drastic variation in their electrical, mechanical as well as in optical properties.

Figure 1(a,b,c) shows the XRD pattern of undoped $ZrO_2$ as well as Ti and Ce doped phosphors prepared by solid state reaction method, the concentration of doping element kept constant (i.e 0.5 mol %). Pattern shows the formation of monoclinic phase in the sample, which is conformed through Crystallography Open database (COD) file no. 9007485 with sub group P1 21/ C1 (14). In figure all the identical peaks are looks similar and having the same position i.e the 2θ values are same. Peak centered at 28.19, 31.49, and 34.23 degree are due to (-1,-1,1), (-1,-1,-1) and (0,0,-2) reflections planes.

Table 1 shows the calculated grain size for undoped as well as doped $ZrO_2$ phosphors. From the observed values we conclude that the average grain size of the prepared phosphor depends on the ionic radii of the doped material also. If the ionic radius of the doped element is lower ($Ti^{3+}$ = 81 pm) than the host radii ($Zr^{4+}$ = 86 pm) then the average grain size of the prepared phosphors ($ZrO_2$:Ti ) is large as compared to the undoped $ZrO_2$ phosphors. If the ionic radius of the doped element is larger ( $Ce^{3+}$ = 115 pm) than the host radii ($Zr^{4+}$ = 86 pm ) then the average grain size of the prepared phosphors ($ZrO_2$: Ce) is lower as compared to the undoped $ZrO_2$ phosphors. M. Hartmanova et al reported that "The growth of zirconia crystals stabilized with rare earth oxides of large ionic radii (Pr,Nd & Ce) leads to a significant decrease of the crystal size"[15].

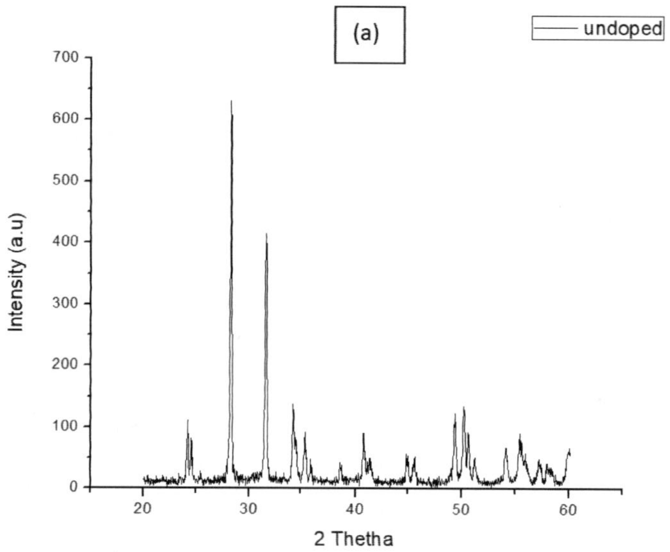

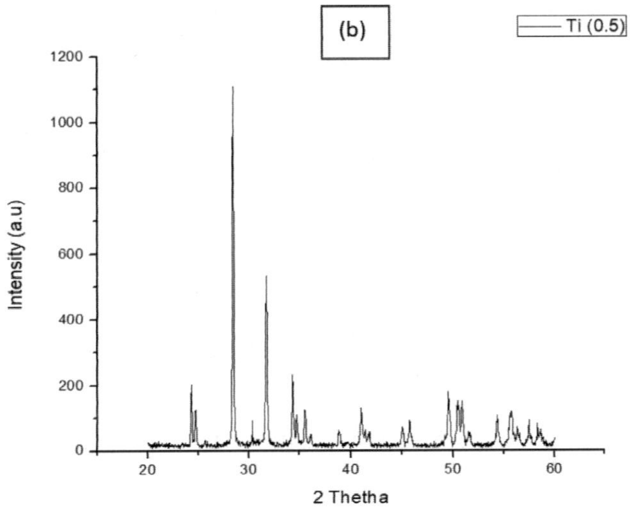


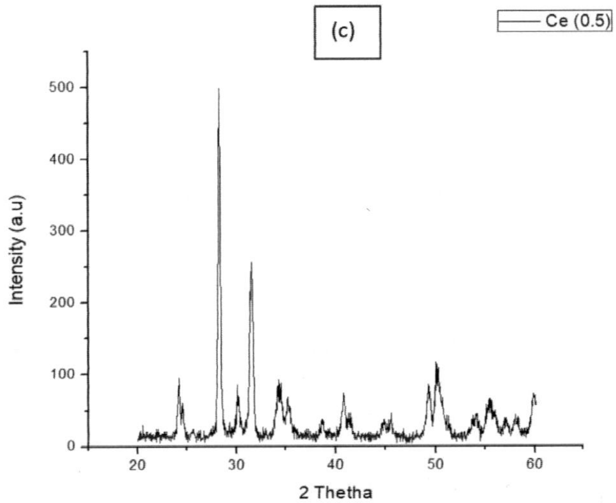


**Figure 1.** (a,b,c) XRD pattern of undoped, Ti and Ce doped $ZrO_2$ phosphors.

**Table 1.** Average crystalline size for undoped, Ti and Ce doped $ZrO_2$ nanophosphors calculated through Diffac.eva software.

| Sample | Ionic radii of host material $Zr^{4+}$ (pm) | Ionic radii of doping material (pm) | FWHM | Calculated grain size (A°) | Size (nm) |
|---|---|---|---|---|---|
| **$ZrO_2$ undoped** | 86 | --- | 0.164 | 555.7 | 55.5 |
| **$ZrO_2$:Ti (0.5 mol %)** | 86 | $Ti^{3+}$ = 81 | 0.158 | 577.1 | 57.7 |
| **$ZrO_2$:Ce (0.5 mol %)** | 86 | $Ce^{3+}$ = 115.0 | 0.256 | 355.1 | 35.5 |

*3.2 Fourier Transform Infrared spectroscopy (FTIR)*

FTIR is one of the most important and emerging tool through which we can analyze and identify different functional groups and conventional impurities (such as NO3, OH-) present in the prepared sample. Figure 2 shows the FTIR spectra of undoped as well as Ti and Ce doped $ZrO_2$ phosphors. For taking spectra the sample is mixed with KBr powder because KBr is inert and it does not react with the sample. KBr is optically transparent over the range of wavelengths typically used in an IR analysis, and also it helps for better resolution. Infrared (IR) spectra of samples prepared by Solid State Reaction method show the weak absorption at 3054 $cm^{-1}$ caused by O-H single bond stretching vibration of Zr-OH or absorbed $H_2O$, indicates that the water molecules or moisture contents have low concentration due to high temperature synthesis [16]. A sharp and intense peak at 2355 $cm^{-1}$ corresponds to C-H stretching vibrations of the alkyl groups. Peaks around 1550-1300 $cm^{-1}$ can be respectively assigned to the anti symmetric and symmetric vibrations of carboxylate groups $\upsilon_{as}(COO^-)$ and $\upsilon_a$ $(COO^-)$ and below 1000$cm^{-1}$ peaks are the characteristic peaks corresponds to Zr-O-Zr vibrations. This confirms the formation of –Zr-O-Zr- networks.[17].

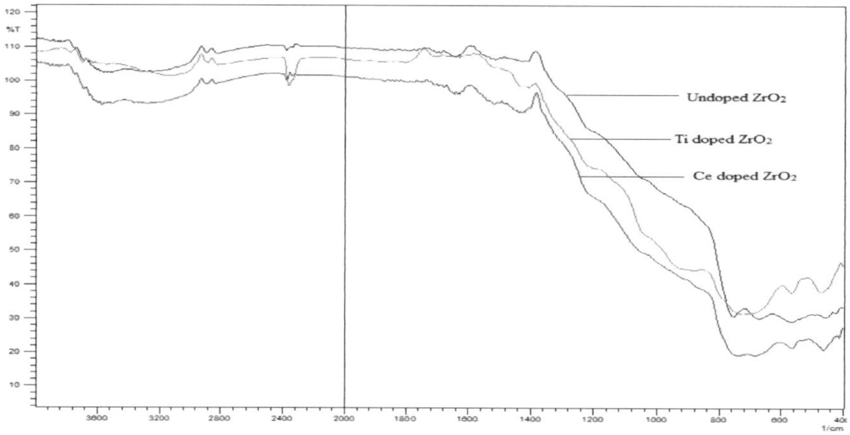


**Figure 2.** shows the FTIR spectra of Undoped as well as Ti, Ce doped $ZrO_2$ phosphors prepared by Solid State Reaction method.

### 3.3 *Surface Morphology*

The morphology of the synthesized undoped $ZrO_2$ and Ti, Ce doped $ZrO_2$ was studied by Field emission electron microscope (FESEM). From the FESEM images shown in the figure 3. We conclude that the particles are in almost irregular in shape and highly agglomerated.

**Figure 3.** FESEM images of (a) Undoped $ZrO_2$, (b) Ti doped $ZrO_2$, (c) Ce doped $ZrO_2$.

*3.4 EDX Study*

Energy –Dispersive Spectrum analysis is performed to investigate the elemental composition of undoped $ZrO_2$ and Ti, Ce doped phosphors. Figure 4 shows the EDX images of undoped and doped phosphors, analysis results confirms the presence of oxygen ions and zirconium ions as well as doping elements in EDX profile.

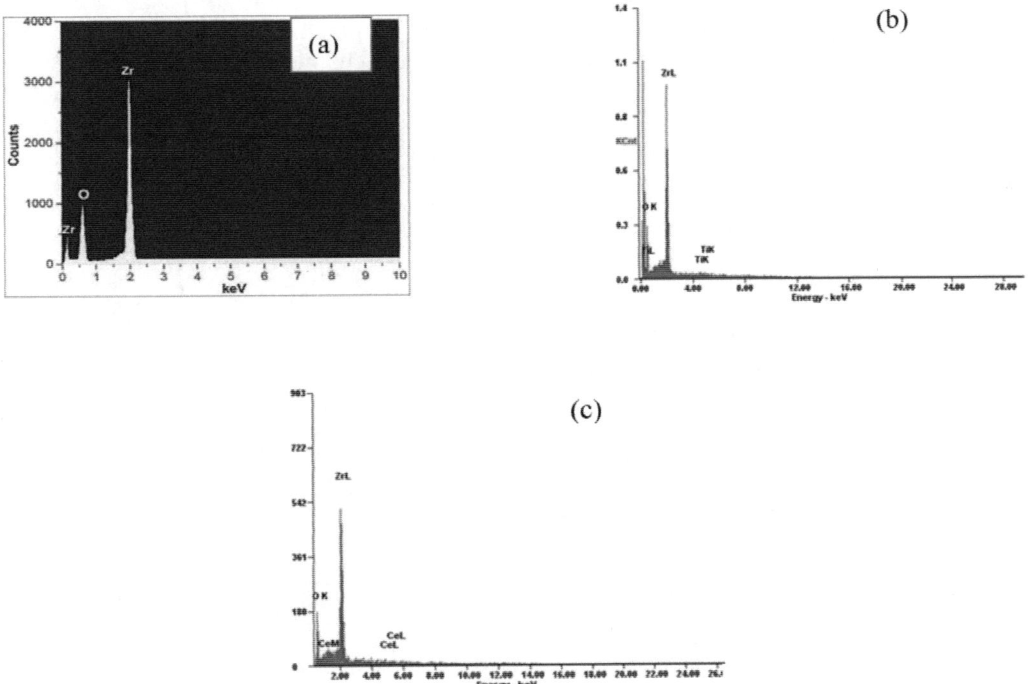


**Figure 4.** EDX profile of (a) Undoped $ZrO_2$, (b) Ti doped $ZrO_2$, (c) Ce doped $ZrO_2$.

**Table 2**. Elemental weight percentage concentration of various elements in the synthesized phosphors.

| Sample Name | Elements | | | | | | | |
|---|---|---|---|---|---|---|---|---|
| | O K | | Zr L | | Ti K | | Ce K | |
| | Wt % | At % | Wt % | At % | Wt % | At % | Wt % | At % |
| **Undoped $ZrO_2$** | 36.75 | 73.99 | 63.25 | 26.01 | Nil | Nil | Nil | Nil |
| **$ZrO_2$:Ti (0.5 mol%)** | 36.94 | 76.65 | 61.86 | 22.51 | 01.20 | 00.83 | Nil | Nil |
| **$ZrO_2$:Ce (0.5 mol%)** | 35.50 | 76.29 | 59.94 | 22.59 | Nil | Nil | 04.56 | 01.12 |

## 3.5 Thermoluminescence Studies

*3.5.1 Thermoluminescence glow curve of Undoped $ZrO_2$ phosphors*

Figure 5 shows the glow curve of undoped $ZrO_2$ phosphors for different UV radiation dose i.e each 5 gm sample is radiated through UV light of 254 nm for 5, 10, 15, 20, 25, and 30 min time and then warmed at uniform heating rate of 10°C/S.

For 5 min irradiation we get single broad peak, when we deconvoluted the peak through Origin 8.1 software we observed that these broad peak is actually constituted with two distinct peak. First peak observed at 107.8 °C/S and second peak at 163.7 °C/S. It indicates that there are at least two types of traps present in the undoped $ZrO_2$ phosphors.

As we increase the UV irradiation time i.e 10 min, 15 min, 20 min, 25 min, and 30 min we observe separation in broad peak and the position of first peak observed at 94.8, 96.6, 111.2, 95.36, 93.4 °C/S and second peak at 168.5, 170.03, 173.4, 165.2, 173.4 °C/S respectively From the peak position we get clear indication that the TL peak position remains constant only its intensity slightly varies. Hence we say that the chemical nature of defects related to these peaks remains unchanged [18]. Activation energy and frequency factor for both deconculuted peak was calculated through Chen's method and it was shown in table 5.

**Table 3.** Kinetic parameters for undoped $ZrO_2$.

| UV dose | Peak | $T_1$ °K | $T_m$ °K | $T_2$ °K | τ | δ | ω | μ =δ/ω | Activation Energy E by Chen's method | Frequency factor S |
|---|---|---|---|---|---|---|---|---|---|---|
| **05 min** | $I^{st}$ | 346.12 | 380.85 | 415.57 | 34.73 | 34.72 | 69.45 | 0.49 | 0.51 | $6.7x10^7$ |
| | $II^{nd}$ | 405.39 | 435.42 | 465.45 | 30.03 | 30.03 | 60.06 | 0.5 | 0.83 | $6.09x10^{10}$ |
| **10 min** | $I^{st}$ | 332.19 | 367.85 | 402.58 | 35.66 | 34.73 | 70.39 | 0.49 | 0.47 | $3.2x10^7$ |
| | $II^{nd}$ | 407.12 | 440.91 | 472.82 | 33.79 | 31.91 | 65.7 | 0.48 | 0.72 | $2.15x10^9$ |
| **15 min** | $I^{st}$ | 333.35 | 369.95 | 404.67 | 36.6 | 34.72 | 71.32 | 0.48 | 0.45 | $1.5x10^7$ |
| | $II^{nd}$ | 411.24 | 444.09 | 475.99 | 32.85 | 31.9 | 64.75 | 0.49 | 0.77 | $7.33x10^9$ |
| **20 min** | $I^{st}$ | 356.52 | 384.67 | 413.77 | 28.15 | 29.1 | 57.25 | 0.5 | 0.67 | $9.3x10^9$ |
| | $II^{nd}$ | 413.77 | 445.67 | 477.58 | 31.9 | 31.91 | 63.81 | 0.5 | 0.81 | $2.02x10^{10}$ |
| **25 min** | $I^{st}$ | 328.29 | 365.83 | 401.49 | 37.54 | 35.66 | 73.2 | 0.48 | 0.42 | $6x10^6$ |
| | $II^{nd}$ | 406.98 | 439.83 | 471.73 | 32.85 | 31.9 | 64.75 | 0.49 | 0.75 | $5.21x10^9$ |
| **30 min** | $I^{st}$ | 330.82 | 366.48 | 400.27 | 35.66 | 33.79 | 69.45 | 0.48 | 0.45 | $1.8x10^7$ |
| | $II^{nd}$ | 411.67 | 446.4 | 479.24 | 34.73 | 32.84 | 67.57 | 0.48 | 0.71 | $1.26x10^9$ |

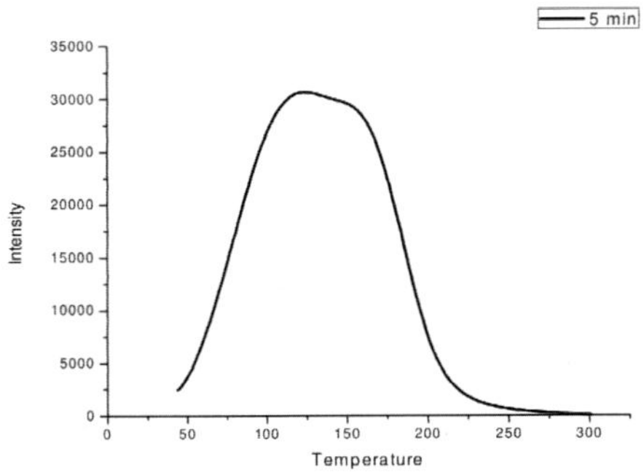
5 min
Intensity
Temperature

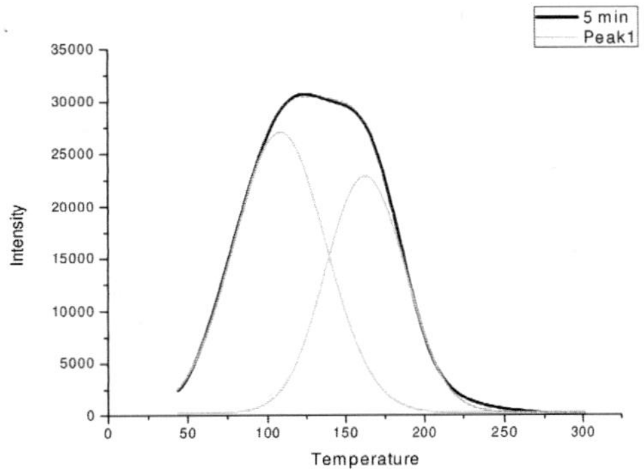
5 min
Peak1
Intensity
Temperature

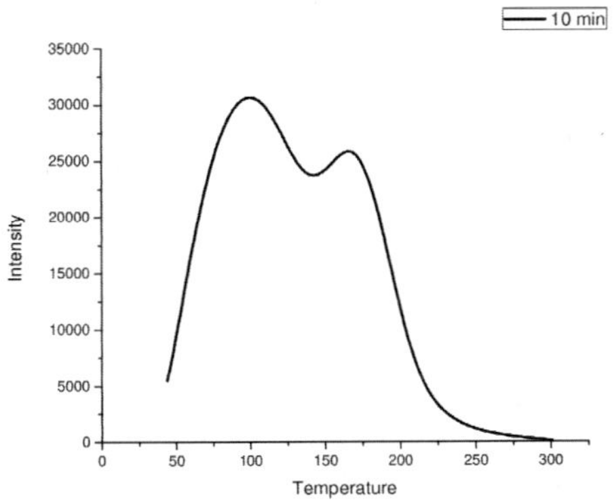
10 min
Intensity
Temperature

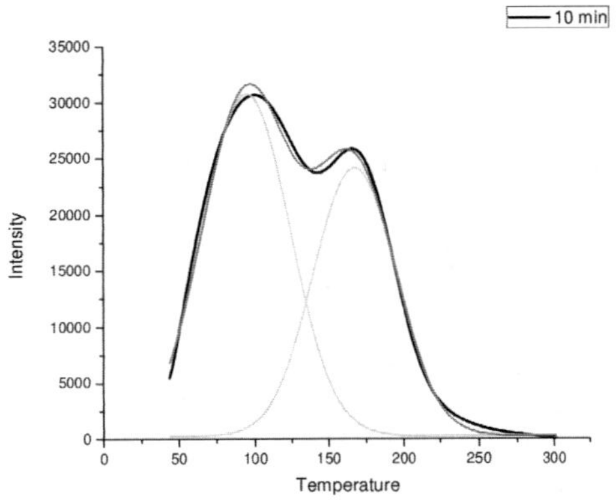
10 min
Intensity
Temperature

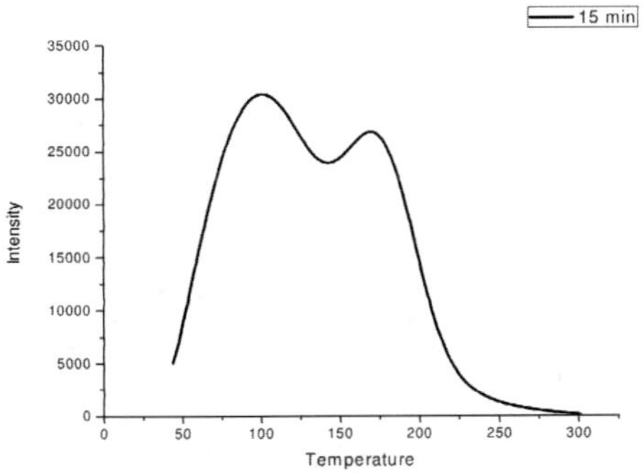
15 min
Intensity
Temperature

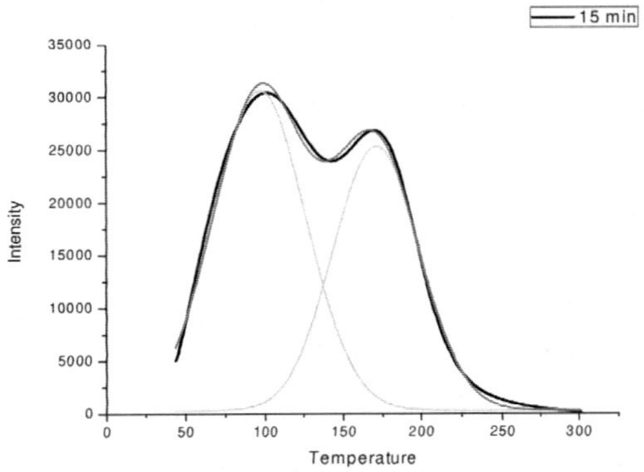
15 min
Intensity
Temperature

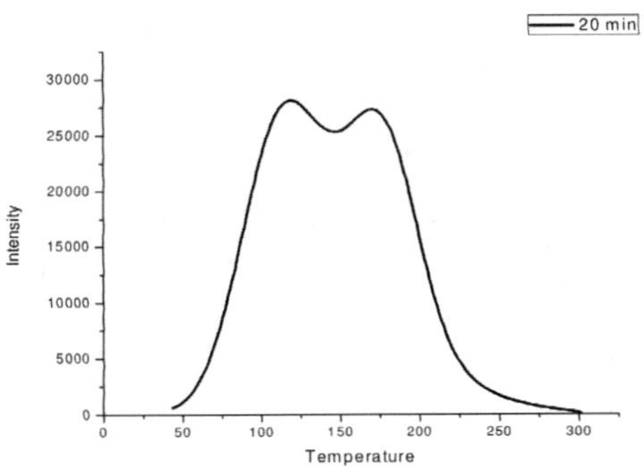
20 min
Intensity
Temperature

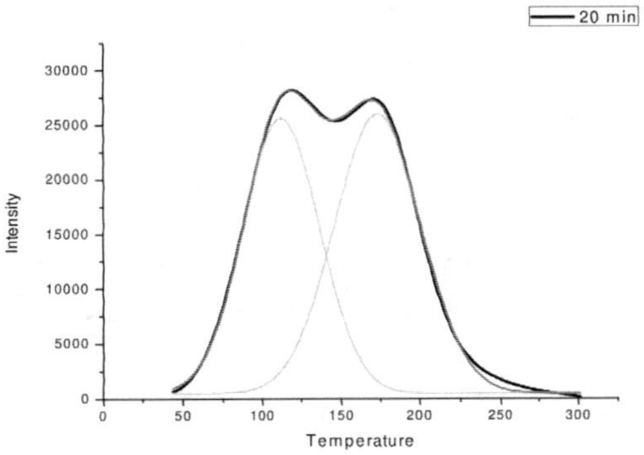
20 min
Intensity
Temperature

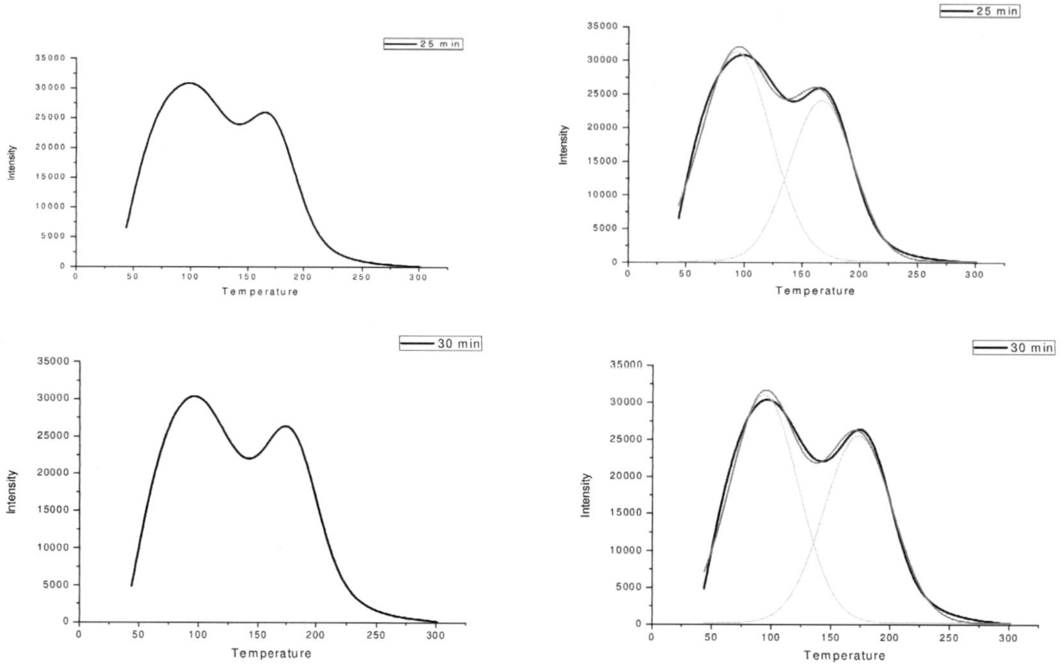


**Figure 5.** TL glow curve for undoped $ZrO_2$ nano phosphors at different UV radiation time.

*3.5.2 Effect of Ti Concentration on Thermoluminescence property of $ZrO_2$ phosphors*

To observe the concentration effect of Ti doping on $ZrO_2$ phosphors, highest intense TL peak at 25 min UV exposure (shown in figure 6 (a)) is taken as raw data and the observed intensity is plotted against the different concentration (ranges from 0.1 – 0.5 mol%) of Ti (shown in figure 6 (b)). For the measurement the heating rate ($\beta$= 10°C/S) and the quantity of the sample (5 mg.) is kept constant. Single prominent peak at 134.61°C, 150.41°C, 147.87 °C, 144.74 °C, 143.48°C for 0.1 mol %, 0.2 mol %, 0.3 mol %, 0.4 mol %, and 0.5 mol % respectively. From observed results TL intensity initially increases for increasing the concentration attains a maxima at 0.4 mol % Ti value and then the intensity is decreased this may be due to the concentration quenching effect i.e when the concentration of an impurity ions is higher than an appropriate value, the intensity of the phosphor material decreases. Activation energy and frequency factor for the peak was calculated through Chen's method and it was shown in Table 4.

**Table 4.** Kinetic parameters of $ZrO_2$: Ti (0.1-0.5 mol %) for constant 25 min UV exposures.

| Ti Concentration | $T_1$ °K | $T_m$ °K | $T_2$ °K | $\tau$ | $\Delta$ | $\omega$ | $\mu = \delta/\omega$ | Activation Energy E by Chen's method | Frequency factor S |
|---|---|---|---|---|---|---|---|---|---|
| **0.1 mol%** | 377.67 | 406.76 | 434.92 | 29.09 | 28.16 | 57.25 | 0.49 | 0.73 | $1.68 \times 10^{10}$ |
| **0.2 mol%** | 385.9 | 422.5 | 455.35 | 36.6 | 32.85 | 69.45 | 0.47 | 0.59 | $1.23 \times 10^{8}$ |
| **0.3 mol%** | 384.1 | 419.88 | 454.68 | 35.78 | 34.8 | 70.58 | 0.49 | 0.62 | $3.31 \times 10^{8}$ |
| **0.4 mol%** | 383.96 | 415.86 | 444.87 | 31.9 | 29.01 | 60.91 | 0.47 | 0.66 | $1.3 \times 10^{9}$ |
| **0.5 mol%** | 386.04 | 417.01 | 445.17 | 30.97 | 28.16 | 59.13 | 0.47 | 0.69 | $2.96 \times 10^{9}$ |

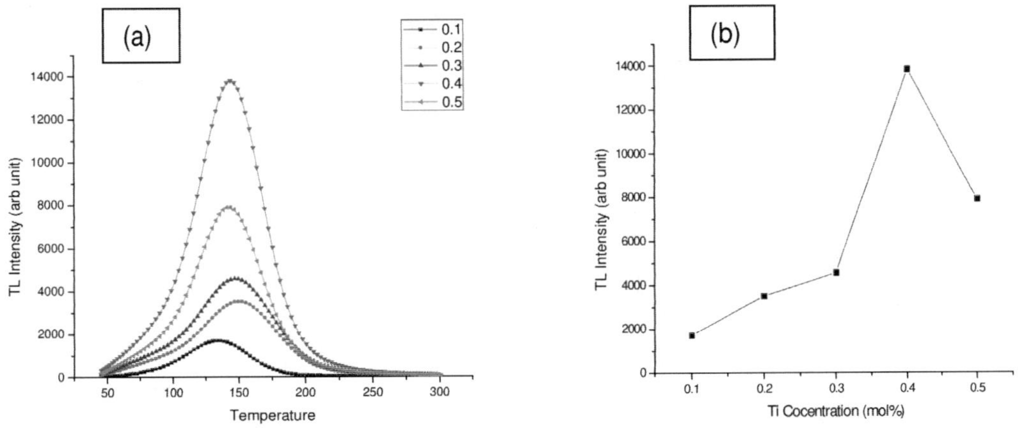


**Figure 6.** (a) TL glow curve of $ZrO_2$: Ti (0.1-0.5 mol %) synthesized by Solid State Reaction method for constant 25 min UV exposures. (b) Concentration Quenching curve for Ti .

*3.5.3 Effect of Ce Concentration on Thermoluminescence property of $ZrO_2$ phosphors*

Figure 7 shows the effect of Ce concentration on $ZrO_2$ phosphors prepared by Solid State reaction method. For the observation of concentration effect different amount of Ce (0.1, 0.2, 0.0.3, 0.4, 0.5 mol %) is mixed on $ZrO_2$ , the amount is calculated according to stoichiometry calculation. From the prepared phosphors taking fixed quantity (= 5 mg) and then irradiated to UV exposure for 25 min and then take the TL glow curve (shown in figure 7(a)). From observed result the intensity is found maximum for 0.2 mol % of Ce doping after that the intensity become saturated and then slightly decreased (shown in figure 7(b)). Activation energy and frequency factor for the peak was calculated through Chen's method and it was shown in table 5.

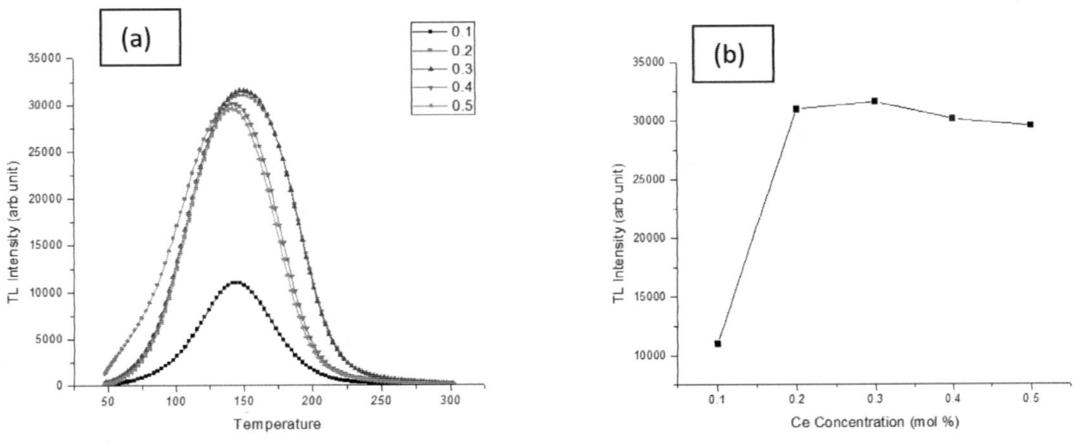


**Figure 7.** TL glow curve of $ZrO_2$: Ce (0.1-0.5 mol %) synthesized by Solid State Reaction method for constant 25 min UV exposures. (b) Concentration Quenching curve for Ce .

**Table 5.** Kinetic parameters of $ZrO_2$: Ce (0.1-0.5mol %) for constant 25 min UV exposures.

| Ce Concentration | $T_1$ °K | $T_m$ °K | $T_2$ °K | τ | Δ | ω | μ =δ/ω | Activation Energy E by Chen's method | Frequency factor S |
|---|---|---|---|---|---|---|---|---|---|
| **0.1 mol%** | 388.35 | 418.39 | 448.42 | 30.04 | 30.03 | 60.07 | 0.49 | 0.73 | $8.88 \times 10^9$ |
| **0.2 mol%** | 370.64 | 421.89 | 465.39 | 51.25 | 43.5 | 94.75 | 0.45 | 0.38 | $0.24 \times 10^6$ |
| **0.3 mol%** | 378.08 | 422.55 | 465.09 | 44.47 | 42.54 | 87.01 | 0.48 | 0.48 | $0.48 \times 10^6$ |
| **0.4 mol%** | 377.49 | 415.19 | 451.93 | 37.7 | 36.74 | 74.44 | 0.49 | 0.57 | $9.3 \times 10^7$ |
| **0.5 mol%** | 378.45 | 415.19 | 450.96 | 36.74 | 35.77 | 72.51 | 0.49 | 0.58 | $1.3 \times 10^8$ |

### *3.6 Photoluminescence Studies*

#### *3.6.1Photoluminescence glow curve of undoped and doped $ZrO_2$ phosphors*

Photoluminescence properties of the sample depends not only the surface quality of the material but also depends on the nature of the optical excitation. Timothy H. Gfroerer reported that when surface structure is poor, the PL spectrum changes with increasing excitation energy whereas the PL spectrum of a high-quality surface is independent of excitation energy. From the PL spectra of undoped $ZrO_2$ ( shown in figure 8) and Ti doped $ZrO_2$ ( shown in figure 9 a, b, c and d) it is clear that the emission and excitation spectra were essentially little changed and we get shift in both the emission and excitation spectra such that for undoped we get two excitation peaks at 292 nm and 311 nm observed under 470 nm emission. Whereas for Ti doped phosphors we obtain one single broad peak at 275 nm under same emission 470 nm . The emission spectra were taken under 295 nm excitation for both the sample. The results shows that the broad peak at 490nm (blue emission) for undoped phosphor and for Ti doped phosphor we observed broad peak at 495 nm (blue emission) and also the PL intensity becomes stronger for doped phosphors. Cong et al [19] reported that when $TiO_2$ dissolved into $ZrO_2$ mixed valence states of $Ti^{4+}$ and $Ti^{3+}$ exist and the substitution of two $Ti^{3+}$ ions for a $Zr^{4+}$ ion produces one anion vacancy for charge compensation and hence concentration of anion vacancies in the $ZrO_2$:Ti is higher than that in the undoped $ZrO_2$ to enhance the emission and excitation intensity [20].

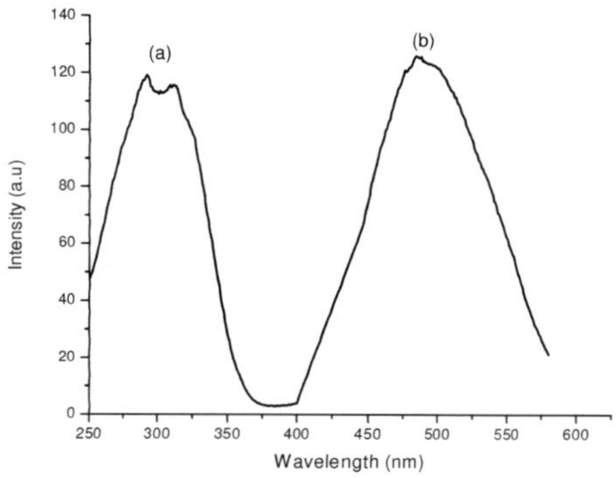


**Figure 8.** PL spectra of undoped $ZrO_2$: (a) excitation spectrum of $ZrO_2$ ($\lambda_{em}$ = 470 nm), (b) emission pectrum of $ZrO_2$ ($\lambda_{exe}$ = 295 nm).

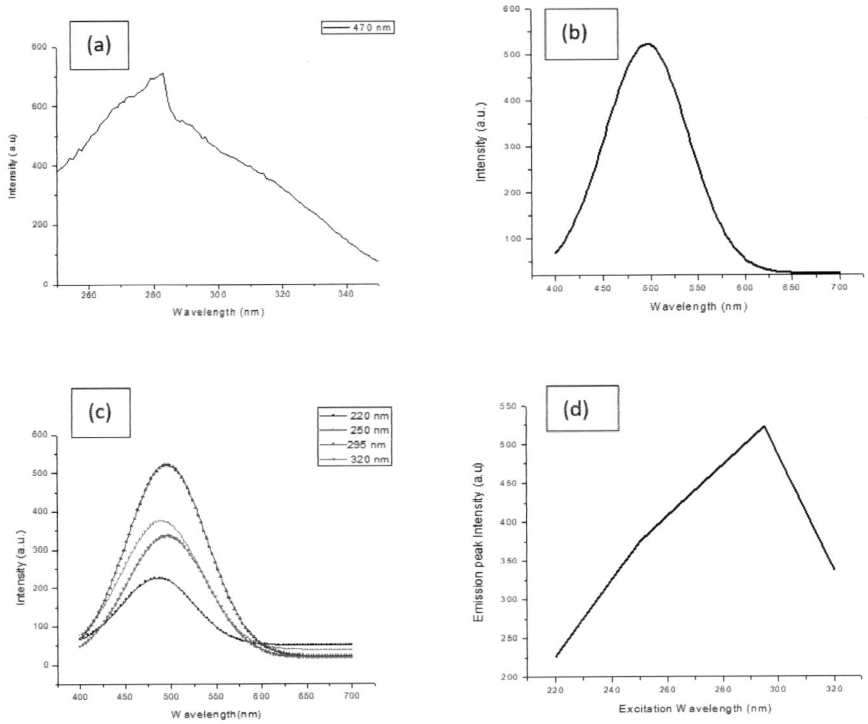


**Figure 9.** PL spectra of Ti doped $ZrO_2$: (a) excitation spectrum of Ti doped $ZrO_2$ ($\lambda_{em}$ = 470 nm), (b) emission spectrum of Ti doped $ZrO_2$ ($\lambda_{exe}$ = 295 nm), (c) emission spectrum of Ti doped $ZrO_2$ under different excitation wavelength i.e ($\lambda_{exe}$ = 220 nm, 250 nm,295 nm and 320 nm), (d) variation of intensity with excitation wavelength.

Figure 10 (a) presents the excitation spectrum for Ce (0.1 mol ) doped $ZrO_2$ sample observed at $\lambda_{emi}$ = 470 nm we get broad peak at 275 nm. Figure 10 (b) shows the emission spectra of Ce (0.1 mol) doped $ZrO_2$ phosphors taken at $\lambda_{exe}$ = 295nm, we observed a sharp emission peak at lower wavelength region at 359 nm (violet region) and a broad peak at 485nm (green region). For different excitation i.e at $\lambda_{exe}$ = 220 nm, 250 nm, 295 nm and 320nm shown in figure 10 (c) .We observed constant sharp emission band at $\lambda_{emi}$ = 359 nm for all excitation also we observed second broad peak and 475 nm, 480nm, 485 nm, 489 nm respectively. Hence we conclude that as we increase excitation wavelength i.e from 220 nm to 320 nm the $Ce^{3+}$ emission band is shifted towards the higher wavelength region or lower energy side, and peak becomes broadened [21]. Figure 10 (d) shows the variation of intensity with excitation wavelength. The electron configuration of $Ce^{3+}$ ion is very simple at ground and excited state $4f^1$ and $5d^1$ respectively. The $4f^1$ ground state split up in a doublet $2F_{7/2}$ and $2F_{5/2}$ levels. The PL properties of $Ce^{3+}$ are due to their fully allowed 4f $\rightarrow$5d transitions [22].

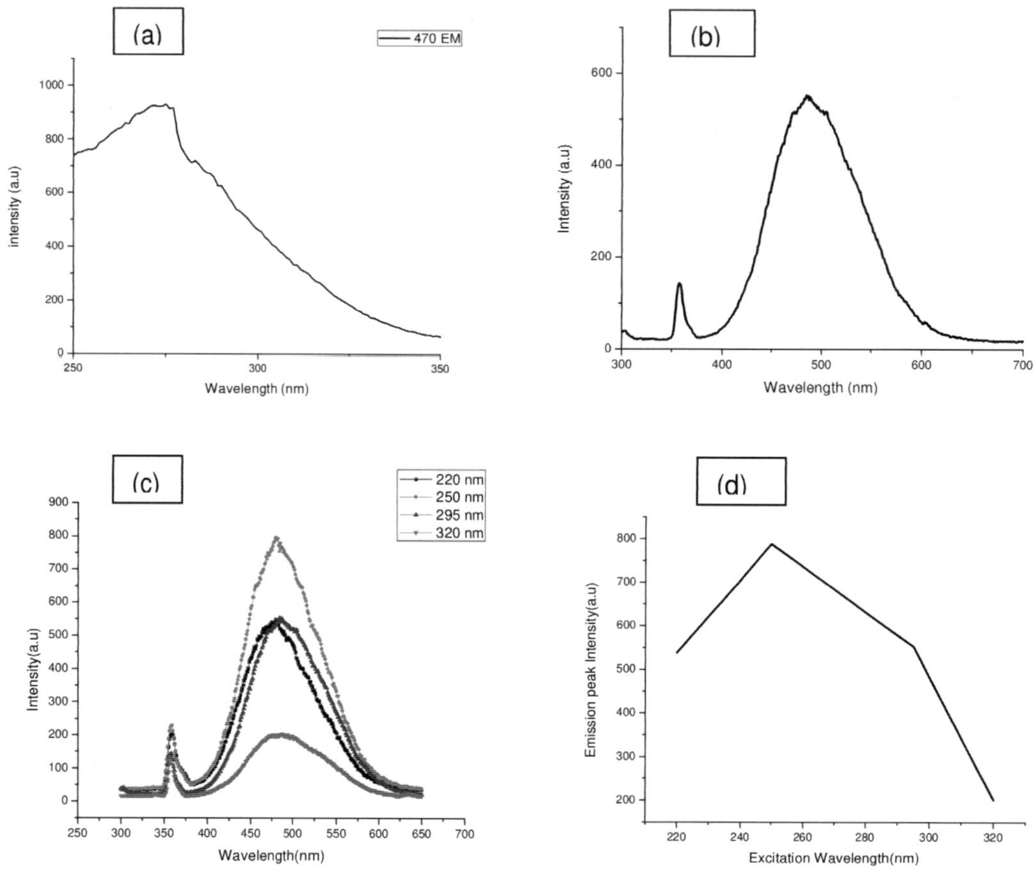


**Figure 10.** PL spectra of Ce doped $ZrO_2$: (a) excitation spectrum of Ce doped $ZrO_2$ ($\lambda_{em}$ = 470 nm), (b) emission spectrum of Ce doped $ZrO_2$ ($\lambda_{exe}$ = 295 nm), (c) emission spectrum of Ce doped $ZrO_2$ under different excitation wavelength i.e ($\lambda_{exe}$ = 220 nm, 250 nm,295 nm and 320 nm), (d) variation of intensity with excitation wavelength.

## 4. Conclusion

In present work we have studied the TL and PL properties of undoped and doped (Ti,Ce) Zirconium oxide phosphors. XRD pattern confirms the monoclinic formation and also reveals that the average grain size of the prepared phosphors depends on the ionic radii of the dopened. From TL glow curve results we conclude that the doping enhance the intensity of the synthesized phosphors and also the value of kinetic energy and frequency factor for each glow curve is presented in the table. From observed PL spectra it is clear that the emission spectra of undoped $ZrO_2$ and doped with Ti & Ce has similar characteristic of broad band. Hence we conclude, emission is strongly related to the crystal structure of host lattice, which makes it possible to tailor and modify the luminescent properties by selecting different host materials.

## References

[1] Hyppanen I, Holsa J, Kankare J, Lastusaari M, Malkamaki M and Pihlgren L 2009 *J. Lumin* **129** 1739
[2] Stojadinovic S, Tadic N and Vasilic R 2019 *J. Lumin.* **208** 296
[3] Rozo C, Fonseca L F, Jaque D and Garcia J S 2009 *J. Lumin.* **129** 696

[4] Ramos A I, Sanchez M, Guzman J,Torres J G, Garcia M and Riverra T 2019 *J. Phy. Conf. Ser.* **1221** 0120021
[5] Cong Y, Li B, Yue S and Fan D , Wang X J 2009 *J. Phys Chem. C* **113** 13974
[6] De la Rosa Cruz E, Diaz-Toprres L A, Salas P, Rodriguez R A and Kumar G A 2003 *J. Appl. Phys* **94** 5
[7] Phatak G M, Gangadharan K, Pal H and Mittal J.P. 1994 *Bull. Mater Sci.* **17** 163
[8] Srigurunathan K, Mecnambal R, Guleria A, Kumar D, Maria J, da Foute Ferreira and Kannan S 2019 *ACS Biomater. Sci. Eng.* **5** 1725
[9] Keiteb A S, Saion E, Zakaria A and Soltani N 2016 *J. Nanomater* Article ID 1913609
[10] Pal M, Pal U, Gracia J, Jimenez Y and Rodriguez F P 2012 *Nanoscale Res. Lett.* **7**
[11] Mari B, Singh K C, Sahal M and Khatkar S P 2010 *J. Lumin.* **130** 2128
[12] Cabello G, Lillo L, Caro C, Buono-Core G E, Chornik B and Soto M A 2008 *J. Non-Cryst. Solids* **354** 3919
[13] Bhagyaraj S M, Oluwafemi O S, Kalarikkal N and Thomas S 2018 (eds) *Micro & Nano Technology Series* (Elsevier)
[14] Ekambaram S, Patil K C and Maazaz M, 2005 *J. Alloy Compd.* **393** 81
[15] Hartmanova M, Lomonova E E, Kubel F, Schneider J, Bursikova V, Jergel M, Navra V and Kundracik F 2009 *J. Mater. Sci.* **44** 234
[16] Mizuno M, Sasaki Y, Lee S and Katakura H 2006 *Langmuir(ACS)* **22** 7137
[17] Nakamoto K 1978 *Handbook of Vibrational Spectroscopy, John Wiley & Sons*
[18] Cong Y, Li B, Wang X J, Leia B and Li W 2008 *J. Electrochem Soc.* **155**K 195
[19] Cong Y, Li B, Leia B and Li W 2007 *J. Lumin* **126** 822
[20] Nakayama S and Sakamoto M 2016 *J. Mater Res and Technol* **5** 289
[21] Malchukova E and Boizot B 2014 *J. Rare Earth* **32** 217
[22] Dierre B, Xie R J, Hirosaki N and Sekiguchi T 2007 *J.Mater. Res.* **22** 7

IOP Conf. Series: Materials Science and Engineering 1120 (2021) 012003 doi:10.1088/1757-899X/1120/1/012003

# Synthesis, characterization and optical properties of rare earth doped barium hexaferrite nano particles

**Jayashri Mahapatro and *Sadhana Agrawal**

Department of Physics, National Institute of Technology, Raipur-492010, India

*Corresponding author's e-mail address: sagrawal.phy@nitrr.ac.in

**Abstract.** In this work we have prepared a series of $BaFe_{12}O_{19}$ :$xEu^{3+}$ (x=0 to 2 mol %) samples by combustion method. The prepared samples were characterized by PXRD, TEM and FTIR. The structural analysis of PXRD pattern affirms that the specimens have space group $p6_3/mmc$ and hexagonal structure. The TEM micrographs show hexagonal plate like structure and the FTIR spectra show vibration between the $4f_1$ and $4f_2$ crystallographic sites of the ferric ions. The absorption spectra of the prepared specimen were obtained from UV-vis spectroscopy. The calculated band gap values range from 3.99 eV to 2.29eV.

## 1. Introduction

Ferrite materials have attracted the attention of technologists, physicists and chemists because of their fascinating magnetic and electromagnetic properties [1]. Ferrites are further sub divided into three types such as spinel ferrite, garnet, and magnetoplumbite/hexagonal ferrite depending on their crystal structure.

Now a day's hexagonal ferrites (hexaferrites) grabbing greater attention due to their high Curie temperature, high saturation magnetization, large uniaxial magnetic anisotropy, high coercivity, low dielectric loss and good chemical stability. Barium hexaferrites ($BaFe_{12}O_{19}$/BaM) are widely used in various devices such as multimeter devices, sound devices, perpendicular magnetic recording devices etc [2-9]. The crystal structure of BaM is described by the space group $p6_3/mmc$. The unit cell of BaM consists of two structural blocks 'R' and 'S' with stacking sequence RSR*S*, ordered along the direction of hexagonal c-axis. The S and R blocks have spinal and hexagonal structure respectively. The S* and R* are the rotated blocks obtained by 180° rotation of S and R block along the c-axis respectively [10]. The synthesis technique of $BaFe_{12}O_{19}$ plays a vital role in determining the structural, electrical and optical properties. Various synthesis methods were reported for the synthesis of high purity $BaFe_{12}O_{19}$ powders such as sol-gel method [11], hydrothermal synthesis [12], combustion method [13] etc.

In this work undoped and europium doped barium hexaferrite powders were prepared using combustion synthesis technique due to its low cost and less time consumption.

## 2. Synthesis Technique

The $BaFe_{12-x}Eu_xO_{19}$(x=0, 1, 1.25, 1.5, 1.75, 2) samples were prepared by combustion technique. Iron nitrate, europium nitrate, barium nitrate and citric acid were used as initial reactants. These reactants were taken according to their stoichiometric ratio and mixed in distilled water. The mixture was placed on a magnetic stirrer, when the temperature of the stirrer reaches $85^0$c combustion process occurs and after that we get the combusted powder. The combusted powder was finely crushed and calcined for 3 hour at $1250^0$c in a furnace. After calcination the final product was again ground finely and then subjected to various characterizations.

## 3. Result and Discussion

*3.1 Structural Analysis*

Figure 1 represents PXRD pattern for the $BaFe_{12}O_{19}:xEu^{3+}$ (x=2 mol%) sample which was recorded using the PANalytical X-ray diffractometer. PXRD pattern analysis confirms formation of hexagonal phase of prepared powder sample. The diffraction peaks present in the PXRD pattern were indexed with respect to their hkl value. The crystallite size was calculated by Debye Scherrer formula $D=\frac{k\lambda}{\beta\cos(\theta)}$ where λ is X – ray wave length, θ is the Bragg's diffraction angle and k is constant, and is found to be 38.15nm approximately.

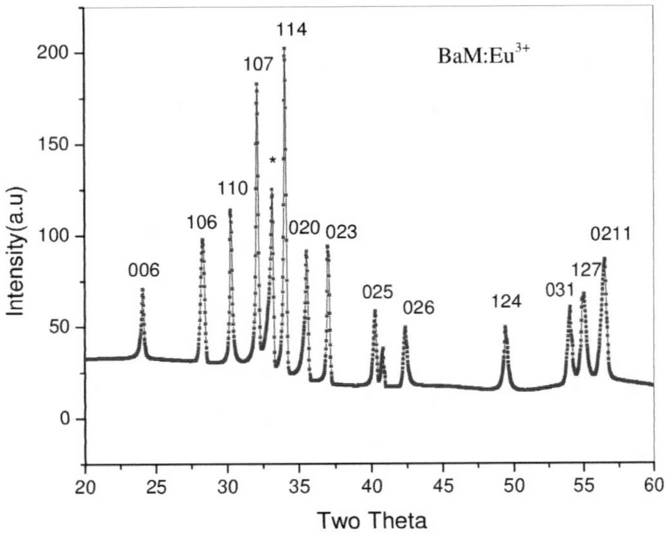


**Figure 1.** PXRD pattern for the $BaFe_{12}O_{19}:xEu^{3+}$ (x=2 mol%) sample.

*3. 2 TEM Analysis*

TEM micrographs of the 2 mol % $Eu^{3+}$ doped hexaferrite sample were recorded using Hitachi H- 600 TEM Analyzer. Figure 2a and 2b show the TEM micrographs of the hexaferrite sample which indicate that the prepared sample shows plate like structure [14].

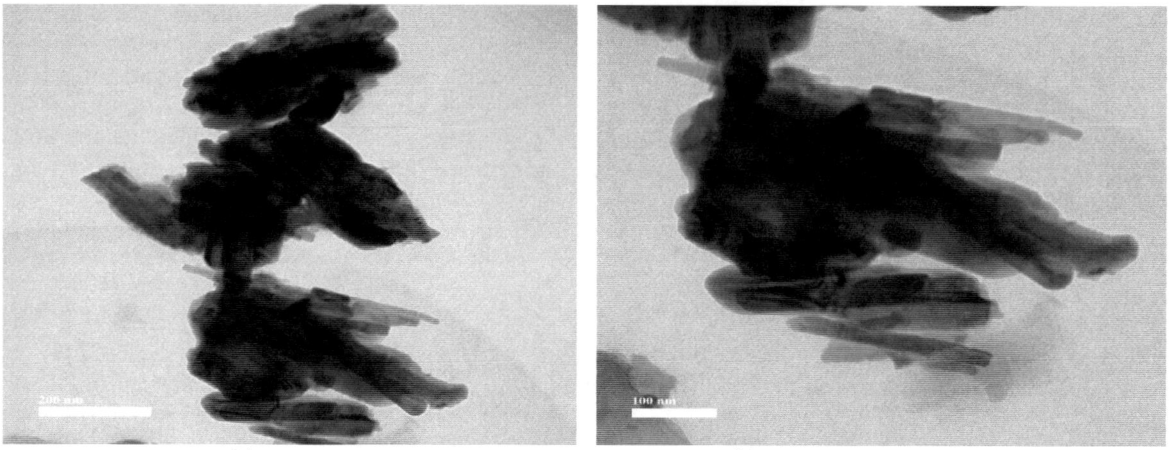

(a) (b)

**Figure 2.** TEM micrographs for the $BaFe_{12}O_{19}$:$xEu^{3+}$ (x=2 mol%) sample.

*3.3 FTIR Analysis*

FTIR analysis was performed to get information about the molecular bands, functional groups etc. Figure 6 shows FTIR spectra of the prepared hexaferrite samples in the range from 400-500 $cm^{-1}$. The strong absorption between 420 $cm^{-1}$ to 480 $cm^{-1}$ indicates the vibration of the $4f_1$ and $4f_2$ crystallographic sites of the ferric ion corresponding to the octahedral vibration mode ($A_{2u}$) [15].

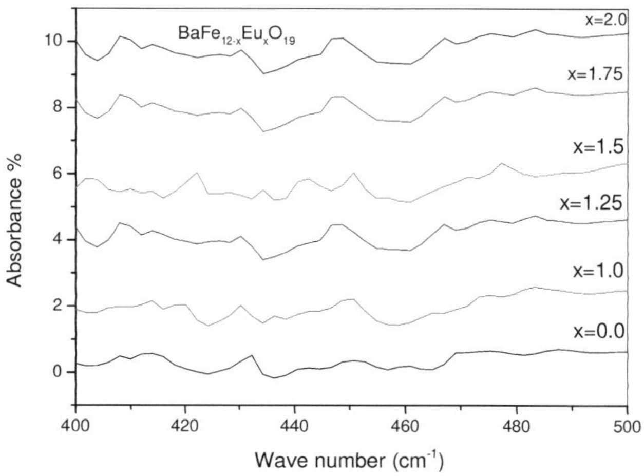


**Figure 3.** FTIR spectra of the prepared hexaferrite samples.

*3.4 UV-Visible Spectroscopy*

The UV visible spectra of $BaFe_{12-x}Eu_xO_{19}$(x= 0 to 2) samples were shown in figure 4 (a-f). Due to the transition between CB (conduction band) and VB(valence band) we get UV-visible spectrum of the samples. The band gap of the hexaferrite samples were calculated by plotting Tauc plot using the equation $\alpha h\nu = A(h\gamma - E_g)^n$ [16]

where hv is the photon energy, $E_g$ is the energy band gap, α is the absorption coefficient, A is a constant and n is the characteristics constant for different transitions. For undoped and doped samples the observed values are ranging from 3.99eV to 2.29 eV. From the figures it is clear that with increase in europium ions concentration, the band gap value decreases. As the europium ions concentration increases the number of lattice defects increases which results in formation of intermediate electronic levels in the forbidden band gap therefore resulting the decrease in the band gap value [17].

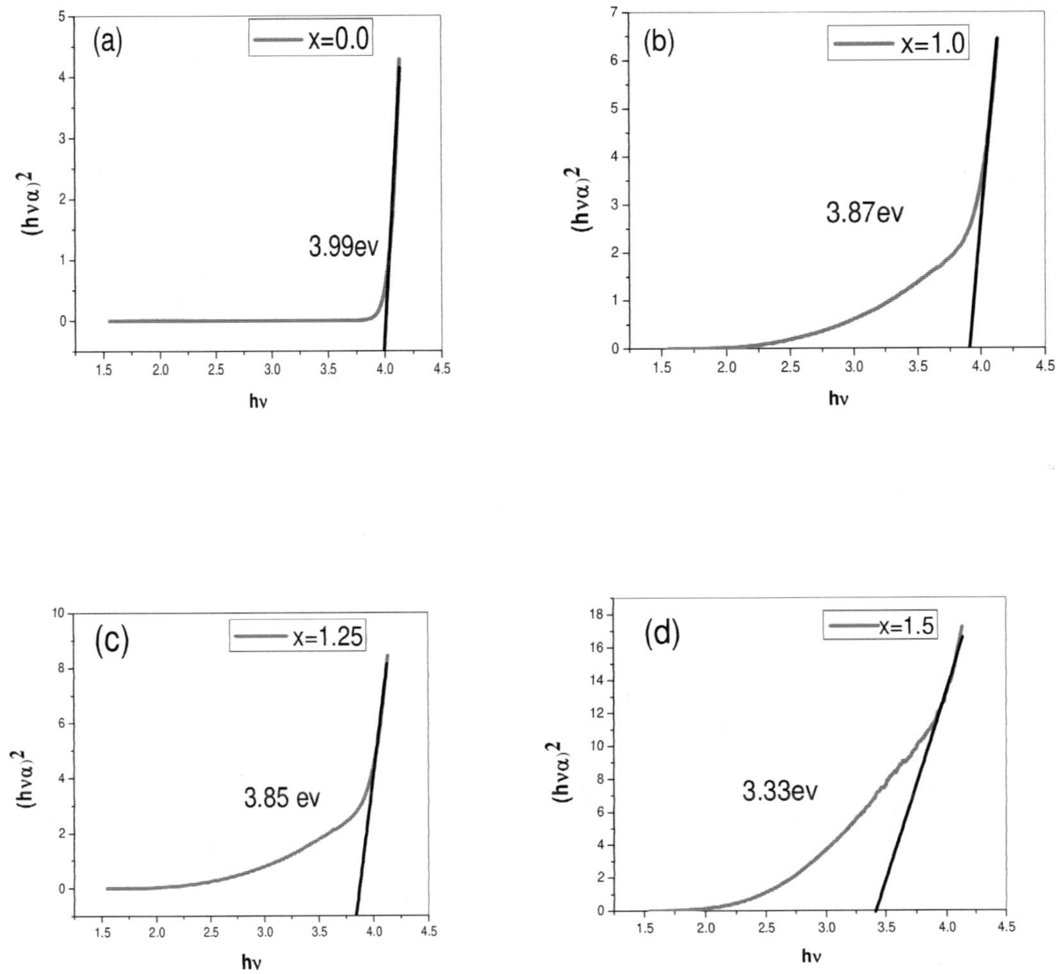

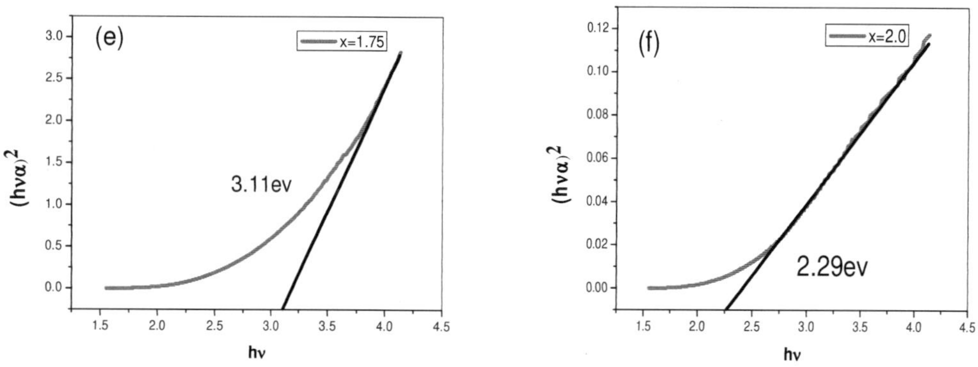


**Figure 4.** UV-visible spectra for the $BaFe_{12-x}Eu_xO_{19}$ (x= 0 to 2 mol %) samples.

## 4. Conclusion

Europium ions doped and undoped barium ferrite nano particles were prepared by combustion synthesis technique and the structure of the prepared nano particles is hexagonal as confirmed by PXRD pattern analysis. TEM micrographs confirm the formation of plate like structure. Vibration between the crystallographic sites ($4f_1$ and $4f_2$ ) of ferric ions of the prepared samples were observed in FTIR analysis. The prepared samples show semiconducting behaviour as confirmed by Uv-vis analysis.

## References

[1] Dawar N, Chitkara M, Singh I et al 2016 *Conjent physics* **3** 1208450
[2] Sharma M and Kashyap S C 2019 *Ceram. Int.* **45** 11226
[3] Pullar R C 2012 *Prog. Mater.* **57** 1191
[4] Afghahi S S S, Jafarian M and Atassi Y 2016 *J. Magn. Magn. Mater.* **419** 62
[5] Kojima H 1982 *Hand book of Ferromagnetic Materials* **3** 305
[6] Pfeiffer H, Chantrell R.W et.al 1993 *J. Magn. Magn. Mater.* **125** 373
[7] Awadallah A, Mahmood S H, Maswadeh Y et al 2016 *Mater.Res.Bull.* **74** 192
[8] Tang X, Hong , Feng W and Badami D 2013 *J.Alloy.Compd.* **562** 211
[9] Gairola S, Verma V, Singh A, Purohit L and Kotnala R 2010 *Solid State Commun.* **150** 147
[10] Rashad, M M and Ibrahim I A 2011 *J. Magn. Magn. Mater.* **323** 2158
[11] Dong L, Han Z, Zhang Y, Wu Z and Zhang X 2006 *Rare Met.* **25** 605
[12] Liu X, Wang J, Gan L and Ng S, 1999 *J. Magn. Magn Mater.* **195** 452
[13] Koga N and Tsutaoka T 2007 *J. Magn. Magn Mater.* **313** 168
[14] Nabil M M, Qurashi A, Assadi A A et al *2015 Nanoscale Res. Lett.* **10** 215
[15] Song F, Shen X, Xiang J and Song H 2010 *Mater. Chem. Phys.* **120** 213
[15] Karmakar M, Mondal B, Pal M and Mukherjee K 2014 *Sens. Actuators B Chem.* **190** 627
[16] Ganguly M, Rout S K, Park H Y, Ahn C W and Kim I W 2013 *Phys. Express* **3** 1

IOP Conf. Series: Materials Science and Engineering 1120 (2021) 012004 doi:10.1088/1757-899X/1120/1/012004

# Mechanoluminescence behaviour on $Eu^{2+}/Dy^{3+}$ activated $SrAl_2O_4$ phosphor

**Manish Kumar Mishra[1, 2, *], Vikas Dubey[3], Pushya Mitra Mishra[4] and Taranjeet Sachdev[5]**

[1] Department of Mechanical Engineering, School of Engineering and Technology, Career Point University, Kota, Rajasthan, India, 324005
[2] Department of Mechanical Engineering, Bhilai Institute of Technology Raipur, India, 493661
[3] Department of Physics, Bhilai Institute of Technology Raipur, India, 493661
[4] Department of Mechanical Engineering, MANIT Bhopal, India, 462003
[5] Department of Mechanical Engineering, SSIPMT Raipur, CG, India, 492015

*Corresponding author's e-mail address: mishramanish68@gmail.com

**Abstract.** Synthesis of Eu/Dy rare earth doped strontium aluminate phosphor or material with varying concentration of co-dopant ($Dy^{3+}$) and fixed concentration of dopant ($Eu^{2+}$) is reported inside the manuscript. Phosphor are prepared and synthesized by using the technique of solid-state reaction in reducing atmosphere and activated charcoal was used to create the atmosphere. Mechanoluminescence pattern were recorded for variable concentration of doping / co-doping ions and it is found with the increment in the concentration of co-dopant up-to 1 mol%, the ML intensity increases and after that the ML intensity decrement occurs due to the phenomenon of concentration quenching. Similarly, for the optimized concentration the ML pattern obtained for gamma – irradiated phosphor and it shows linear response with dose so our study supports the ML dosimeter application to detect the radiation via mechanoluminescence study. The suitable models are compared herein, which are already reported by so many authors.

## 1. Introduction

The concept of occurrence of mechanoluminescence behaviour, in which the light is emitted by phosphors when they undergo with mechanical deformation like rubbing, friction, cutting, collision, grinding and striking is reported by so many authors [1-3]. Various reports say that on around 50% of organic molecular solids and inorganic type of substances possesses the mechanoluminescence behaviour when deformed mechanically. Some researchers [4, 5] have got focus on the application of this mechanoluminescence phenomenon for the deformation detection or for finding the mechanical stresses generated from a far or remote location in which they use to convert the mechanical energy developed in to the visible source of lights. Earlier the detection of the developed mechanical energy is carried out with the use of piezoelectric substances, but this piezoelectric kind detection technique involves the use of electrodes to be in physical contacts with the material required testing and analysis for conversion of the mechanical energy in to the electrical pulses or signals.

The sensing of developed mechanical stresses can be carried out from a far or remote location only when the emission induced by mechanoluminescence is very strong, but the drawback associated with this is that, the intensity produced in these occasions are normally very low due to which the real-life practical employability of this phenomenon becomes too much burdensome.

Till date, only a very few numbers of studies were reported for the non-destructive kind of mechanoluminescence development in substances. The phosphors based on $SrAl_2O_4$, when they are produced by codoping with the rare earth substances like $Eu^{2+}$ and $Dy^{3+}$ exhibits grater qualities like either no radiation or very less radiation, long lasting time for light emitting, higher side brightness and a very good quality of photo resistance. These newly developed phosphors have got a tremendous usages or applications in variety of fields of sensing and actuating [6]. A lot of studies are reflecting that the mechanoluminescence behaviour is observed when the holes are got released due to the occurrence of deformation/stresses on the crystalline phosphors and then these holes are got associated with the metastable $Eu^{+1}$ [6]

The study of luminescent phosphors like $SrAl_2O_4$:Ce, SAO:Ce and $SrAl_2O_4$:Ce,Ho, SAOHoCe, which are emitting the ultraviolet mechanoluminescence were reported by Zhang et al. [7]. The possibility behind the use of phosphors $SrAl_2O_4:Eu^{2+}$, $Dy^{3+}$ (SAO) is of great interest as these materials can be utilized for formation of the non-destructive and reproducible kind of load sensors or stress sensors has been reported by Sohn et al. [8].

The concept of impulsive excitation in the mechanoluminescent materials like $SrAl_2O_4$ have been reported by Jha et. al [9]. The authors have also reported that with the successive and consecutive impacts & collisions provided by loads on to the prepared phosphors the mechanoluminescence intensity have got decreased and the declined & reduced intensity of mechanoluminescence can roughly be recovered by using the UV irradiation [9].

The researchers Jha and Chandra [10] concluded about the presence of impulsive excitation of the mechanoluminescence phenomenon, in the prepared phosphors $SrAl_2O_4:Eu^{2+}$, $Dy^{3+}$ (SAO) when developed in the reducing environment defined by 95% Ar & 5% $H_2$ with using the solid-state reaction technique.

In this manuscript time resolved spectra shows the decay kinetics and it has persistence luminescence behaviour of prepared phosphor. The mechanoluminescence intensity of the prepared phosphor will increase with time because of the impulsive excitation produced due to the impact of load, gradually it increases and the it gets decreased after attaining the maximum value. When a graph has been plotted between the time and mechanoluminescence intensity, it has been observed that the peaks go on increasing and afterward they shift towards the smaller time values with the provision of increment in the impact velocities. The mechanoluminescence intensity which is represented by $I_m$, related and calculated on the basis of a graph plotted between the time and the mechanoluminescence intensity, has a tendency to increase linearly with the impact velocity. Also, the variation of the time $t_m$ is found linear and related with the factor $1000/V_0$. The study of formation of mechanoluminescence, due to the impulsive excitation in the synthesized phosphors, have got a tremendous application on crack sensing as it is a very useful phenomenon in grasping the crack formation.

Sik kim et al. [11] monitored about the mechanoluminescence behaviour of $SrAl_2O_4:Eu^{2+}$, $Dy^{3+}$ materials & phosphors and they have reported that mechanoluminescence of $SrAl_2O_4:Eu^{2+}$, $Dy^{3+}$ phosphors are strongly dependent and are a functional factor of their expeditious loading rate.

They had reported about the systematic investigation of ML of $SrAl_2O_4:Eu^{2+}$, $Dy^{3+}$ phosphors using rate equations, which are almost same as experimental data. They have claimed that the ML of $SrAl_2O_4:Eu^{2+}$, $Dy^{3+}$ materials/phosphors is developed with the variation in load instead of static load, hence the rate of loading gives the profile of ML versus time curve.

Figure 1 represents the line diagram of the mechanism of the mechanoluminescence phenomenon in the prepared or synthesized $SrAl_2O_4:Eu^{2+}$ crystals. Here the step 1 represents the excitation of rare earth $Eu^{2+}$, step 2 represents the movement of electron in Conduction band, step 3 shows about the trapping of the electron, step 4 reveals about the release of the electrons, step 5 explained about the movement of the electron inside of the Conduction Band, step 6 shows the electron-capture at 5d of $Eu^{2+}$, and the step 7 shows the emission of the light. By using this representation, the locations of the stages or band or levels of rare earth $Eu^{2+}$ and the traps in the band-gap were explained on the basis of data reported by Clabau et al. [12].

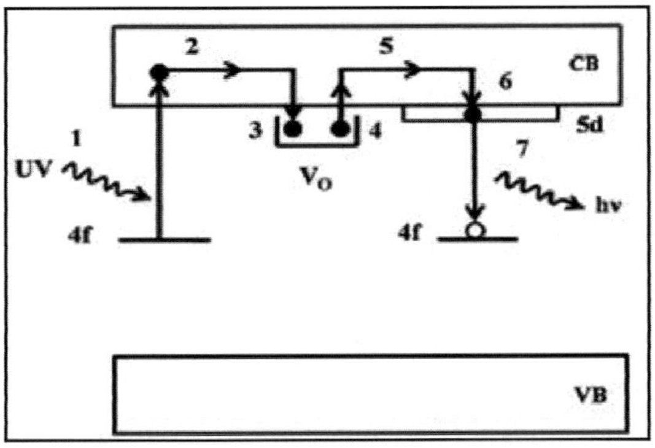


**Figure 1.** Representation diagram of the ML emission in $Eu^{2+}$ doped phosphor [13].

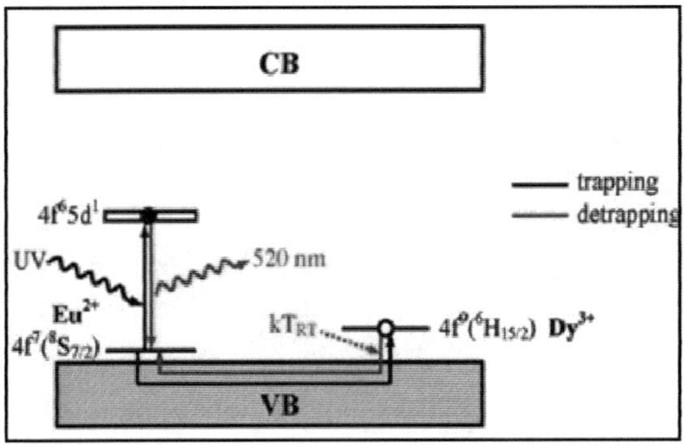


**Figure 2.** The technique of phosphorescence mechanism projected by Matsuzawa et al. for $SrAl_2O_4:Eu^{2+}$, $Dy^{3+}$ phosphors [12].

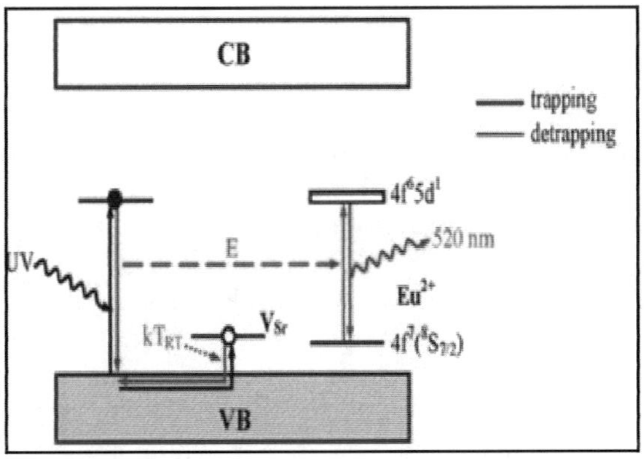


**Figure 3.** The technique of phosphorescence mechanism projected by Beauger for $SrAl_2O_4:Eu^{2+}$ phosphors [12].

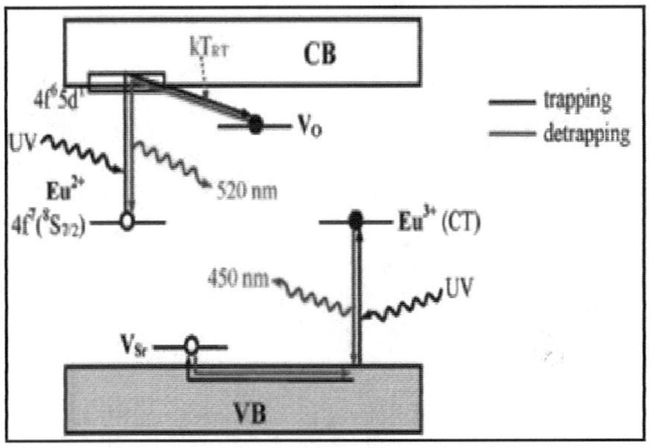


**Figure 4.** The technique of phosphorescence mechanism projected by Clabau for $SrAl_2O_4:Eu^{2+}$, $Dy^{3+}$, $B^{3+}$ [12].

Figure 2, 3 and 4 shows the schematic diagram for the Phosphorescence mechanism proposed by Matsuzawa et al. for $SrAl_2O_4:Eu^{2+}$, $Dy^{3+}$ phosphors, Phosphorescence mechanism proposed by Beauger for $SrAl_2O_4:Eu^{2+}$ phosphors and Phosphorescence mechanism proposed by Clabau for $SrAl_2O_4:Eu^{2+}$, $Dy^{3+}$, $B^{3+}$ [12].

## 2. The constituent materials and the methods involved

### *2.1. Ingredient Materials utilized*

Below mentioned table 1 contains the list of materials or chemicals utilized for making or synthesis of desired phosphors including their chemical formula, molecular weight in gm/mol and density in gm/cm$^3$.

**Table 1.** List of Chemicals with their molecular weight used for the synthesis of phosphors.

| S. No. | Chemical Formula | Name of Chemical | Molecular Weight (gm/mol) | Density (gm/cm$^3$) |
|---|---|---|---|---|
| 1. | $SrCO_3$ | Strontium Carbonate | 147.63 | 3.5 |
| 2. | $Al_2O_3$ | Aluminium Oxide | 101.96 | 3.95 |
| 3. | $H_2BO_3$ | Boric Acid | 61.83 | 1.435 |
| 4. | $ZrO_2$ | Zirconium Dioxide | 123.22 | 5.8 |
| 5. | $TiO_2$ | Titanium Dioxide | 79.87 | 4.23 |
| 6. | $Eu_2O_3$ | Europium Oxide | 351.926 | 7.4 |
| 7. | $Dy_2O_3$ | Dysprosium Oxide | 372.998 | 7.8 |
| 8. | $(CH_3)_2CO$ | Acetone | 58.08 | 0.7845 |
| 9. | C | Activated charcoal | 12.011 | 2.1 |

### *2.2. Preparation of samples*

Many researchers have reported and concluded about the processes of synthesis and characterization of rare earth doped luminescent materials and phosphors [14-17]. The reports included about the combustion technique and solid-state reaction technique for synthesis. The several characterization techniques are XRD, SEM, TEM, EDS and FTIR, which provides the details of the synthesized phosphors.

To prepare the rare earth (europium and dysprosium) doped $SrAl_2O_4$ phosphor of different concentration, the stoichiometrically calculated (in mol% criteria of total weight of chemicals required for preparation of samples) quantity of ingredient chemicals are mixed to prepare a 5-gram sample and it is kept in alumina crucible and allowed for calcination (heated in muffle furnace for a time period of 2 hours) for a temperature of around 1000°C. After the calcination is done for 2 hours the prepared samples are allowed to cool inside the furnace and allowed to reach the room temperature. The samples were then taken out and successive intermediate grinding of calcinated samples is carried out using the mortar and pestle manually. The technique adopted for the preparation of rare earth $Dy^{3+}$ and $Eu^{2+}$ activated phosphor samples is a solid-state reaction/diffusion method. The constituent chemicals utilized for the preparation of desired phosphor are Strontium Carbonate ($SrCO_3$), Aluminium Oxide ($Al_2O_3$), Europium Oxide ($Eu_2O_3$), Dysprosium Oxide ($Dy_2O_3$) and Boric Acid ($H_2BO_3$) as flux.

For making the homogeneous powder mixture of samples, the mixture of reagent ingredients are grinded firmly using the agate mortar and pestle. The time required for making the firmly homogenous mixture is around 45 minutes by grinding process. After the intermediate grinding of samples, they were again transferred in to the alumina crucible for sintering process. These properly mixed and grinded samples are again kept in alumina crucible and heated in the muffle furnace up-to a temperature of 1250°C for a time period of 3 hours, while keeping the reducing environment inside the furnace, with placement of a couple of activated charcoal filled crucibles along with the crucible of phosphor samples [18]. Then after the desired and prepared samples are allowed to cool naturally inside the furnace up to the room temperature.

The chemical reaction take places is:

$$SrCO_3 + Al_2O_3 = SrAl_2O_4 + CO_2 \tag{1}$$

Below mentioned table 2 and table 3 contains the weight of ingredient chemicals required for the preparation and synthesis of phosphors when the concentration of rare earth $Eu^{2+}$ and $Dy^{3+}$ in varying in mol% from 0.1, 0.2, 0.5, 1.0, 1.5, 2.0 and 2.5.

**Table 2.** Calculation of weight for ingredients in 5 grams of mixture keeping 1 mol% of $Eu_2O_3$ and variation of $Dy_2O_3$ from 1 mol% to 2.5 mol%.

| $Dy_2O_3$ Mol % | 0.1 | 0.2 | 0.5 | 1.0 | 1.5 | 2.0 | 2.5 |
|---|---|---|---|---|---|---|---|
| Weight (in grams) | | | | | | | |
| $SrCO_3$ | 2.9120 | 2.9078 | 2.8950 | 2.8740 | 2.8533 | 2.8328 | 2.8127 |
| $Al_2O_3$ | 2.0112 | 2.0082 | 1.9994 | 1.9849 | 1.9706 | 1.9565 | 1.9426 |
| $Eu_2O_3$ | 0.0694 | 0.0693 | 0.0690 | 0.0685 | 0.0680 | 0.0675 | 0.0671 |
| $Dy_2O_3$ | 0.0074 | 0.0147 | 0.0366 | 0.0726 | 0.1081 | 0.1431 | 0.1777 |
| Total Weight | 5.0000 | 5.0000 | 5.0000 | 5.0000 | 5.0000 | 5.0000 | 5.0000 |

Table no. 2 shows the calculated weight of ingredients in 5 grams of mixture while keeping 1 mol% of $Eu_2O_3$ and variation of $Dy_2O_3$ from 1 mol% to 2.5 mol%. Similarly, the table no. 3 shows the calculated weight of ingredients in 5 grams of mixture while keeping 1 mol% of $Dy_2O_3$ and variation of $Eu_2O_3$ from 1 mol% to 2.5 mol%. For proper and accurate weighing of ingredients in the mixture an electronic digital balance is utilized.

**Table 3.** Calculation of weight for ingredients in 5 grams of mixture keeping 1 mol% of $Dy_2O_3$ and variation of $Eu_2O_3$ from 1 mol% to 2.5 mol%.

| $Eu_2O_3$ Mol % | 0.1 | 0.2 | 0.5 | 1.0 | 1.5 | 2.0 | 2.5 |
|---|---|---|---|---|---|---|---|
| Weight (in grams) | | | | | | | |
| $SrCO_3$ | 2.9099 | 2.9058 | 2.8938 | 2.8740 | 2.8544 | 2.8351 | 2.8161 |
| $Al_2O_3$ | 2.0097 | 2.0069 | 1.9986 | 1.9849 | 1.9714 | 1.9581 | 1.9449 |
| $Dy_2O_3$ | 0.0735 | 0.0734 | 0.0731 | 0.0726 | 0.0721 | 0.0716 | 0.0712 |
| $Eu_2O_3$ | 0.0069 | 0.0139 | 0.0345 | 0.0685 | 0.1021 | 0.1352 | 0.1678 |
| Total Weight | 5.0000 | 5.0000 | 5.0000 | 5.0000 | 5.0000 | 5.0000 | 5.0000 |

Figure 5 and 6 shows the alumina crucible contains the 5 gram of phosphors in varying concentration of $Eu^{2+}$ and $Dy^{3+}$ rare earth and the sequence and pattern followed for keeping of samples in the muffle furnace as shown in figure 7, for the calcination and sintering process. After the calcination and sintering the samples were kept in LDPE bottles to keep them protected from any impurity and moisture. The coding was adopted and mentioned on bottles for each of the sample with a specific concentration of $Eu^{2+}$ and $Dy^{3+}$ rare earth.

**Figure 5.** Photograph of synthesis (sintering) of phosphors in alumina crucible using muffle Furnace

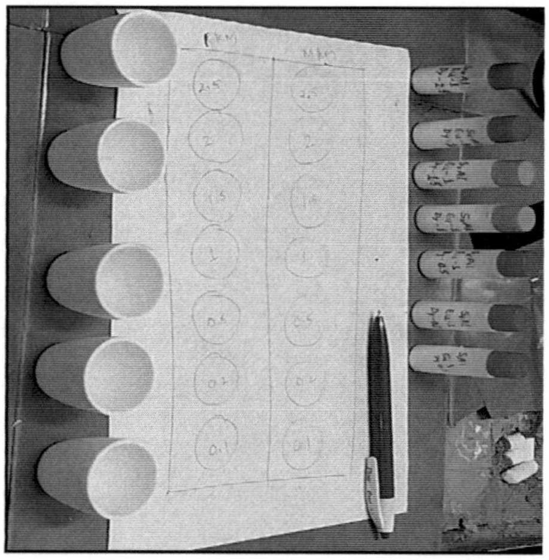

**Figure 6.** Keeping the prepared phosphors in LDPE screw cap bottles with reference code on bottles and poly-pack for further analysis and testing.

*2.3. Theoretical approach related to mechanism of the mechanoluminescence in the prepared phosphor SAO doped with RE ($Eu^{2+}$, $Dy^{3+}$)*

Some of the researchers have reported that for providing a good explanation of mechanoluminescence property of prepared samples one of the finest ways is electron detrapping model with a piezo-electric induction. As Jha and Chandra [10] have presented that, as in presence of Mn, the ZnS:Mn crystalline phosphor will emits the Elastico-mechanoluminescence and in presence of Ti, the $ZrO_2$ :Ti crystals are possessing the Elastico-mechanoluminescence, the prepared desired samples of SAO doped with RE ($Eu^{2+}$, $Dy^{3+}$) will show the Elastico-mechanoluminescence phenomenon. With the above study it is clear that the property of piezoelectric constant is in more amount in the Elastico-mechanoluminescence phosphors near to the region of distortion at the activator ions. Following are some of the basic stages involved in the Elastico-mechanoluminescence emission from the persistent luminescent crystals:

- The characteristic of light emission from the excited ions may gets increased due to the de-excitation of the excited activator ions.
- The capturing of electrons, in excited state while moving in the conduction band, by the activator ions, which are placed at the bottom of the conduction band may occur where there is possibility of emission or release of excited $Eu^{+2}$ ions [19-22].
- Their may be chances of occurrence of thermal detrapping of the carriers due to the decrease of trap depth of carriers.

*2.4. Experimental work*

The solid-state diffusion technique, involved in the preparation of phosphors and samples, is one of the commonly practiced technique for the development of multicomponent solid ingredient chemical materials carried out at the higher temperatures. Due to the barrier faced by solids, that they can not interact with each other at the common room temperature, even when the thermodynamics favours, a high temperature involvement with a controlled environment is required to get the desired reaction between the reagent chemicals. The better feature associated with the solid-state reaction/diffusion technique is that it is a low-cost technique at the commercial level and there will be availability of ample of precursors [23]. The below figure 7 shows the muffle furnace utilized in this work for the calcination and the synthesis of desired phosphors doped with rare earth materials.

Before going to record the ML pattern of prepared phosphor samples, they were irradiated by $Co^{60}$ gamma rays. The magnitude of load applied or dropped is 500gm in each attempt and the quantity of sample used is 2mg for the recording of ML pattern. This process is carried out at RTM university Nagpur.

**Figure 7.** Photograph of Muffle Furnace utilized for synthesis of phosphors

*2.5. Procedural analysis for Mechanoluminescence and Discussions*

The below Figures 8, 9 and 10 shows the experimental setup of mechanoluminescence device utilized for the measurement of ML in the synthesized crystalline phosphors.

**Figure 8.** Experimental setup ML Device with stake height of 1 meter used for the measurement of ML Property in synthesized phosphors on time dependent basis.

**Figure 9.** The wooden block, arrangement with transparent lucite plate and connections.

**Figure 10.** The photomultiplier tube arrangement with connections to oscilloscope of ML device.

It has been found that the developed phosphor is possessing only one curve of glow with only one peak for the mechanoluminescence. Below Figure 11 depicts the time vulnerability of mechanoluminescence strength of ϒ- ray irradiated ($Eu^{2+}$, $Dy^{3+}$) doped SAO sample phosphor for the varying concentrations of different impurity $Dy^{3+}$.

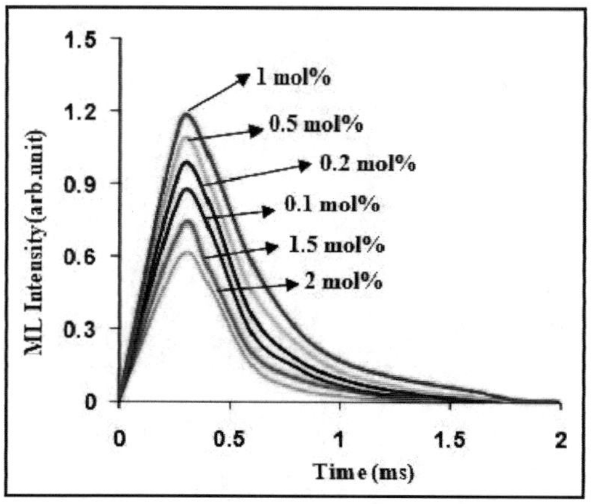


**Figure 11.** Time dependence of ML intensity of SAO:Eu, Dy (0.1 to 2 mol%) samples for the different concentrations of impurity.

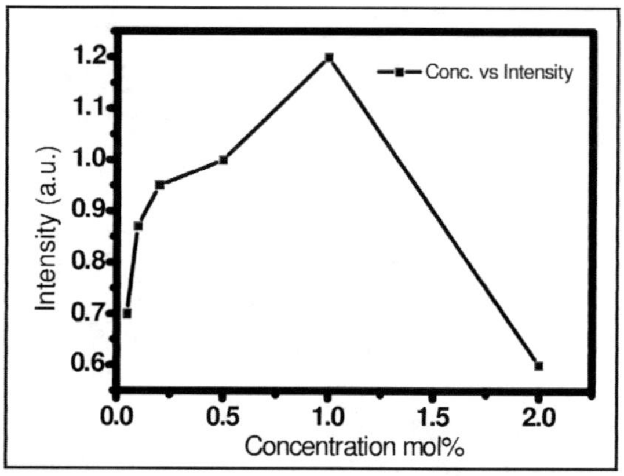


**Figure 12.** Dopant concentrations versus intensity plot for SAO phosphor doped with $Dy^{3+}$.

It has been observed that there will be an increment in mechanoluminescence intensity with the increment in the concentration of different impurity of dopants in the prepared desired phosphors. There will be an indication of single emission centre by the occurrence of only one single peak and hence the first observed peak can be compared with the time required for detrapping of the trapped electrons from the traps. Optimized concentration for intense ML is 1mol% for Eu and Dy doped SAO phosphor. Figure 12 shows the dopant concentrations versus intensity plot for SAO phosphor doped with $Dy^{3+}$, in which it has been observed that with the increase in concentration the intensity increases for up to 1mol% and after that the intensity decreases, for concentration of 1 mol % of $Dy^{3+}$, the maximum intensity is achieved.

Figure 13 depicts the glow curves for the mechanoluminescence of ϒ- ray irradiated ($Eu^{2+}$, $Dy^{3+}$) doped SAO samples, involving the varying impact velocities of the moving piston. The effect of increment in impact velocity governs the increment in the ML intensity. Also, it has been observed that there is a shifting of amount of time required related to the ML peak is towards the lower side while there will be an increase in the impact velocity.

When the experimental work of, dropping of a small mass from a particular height has been carried out, it has been observed that for the samples based on aluminium ingredients, the intensity of mechanoluminescence is increased for initial period, it increases and goes up to the peak value and then time dependent decrement is there. While comparing the appearance of the mechanoluminescence and its decay, it has found the quantities are lower in above case as compared to the amount developed in case of pressure application. The piezoelectric induces electron detrapping model is capable of explaining the mechanoluminescence behaviour in aluminium based phosphors. In these samples the trap depth has been reduced by piezoelectric field near the impurities of $Eu^{2+}$ & $Dy^{3+}$. Hence there will be occurrence of detrapping of the electron traps with the release of energy in non-radiative mode and obviously the mechanoluminescence has been observed.

IOP Conf. Series: Materials Science and Engineering 1120 (2021) 012004 doi:10.1088/1757-899X/1120/1/012004

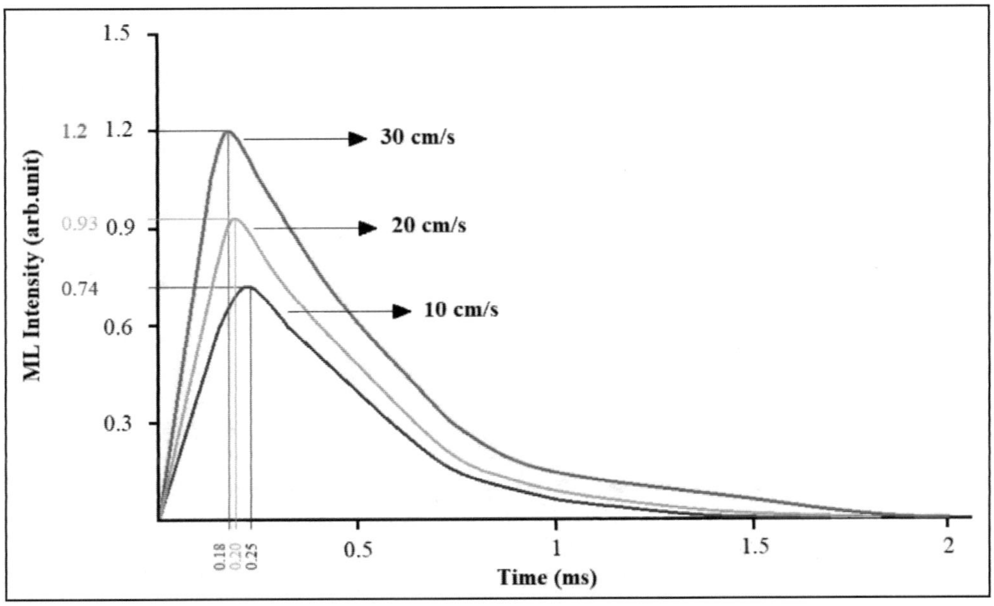


**Figure 13.** Mechanoluminescence curve of glow for ϒ- ray irradiated SAO:Eu/Dy phosphors for various velocities of impact of the piston (for the 1 mol% concentration of $Dy^{3+}$ ).

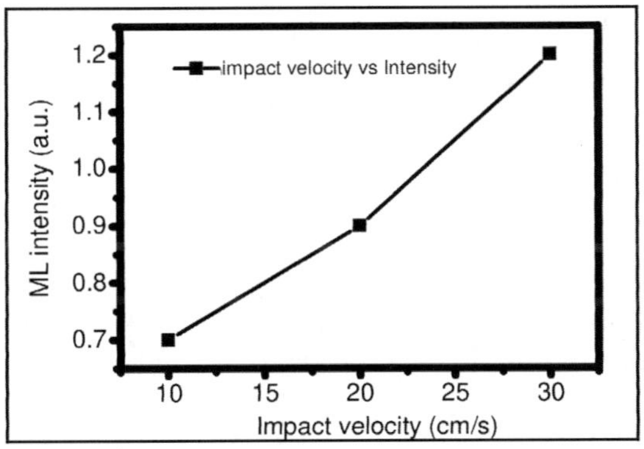


**Figure 14.** Variation of peak obtained of the ML intensity $I_m$ of SAO phosphors doped with RE ($Eu^{2+}$, $Dy^{3+}$) on different impact velocity.

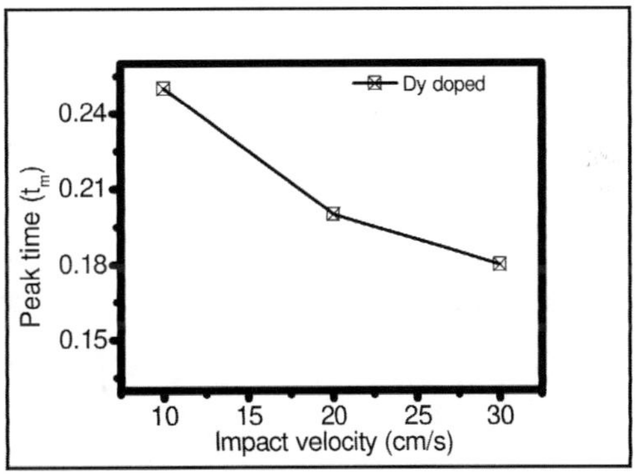


**Figure 15.** Variation of peak time ($t_m$) with respect to the impact velocity ($V_0$) for the SAO doped with ($Eu^{2+}$, $Dy^{3+}$) phosphor.

The curve plot between the impact velocity and ML intensity is shown in Figure 14, which shows that the dependence of peak ML intensity $I_m$ of SAO phosphors doped with RE ($Eu^{2+}$, $Dy^{3+}$) on different impact velocity. For the lower impact velocity, the ML intensity is lower and on increasing the impact velocity the ML intensity is increased. Figure 15 represents the dependence of peak time ($t_m$) on impact velocity ($V_0$) for the SAO doped with ($Eu^{2+}$, $Dy^{3+}$) phosphor which concludes that on increasing the impact velocity the peak time is decreasing.

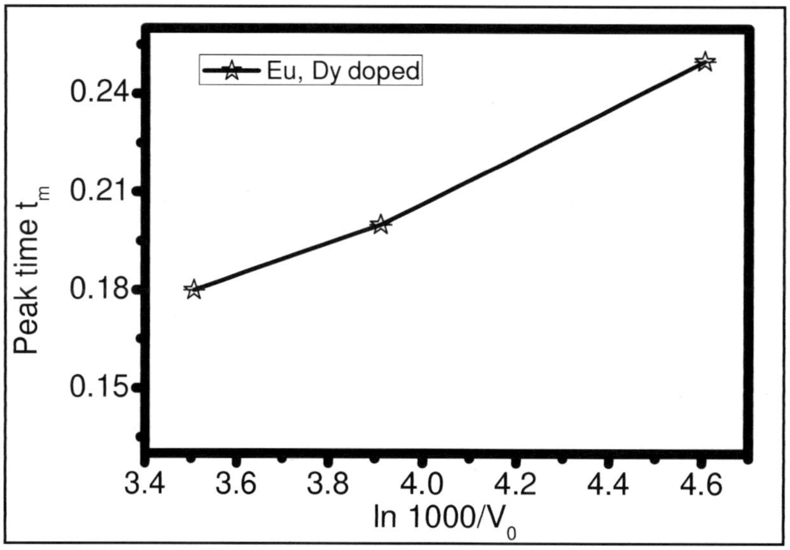


**Figure 16.** Variation of Peak Time with increments in ln(1000/V0 ).

*2.5.1 Time ($t_m$)*

Chandra et. al. [24] have proposes the equation for time $t_m$, as below

$$t_m = H[\delta - \delta^2 V_0]\ln 2 \quad (2)$$

As $\delta$ is a minute quantity for microcrystalline phosphors, it is clear that there will be very minor change in $t_m$ for increment in impact velocity $V_0$, as observed in Figure 16. Sahu et. al. [25, 26, 27] have reported about the variation of time with respect to the increasing impact velocity, for the powder phosphors.

*2.5.2 Peak ML intensity $I_m$*

Chandra et. al. [24] have also presented that the intensity of mechanoluminescence is having proportionality with the quantity of multiplication of rate of change of pressure and the pressure. The relation between impact pressure and impact velocity is studied and presented in various works [28-32].

*2.5.3 Absorption Spectra and behaviour on load application*

Choubey et. al. [33] reported about the Luminescence behaviour of $SrAl_2O_4$ on load application from height that for a time of 10 ms the maximum intensity of ML has been observed and they have also presented the absorption spectra of $SrAl_2O_4$ having absorption edge at 240nm.

*2.5.4 XRD Analysis of Prepared phosphor*

In order to determine the phase structure, powder XRD analysis has been carried out. The typical XRD patterns of $SrAl_2O_4$:$Eu^{2+}$, $Dy^{3+}$ nanophosphor with the standard XRD pattern is shown in Fig. The diffraction pattern depicted much broader and less intense peaks, thereby, giving a strong indication that the prepared nanophosphor has smaller grain size and lower crystallinity.

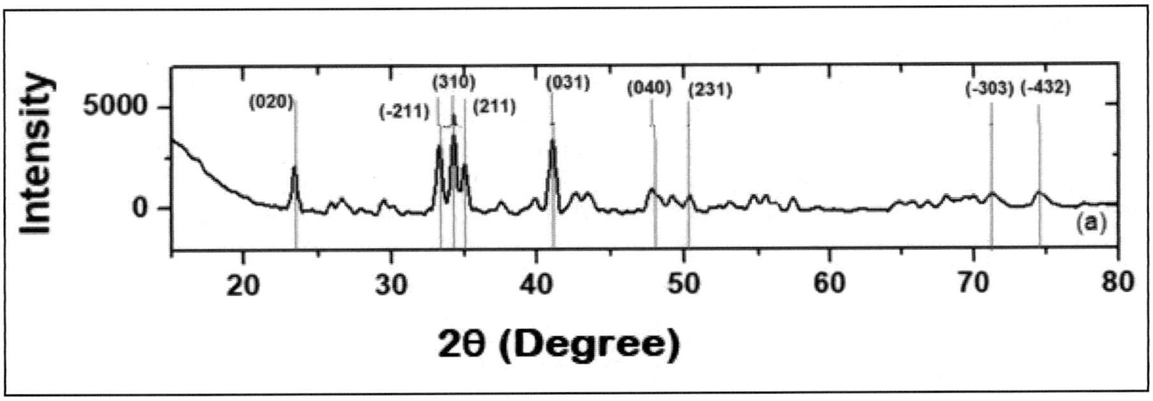


**Figure 17.** XRD pattern of $SrAl_2O_4:Eu^{2+}$, $Dy^{3+}$ nanophosphors.

## 3. Results

The work done in the present study shows that, the mechanoluminescence behaviour is owned by Eu2+ and Dy3+ activated SrAl2O4 phosphor. The prepared phosphor is having a single mechanoluminescence peak with a single glow curve for every sample with variation of mol% in weight. The mol% concentration is varying as 0.1, 0.2, 0.5, 1.0, 1.5, 2.0 and 2.5.

The time dependent ML intensity study of irradiated rare earth (Eu2+, Dy3+) doped SAO phosphor shows that, the ML intensity id maximum for concentration of 1 mol%. The analysis of dopant concentrations versus intensity plot for SAO phosphor doped with Dy3+, indicates that the increase in concentration the intensity increases for up to 1mol% and after that the intensity decreases, for concentration of 1 mol % of Dy3+, the maximum intensity is achieved.

The ML glow curves of ϒ- ray irradiated (Eu2+, Dy3+) doped SAO samples for different impact velocities of the piston indicates the ML intensity increases with increasing impact velocity. For 30cm/s velocity the maximum intensity of ML in arbitrary unit is achieved. Also, as the impact velocity increases the peak time goes on decreasing.

## 4. Conclusion

The developed phosphors can be employed to a wide range of applications, in a non-invasive mode. The synthesized phosphors are capable of providing signals in quite well characterized pattern. They are also able to depict the ability of discrimination of impact energy produced beyond the binary sensing. The recent and trending application of these phosphors may include, the work out for finding of dependency of mechanoluminescence development on the impact energy. By utilizing the mechanoluminescence phenomenon, study about the trapping and detrapping of charge carriers can be carried out. The mechanoluminescence excitation can be achieved by dropping a mass from a prescribed height on to the phosphors, the produced ML intensity gradually increases with the time then reaches to its maximum value and the goes on decreasing.

**Acknowledgments**

Authors are very thankful to TEQIP-III, CSVTU, Bhilai for funding through Collaborative Research Project to PI: MANISH KUMAR MISHRA and Co-PI: TARANJEET SACHDEV, Titled "Luminescence behaviour of rare earth activated $SrAl_2O4$ and $ZrO_2$ nano-phosphors and their applications for Deformation detection techniques with possibilities of their integration with MEMS devices" wide letter NO.CSVTU/TEQIP-III/2019/CRP-Final/3272, Dated : 03/10/2019 with CRP sanction order no. CSVTU/CRP/TEQIP-III/84, Dated : 05/09/2019 . Also, Authors are very thankful to Govt. V.Y.T.P.G. Auto. College Durg for providing facilities of research lab.

IOP Conf. Series: Materials Science and Engineering 1120 (2021) 012004 doi:10.1088/1757-899X/1120/1/012004

## References


[1] Walton A J 1977 *Adv. Phys.* **26** 887

[2] Alzetta G, Chudacek I and Scarmozzino R 1970 *Phys. Status Solidi* A**1** 775

[3] Enomoto Y and Hashimoto H 1990 Nature **346** 641

[4] Xu C N, Akiyama M, Sun P and Watanabe T 1997 *Applied Physics Letters* **70** 1639

[5] Takada N, Hieda S and Sugiyama J 2000 *Synth. Met.* **111** 587

[6] Lin Y H, Dang Z, Deng Y and Nan C W 2003 *Materials Chemistry and Physics* **80** 20

[7] Zhang H, Yamada H and Terasaki N 2007 *Applied Physics Letters* **91** 081905

[8] Sohn K S, Park W B, Timilsina S and Kim J S 2014 *Optics Letters* **39** 6

[9] Jha P 2016 *Luminescence* **31** 1302

[10] Jha P and Chandra B P 2013 *J. Lumin.* **143** 280

[11] Kim J S, Kibble K, Kwon Y N and Sohn K S 2009 *Optics Letters* **34** 1915

[12] Clabau F, Rocquefelte X, Jobic S, Deniard P, Whangbo M H, Garcia A and Mercier T L 2005 *Chem. Mater.* **17** 3904

[13] Chandra V K, Chandra B P and Jha P 2013 *Appl.Phys.Lett.* **102** 241105

[14] Kaur J, Parganiha Y, Dubey V, Singh D and Chandrakar D 2014 *Superlattices and Microstructures* **73** 38

[15] Dubey V, Kaur J, Agrawal S, Suryanarayana N S and KVR Murthy 2013 *Optik – Int. J. Light Electron Opt.* **5** 187

[16] Parganiha Y, Kaur J, Dubey V and Murthy K V R 2015 *Materials Science in Semiconductor Processing* **31** 715

[17] Dubey V, Kaur J and Agrawal S 2015 Materials *Science in Semiconductor Processing* **31** 27

[18] Dubey V, Kaur J, Agrawal S, Suryanarayana N S and Murthy KVR 2013 *Optik – Int. J. Light Electron Opt.* **124** 5585

[19] Clabau F, Rocquefelte X, Mercier T L, Deniard P, Jobic S and Whangbo M 2006 *Chem. Mater.* **18** 3212

[20] Clabau F, Garcia A, Bonville P, Gonbeau D, Mercier T L, Deniard P and Jobic S 2008 J. *Solid State Chem.* **181** 1456

[21] Clabau F, Rocquefelte X, Jobic S, Deniard P, Whangbo M H, Garcia A and Mercier T L 2005 *Chem. Mater.* **17** 3904

[22] Jha P and Chandra B P 2013 *J. Lumin.* **143** 280

[23] Schubert U and Hüsing N 2005 *Synthesis of Inorganic Materials* 2nd Edition WILEYVCH.

[24] Chandra B P and Chandra V K 2012 *J. Lumin.* **132** 858

[25] Sahu I P, Bisen D P, Bramhe N, Tamrakar R K and Shrivastava R 2015 *J Mater Sci: Mater Electron* **26** 9907

[26] Sahu I P, Bisen D P, Tamrakar R K, Murthy K V R and Mohapatra M 2017 *Journal of Science: Advanced Materials and Devices* **2** 59

[27] Sahu I P, Bisen D P, Tamrakar R K, Murthy K V R and Mohapatra M 2017 *Journal of Science: Advanced Materials and Devices* **2** 69

[28] Wilbeck J S and Rand J L 1981 *ASME J. Eng. Power* **103** 725

[29] Zenit R, Hurt M L and Brennen C E 1997 *J. Fluid Mech.* **353** 261

[30] Ahrens T J and Rubin A M 1993 *J. Geophys. Res.* **98** 1185

[31] Klinkov K V and Rein M 2006 *J. Appl. Phys.* **100** 013529

[32] M. Rein and Eur J 1995 *Mech. B/Fluids* **14** 301

[33] Choubey A K, Brahme N, Bisen D P, Sharma R 2011 *The Open Nanoscience Journal* **5** 41

IOP Conf. Series: Materials Science and Engineering 1120 (2021) 012005 doi:10.1088/1757-899X/1120/1/012005

# Photoluminescence studies and synthesis of $KSrPO_4:Ce^{3+}$, $Eu^{3+}$ blue and orange-red emitting phosphor

**Gaurav Rahate[1], V R Panse *[2], S J Dhoble[3], N S Kokode[4] and Khushbu Sharma[5]**

[1,5] Department of basic science & humanities G. H. Raisoni University, Amravati-, India
[2] Department of Physics, Late B. S. Arts, Prof. N. G. Science & A. G. Commerce College Sakharkherda-443202, India
[3] Department of Physics, RTM Nagpur University, Nagpur-India
[4] Nevjabai Hitkarini College, Gondwana University, Gadchiroli-441206, India

*Corresponding author's e-mail address: vshl.panse@gmail.com

**Abstract.** The growing need for leftover fuels and the ecological effects of their use was intended to give life to the traditionally extended global superpower infrastructure. Conventional, reliable fluorescent and incandescent lamps emit energy in the form of either heat or gas discharge. Both principles are strongly owing to the greater energy fluctuations that arise related to higher temperatures with the greater shifts of Stokes concerned. Within this work, we recorded sulphate-based blue & Orange-red emitting phosphors for lamp industry applications prepared using updated wet chemical synthesis methods, because the production of superb materials with high and stable fluorescent and luminescent properties is necessary to achieve high emission and fine clarity in the innovation of display systems. Morphological and structural tests and validation of phase and quality were conducted using XRD and SEM. The photoluminescent emission spectra of $KSrPO_4:Ce^{3+}$, $Eu^{3+}$ phosphors show a blue emission band centered at 440 nm and a orange-red emission band centered at 593 nm and 618 nm. Typically, the structure of the $Ce^{3+}$ ion in the ground state is split into two levels such as $2F_{5/2}$ and $2F_{7/2}$, while the $^5d_1$ excited configuration is divided by the crystal playing field into 2 and 5 elements.

## 1. Introduction

Currently, phosphor-converted white light-emitting diodes (w–LEDs) have been attracted worldwide attention as a new source for the solid-state lighting to next generation due to their special advantages such as long lifetimes, luminous efficiency, absence of mercury, low energy consumption, weak environmental impact, excellent chemical as well as physical stability, availability of final products in different sizes, and so on [1, 2]. A lot of studies have been investigating different luminescent materials to help the growth of suitable phosphors. Phosphors are also useful for producing white light when they are excited by blue or near-ultraviolet (NUV) lights (300–420 nm). It is very important to choose the right compound materials and assure they have outstanding physical and chemical stability for obtaining phosphors with highly proficient emissions.

Such materials are recombined with higher light capitulate, constructive emission wavelength, quicker fluorescence breakdown and temperature tolerance, rendering them desirable for use in high energy physics detectors [3] and even in medical imaging [4]. Thus, with the help of excellent luminescence properties, inorganic materials activated with $Ce^{3+}$ ions are used in the lighting industry, detectors with ionizing radiation display systems [8-9]. A sample ion is therefore considered to measure the 5d energies of another rare earth ion in the similar host lattice material trivalent cerium ion [5]. Thus the study of spectroscopic properties of $Ce^{3+}$ ion in different host material is essential for either actual applications or fundamental science. Wet chemical process is not yet another synthesis procedure recognized as a wet-chemical synthesis method but does not involve constant heating and calcinations. Starting up with the preparation of strontium nitrate, potassium nitrate, potassium phosphate with dual distillation water and cerium nitrate using the principle of propellant chemistry [6-7].

## 2. Experimental

The method of wet chemical synthesis was carried out roughly at a temperature of $100^0$C employing oven and with the aid of Merck analytical grade inorganic materials such as given below, $KNO_3$ (99.99 percent Merck purity), $Sr(NO_3)_2$ (99.99 percent Merck purity), $KPO_4$ (99.99 percent Merck purity), double distillation water and Cerium Nitrate ($Ce(NO_3)_3$, REI 99.9 percent) and $Eu^{3+}$ . The concentration of $Ce^{3+}$ ranges by 1-10 mol percent and $Eu^{3+}$ ranges from 0.1 to 2 mol %.

All the ingredients were arranged as per the stochiometric ratio in the beaker with double distillation water and then lead to the stirring phase with the aid of the magnetic stirrer and then held the homogeneous mixture in oven, the temperature of which was sustained at 100 $^0$C for 24 hours and at the end a pasty solution was formed, wherein the solution was then moved to the silica crucible, and retained inside the muffle furnace, that was maintained at a low temperature of around 100 $^0$C. Foamy powder with flame was developed and the powder was extracted and thereby examined for emission and excitation spectra for photoluminescent evaluation.

## 3. Results and discussion

### *3.1 X-ray diffraction study*

The preparation of the powder compound was defined for the analysis of its phase purity and crystallinity by powder X-ray diffraction (XRD) using the PAN Analytical X'pert Pro diffractometer XRD for the investigation of the XRD pattern of the prepared powder compound was validated using Cu-Kш radiation (1, 54060 nm) with a scanning stage time of 10,3377s with a continuous scan form. This $KSrPO_4:Ce^{3+}$ ion was prepared with respect to the normal JCPDS data source phosphor and found to be in good conformity with JCPDS data source file no.087-1854. The XRD level of the $KSrPO_4$ compound is like as follow. The XRD-pattern indicates that phosphor has a strong crystalline nature. The X-ray diffraction pattern of the $KSrPO_4$ compound as seen in the following figure 1. The material is heated up to 500°C at a steady temperature of 20°C in the argon atmosphere.

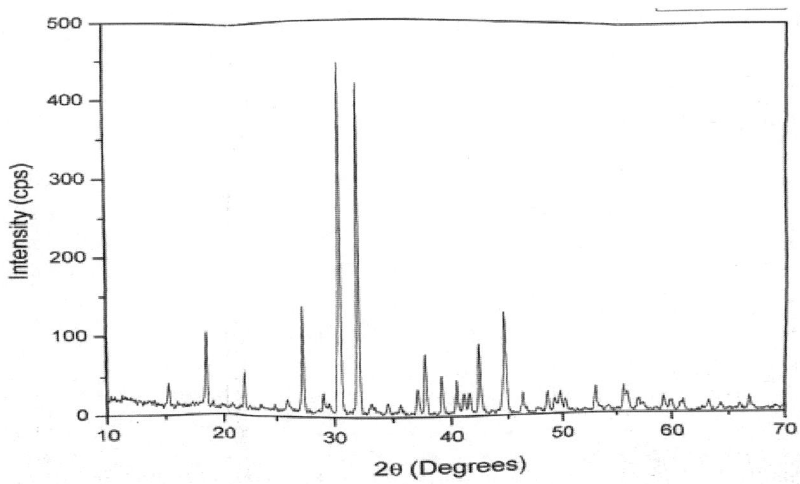


**Figure 1.** XRD-pattern of $KSrPO_4$ phosphor.

*3.2 Scanning electron microscopic study of $KSrPO_4$ phosphor*

The surface morphology analysis S.E.M. and the crystalline dimensions of the synthesized phosphor powder have been investigated. A wet chemical synthesis method conducted out by a heating reaction. This demonstrates that the wet chemical method of mixtures was quite well taken place. The characteristic morphological images are seen in figure 2, for $KSrPO_4$, consequently prepared phosphor materials. It is easily shown that the size of the crystalline micrographs ranges from a few microns to a range of tens of microns. The crystals include intelligent surface morphological representations and involve crystalline grains. Particles have foam-like composition developed from heavily agglomerated crystals. The normal crystalline scale is in the sub-micrometer range as seen in the SEM micrographic images; the crystalline sizes are exactly the same for all the prepared formulations.

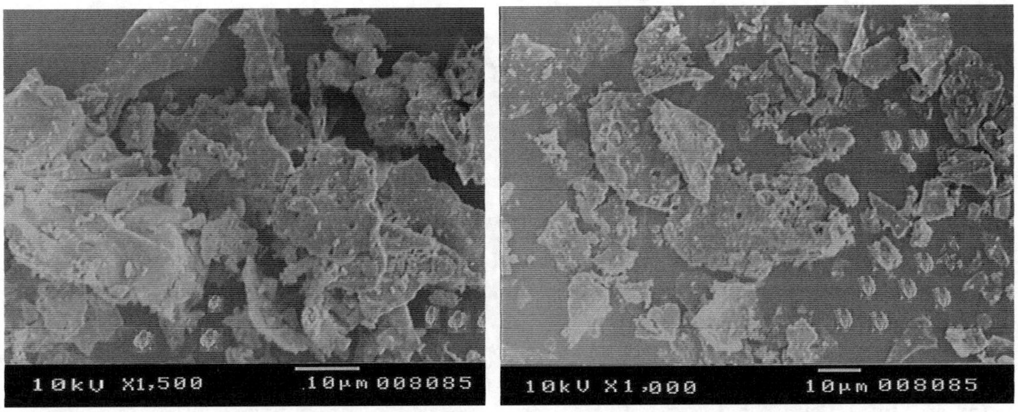


**Figure 2.** SEM morphological images of $KSrPO_4$ phosphor.

### *3.3 Photoluminescence studies of $KSrPO_4$: $Ce^{3+}$ blue emitting phosphor*

Figure 3 and 4 shows photoluminescent excitation and emission spectra of $KSrPO_4$:$Ce^{3+}$ phosphors displaying a broad absorption band in the range of 280 to 380 nm owning to the 4f–5d transformation of $Ce^{3+}$ ions peaking at 329nm. Photoluminescent emission spectra of $KSrPO_4$:$Ce^{3+}$ phosphor as seen in figure 4 exhibit a blue emission band centered at 440 nm. Normally, the structure of $Ce^{3+}$ ion in the ground state is split into 2 levels such as, $2F_{5/2}$ and $2F_{7/2}$, while the $^5d_1$ excited configuration is divided by a crystal playing area of between 2 and 5 elements. Within this research work, the emission spectra of the ready samples exhibit a large blue emission band in the range of 400-650 nm at a maximum of 440 nm. In compliance with the literature analysis [8-9-10], the transition of the electron from the excited state of $^2D_J$ to the $Ce^{3+}$ ion ground states such as, $^2F_{5/2,7/2}$ in the formulated $KSrPO_4$:$Ce^{3+}$ host materials is known, the double feature of trivalent cerium ion never detected in the emission spectra. This photoluminescent emission spectrum arising from the spectral convergence of the various emission peaks associated with trivalent cerium ions have been replaced by five Sr sites in the T segment. As a result, there are multiple 4f–5d absorption bands in the PLE spectra for $Ce^{3+}$, whereas the 5d-4f emission found in the photoluminescent emission spectrum is a typical double-band profile [11] based strongly on the crystallographic field. $Ce^{3+}$ ion emission is mostly in the ultraviolet or blue spectral region, but it is moved to the green to the yellow zone (e.g. $Y_3Al_5O_{12}$:$Ce^{3+}$), through the modulation of the crystallographic environment.

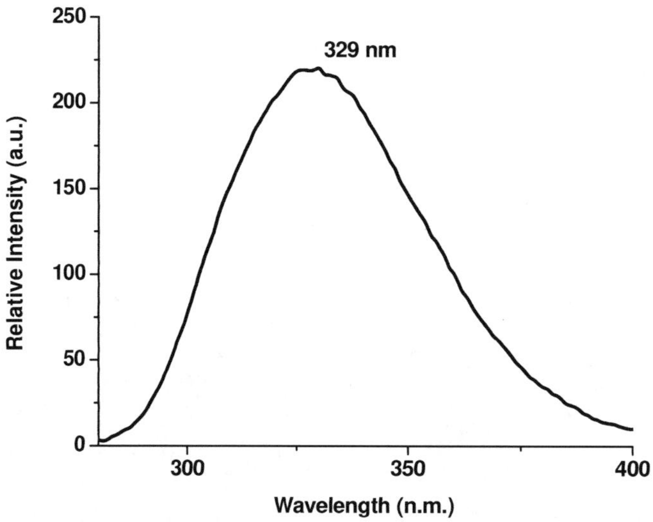


**Figure 3.** Excitation spectra for $KSrPO_4$:$Ce^{3+}$, $\lambda_{em}$:-440 nm.

Often, the characteristic emission of trivalent cerium ions at a particular lattice site occurs as doubling bands with transitions detected attributable to the relaxed low 5d excitation at $^2F_J$ (J=5/2,7/2) spin–orbit splitting 4f ground state. The energy separation of the two bands referring to the typical spin–orbit splitting of the 4f ground state (2000 $cm^{-1}$). The height of excitation, represented by the energy gap of the lowest 5d degree of excitation of trivalent cerium ion (329 nm) in the existing $Sr^{2+}$ host material relative to the free trivalent cerium ion (6.118 eV) in the existing situation as description in figure 4. The emission spectra are distinctly from the same detectable location. Almost no emission band has been found in the emission spectrum suggesting that trivalent cerium ion holds one group of locations in the host material [12].

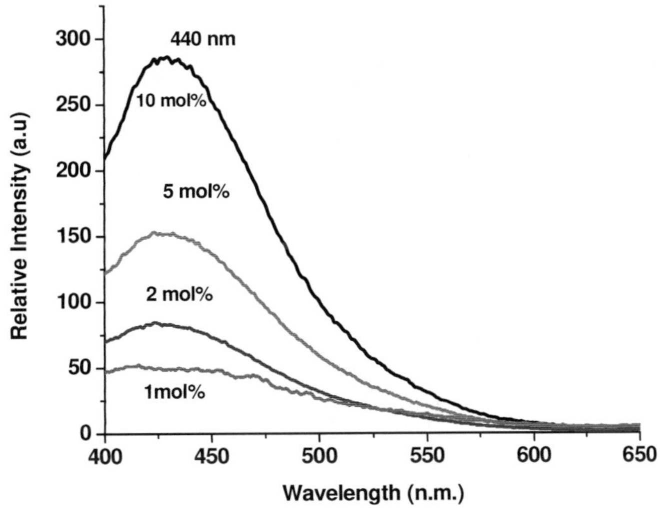


**Figure 4.** Emission spectra for $KSrPO_4:Ce^{3+}$, $\lambda_{ex}$:- 329 nm.

Hence $KSrPO_4:Ce^{3+}$ between blue emission is capable of finding useful uses as a blue emitting lamp phosphor. The substances $KSrPO_4:Ce^{3+}$ were synthesised by updated wet chemical synthesis process and triggered by alkaline earth $Ce^{3+}$ ion, varying from 1 to 10 mol percent with reference to corresponding source content. Crest emission wavelengths of processed fluoride-based aluminates relative to compounds produced at high temperature [13-14] are summarized by emission spectrum graphs. It will be examined that our prepared $KSrPO_4:Ce^{3+}$ blue emitting phosphors emits a mild blue change in emission spectra with commercially available phosphors and has prospective applications for the blue lamp lighting industry.

*3.4 Photoluminescence studies of $KSrPO_4$: $Eu^{3+}$ Orange-Red emitting phosphor*

Figure 5 explains the excitation spectrum of the $KSrPO_4:Eu^{3+}$ phosphor monitored at 618 nm. The excitation spectrum shows the characteristic bands that were located at 370 nm because of intra configurational 4f–4f transitions of $Eu^{3+}$ ions. Figure 6 demonstrate the emission spectrum of $KSrPO_4:Eu^{3+}$ phosphors. Emission spectra shows two emission peaks located at about 593 nm,618 nm which was assigned because of transitions $^5D_0$-$^7F_1$ and $^5D_0$-$^7F_2$ respectively. The maximum emission intensity was noted at 618 nm, recommended that the $Eu^{3+}$ ions engaged a middle of regularity in the $KSrPO_4$ host.

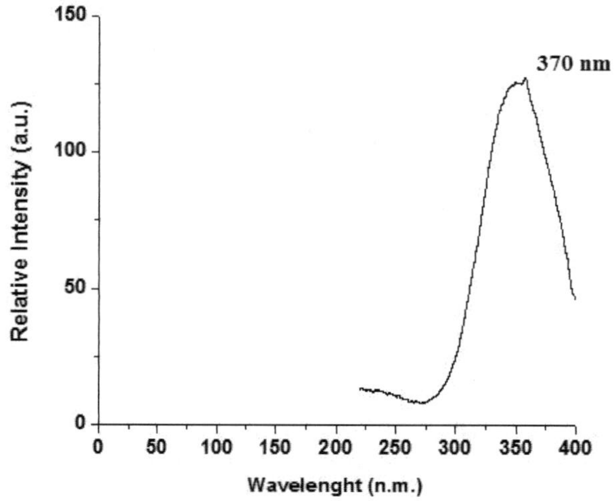


**Figure 5.** Excitation spectra for $KSrPO_4:Eu^{3+}$ phosphor.

It was noted that the as emission intensity improved with increasing $Eu^{3+}$concentration and maximum emission intensity observed at 1 mol%. So, the photoluminescence results obtained concluded that the prepared phosphor may have possible applications in the area of solid-state lighting.

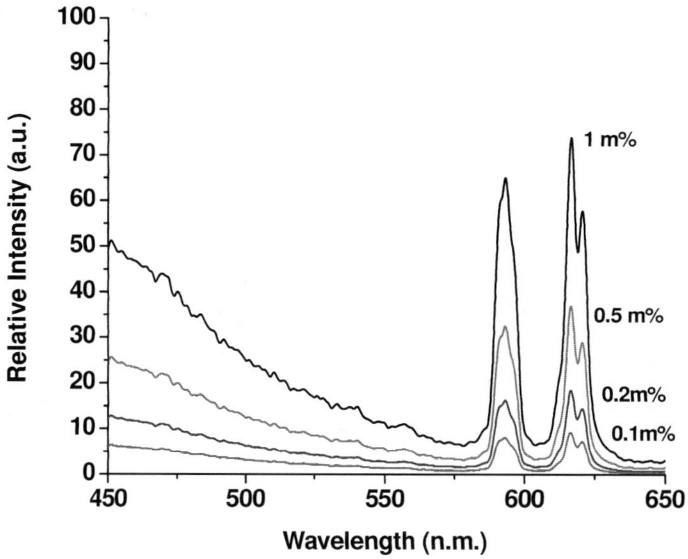


**Figure 6.** Emission spectra for $KSrPO_4:Eu^{3+}$, $\lambda_{ex}$:- 370 nm.

### *3.5 Chromatic properties of $KSrPO_4:Ce^{3+}$ and $Eu^{3+}$ phosphor*

We assume that when $Ce^{3+}$ ion is involved with the host lattice, the amount of energy may be transferred to the activator ion, arising from the distinctive special emission peak of these activator ions. [3-15] It is a well-known reality that the result obtained from the luminescence properties of prepared inorganic phosphors in powder form depending on the concentration of activator ion, hence recognition of doping concentration is necessary [14-15]. Considering the emission spectrum of $Ce^{3+}$ which was situated in the blue zone, the luminescent characteristics of $KSrPO_4:Ce^{3+}$ blue emitting phosphor were selected for further examination and characterization, and the total emission of color was achieved. Here we evaluate the chromaticity coordinate indexed by the $Ce^{3+}$ emission spectra.

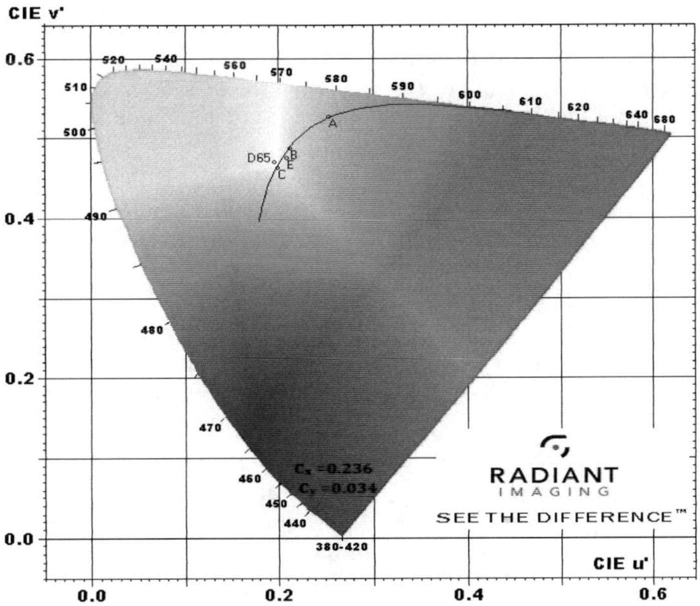


**Figure 7.** CIE chromatic diagram for $KSrPO_4:Ce^{3+}$ phosphor.

Usually lighting means colors that apply to lighting under chromatic color coordinate conditions that recognize the human visualization scheme using 03 major colors: that is blue, red and green [16-14]. CIE chromaticity schematic $KSrPO_4:Ce^{3+}$ blue phosphor emitting diagram seen in Figure 7.The color coordinates of $KSrPO_4:Ce^{3+}$ Phosphor found in the blue region ($C_x$ =0.236 $C_y$ =0.034) Phosphor while in case of $Eu^{3+}$ phosphor found in the orange –red region with the aid of the CIE diagram, it is simple to understand that the $KSrPO_4:Ce^{3+}$ and $Eu^{3+}$ phosphor are quite similar to the CIE graph frame, which is easy to show the highest color transparency of the prepared phosphor material. As a consequence of the emission and excitation process found by connecting these points in the tri-angle structure jointly with white light (0.31 and 0.32)], the go-between arrangement with a exacting percentage of prepared inorganic blue emitting phosphor [17-18]. Due to the growing consequence of $Ce^{3+}$ activated $KSrPO_4$ with desirable blue and orange –red emitting optical properties which make up the prepared inorganic blue emitting phosphor, it has become an outstanding prospective applicant for industrial blue emitting phosphor applications.

## 4.Conclusion

In this research work, $KSrPO_4:Ce^{3+}$ and $Eu^{3+}$ phosphor prepared using a wet chemical synthesis technique. High resolution electron micrographic scanning analysis reveals the presence of various tiny particles within the plate as well as normal crystalline grains in the micron to submicron scale. The emission bands obtained because of the $4f^65d^1 \rightarrow 4f^7$ transformation of $Ce^{3+}$ ion, which belongs to the deep blue spectrum region excited at 329 nm, near uv-excitation and in case of Eu $^{3+}$ ion emission spectra shows two emission peaks located at about 593 nm, 618 nm which was assigned because of transitions $^5D_0$-$^7F_1$ and $^5D_0$-$^7F_2$ respectively. Chromaticity coordinates specify the details location of the color rendering index with the help of emission spectra analysis. As a result, we reached the conclusion that our prepared inorganic phosphores have a prospective application for the Blue and orange-red Lighting Lamp Industry applications.

## References

[1] Zheng J, Guo C, Ding X, Ren Z and Bai J 2012 *J. of Current Applied Physics* **12** 643
[2] Panse V R, Kokode N S, Yerpude A N and Dhoble S J 2016 *Optik* **127** 1603
[3] Sun J, Sun Y, Lai J, Xia Z and Du H 2012 *J.Lumin.* **132** 3048
[4] Yerpude A N and Dhoble S J 2012 *J. Lumin.* **132** 1781
[5] Shinde K N and Dhoble S J 2011 *J. Fluoresc.* **21** 2053
[6] Lin H, Liu X R and Pun E Y B 2002 *J. Opt. Mater.* **18** 397
[7] Xia Z G, Du H Y, J.Y. Sun JY, Chen D M and Wang X F 2010 *Mater. Chem. Phys.* **119** 7
[8] Xiao L, Xie Y, He M, Chen Y, Li W and Yu W 2010 *J. Rare Earth* **28** 225
[9] Panse V R, Kokode N S, Dhoble S J and Yerpude A N 2016 Luminescence **31** 893
[10] Panse V R, Shukla A and Dhoble S J 2017 Optik **130** 539
[11] Panse V R, Kokode N S and Dhoble S J 2013 AIP Conference Proceedings **1569** 372
[12] Sowjanya G, Rama Moorthy L, Basavapoornima C and Jayasankar C K 2017 *J. Spectrochimica Acta Part A: Molecular and Biomolecular Spectroscopy* **170** 206
[13] Yerpude A N and Dhoble S J 2012 *J. Lumin.* **132** 1781
[14] Kumar A, Dhoble S J, Peshwe D R, Bhatt J, Terblans J J and Swart H C 2016 *Ceramics International* **42** 10854
[15] Dhoble S J, Kore B P, Yerpude A N, Kohale R L, Yawalkar P W and Dhoble N S 2015 *Optik* **126** 1527
[16] Yerpude A N and Dhoble S J 2012 *J. Lumin.* **132** 2975
[17] Yerpude A N and Dhoble S J 2013 *Optik* **124** 3567
[18] Panse V R, Kokode N S, Yerpude A N and Dhoble S J 2016 *Inter.J.of Photonics & Optical Tech.* **2** 21

# Structural and thermoluminescence property of $Y_6Ba_4(SiO_4)_6F_2$:Ce,Tb

**Pailendra Kumar Sahu and Sadhana Agrawal***

Department of Physics, National Institute of Technology Raipur, Raipur-492010,India

*Corresponding author's e-mail address: sagrawal.phy@nitrr.ac.in

**Abstract**. Present paper reports the synthesis, structural, morphological and thermoluminescence behavior of $Y_6Ba_4(SiO_4)_6$:Ce,Tb apatite phosphors prepared by solid state reaction method. X-ray diffraction pattern confirms the formation of $Y_6Ba_4(SiO_4)_6$:Ce,Tb into apatite structure. Morphological study was done by scanning electron microscope (SEM) and it reveals that the prepared phosphors have rod-like structure with different orientation. Energy dispersive spectroscope (EDX) confirms the presence of initial reactant. Thermoluminescence glow curve of the prepared phosphors were studied and its frequency factors, activation energy, order of kinetics etc are calculated and reported in the present paper.

## 1. Introduction

Inorganic compounds doped with rare earth ions have wide range of applications in the domain of luminescence like Photoluminescence, Electroluminescence, Cathodoluminescence, thermoluminescence etc. Thermoluminescence(TL) is observed from a material called phosphors [1]. Thermoluminescence arises from a phosphor when it is thermally stimulated subjected to prior irradiation of the phosphor by any form of light energy. Emission of light from the phosphor is due to the trapped charges and energy transfer mechanism. A TL glow curve contains the information regarding the mechanism involved. By analysis of TL glow curve various kinetic parameters are evaluated like order of kinetics (I order, II order or general order), Trap depth, frequency factor etc [2]. For the generation of TL from a phosphor three conditions must be satisfied-(i) Phosphor must be a semiconductor or insulator; metal do not exhibit TL. (ii) Phosphor must have exposed to some form of light energy, which results in the absorbance of this energy. (iii)Finally the TL is observed by heating the phosphors [3,4].

Apatite group of compound have general structural formula $A(I)_{10-x}A(II)_x(ZO4)_6Q2$ with space group $P6_3/m$. There are two distinct crystallographic sites namely 4f and 6h for A(I) and A(II) which are occupied by rare earth, alkali or alkaline. Phosphorous, vanadium, silicon, boron or germanium resides in Z site and Q denotes the anion site. Apatite compound has showed good results for the generation of cathodoluminescence and white LED's [5–7].

In the past few decades , many researchers have worked on the thermoluminescene phophors and have reported several phosphors such as potassium fluoride, $CaF_2$,$K_2YF_5$, $CaSrAl_2SiO_7$, $NaBaBO_3$, $K_2Ca_2(SiO_4)_3$ etc which are used in radiation dosimetry [8–12]. As per our literature survey TL properties of apatite compounds have not been widely explored. This motivated us for the experimental study of the thermoluminescence behaviour of $Y_6Ba_4(SiO_4)_6F_2$.

## 2. Experimental

$Y_6Ba_4(SiO_4)F_2$:0.6Ce,xTb phosphors were synthesized by keeping the ratio of cerium fixed at 0.6mol% whereas doping of terbium ion was taken as 0.2mol%, 0.4mol%,0.6mol%,0.8mol% and 1mol% by solid state reaction method. Initial reactant used for synthesize are $Y_2O_3$, $BaCO_3$, $SiO_2$, $NH_4F$, $CeO_2$ and $Tb_2O_3$. Stoichiometric amount of initial reactant were taken and weighed accordingly with excess amount of $NH_4F$ by 50mol% apart from the stoichiometric ratio to compensate the loss of fluorine ion during the synthesis process. By using acetone as a mixing agent all the initial reactant were thoroughly mixed in an agate mortar and pestle for about 30 minutes. After that the mixture were transferred in alumina crucible and kept in a muffle furnace at 1250°C for 4 hours with a heating rate of 5°c/s. The mixture were air cooled and was ground for 35 minutes by using agate mortar and pestle and again kept in a muffle furnace at 1500°C for 5 hours with a heating rate of 5°c/s and cooled naturally to room temperature. The obtained product was ground to make the final product in powder form.

## 3. Result and discussion

### *3.1 X-Ray diffraction pattern*

"3 kW PANalytical Xpert powder diffractometer" with generator voltage 40kV and tube current 30mA was used to determine the crystal structure and phase purity of the prepared phosphors parameter for XRD graph in 2Θ angle range from 20° to 70° with scan step size of 0.1 were taken. $La_6Ba_4(SiO_4)_6F_2$ with ICDD card number 98-017-0852 was taken as a reference pattern for matching the obtained XRD pattern [13]. The XRD pattern for undoped $Y_6Ba_4(SiO_4)_6F_2$ is shown in figure 1.

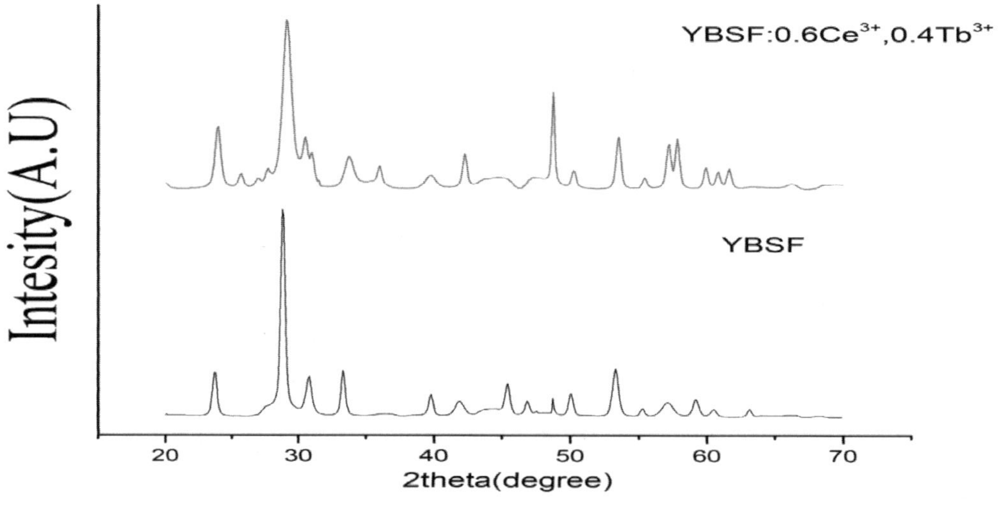


**Figure 1.** XRD graph.

The lattice parameter and unit cell volume was calculated for $Y_6Ba_4(SiO_4)_6F_2$ and $Y_6Ba_4(SiO_4)_6F_2$:0.6Ce,0.4Tb from the obtained graph. The lattice parameter for $Y_6Ba_4(SiO_4)_6F_2$ and $Y_6Ba_4(SiO_4)_6F_2$:0.6Ce,0.4Tb phosphors are a =3.43A & c = 5.445A and a = 4.3A & c = 6.8A. Axial ratio for undoped and doped samples are 1.587A and 1.581A respectively. There is no significant change in the axial ratio of the prepared phosphors and the Change in the lattice parameter is due to the presence of doping ions. The unit cell volume for undoped and doped phosphors are $55.477A^3$ and $108.887A^3$ respectively, which shows an increase in the volume of the unit cell due to the presence of cerium and terbium ions. The average crystalline size of the prepared phosphors was calculated from X-ray diffraction pattern by using debye scherrer formula [14] and was found to be 7.9nm for undoped and 11.08nm for $Y_6Ba_4(SiO_4)_6F_2$:0.6Ce,0.4Tb. Figure 2 shows the unit cell of $Y_6Ba_4(SiO_4)_6F_2$[13,15,16].

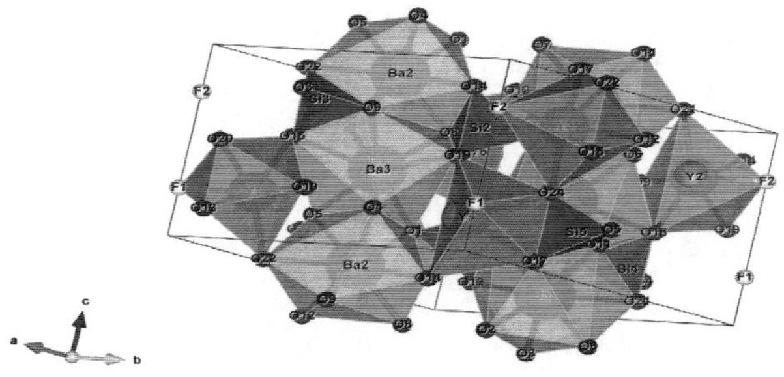

**Figure 2.** Unit cell of $Y_6Ba_4(SiO_4)_6F_2$

### *3.2 Scanning electron microscopy (SEM)*

Zeiss-Oxford SEM analyzer was used to obtain SEM image of the prepared phosphors. The SEM image of the prepared phosphors shows elongated rod-like structure randomly oriented in different direction as shown in figure 3 and the average particle size for $Y_6Ba_4(SiO_4)_6F_2$:0.6Ce,0.4Tb was calculated and is found to be 34μm.

**Figure 3.** SEM image of $Y_6Ba_4(SiO_4)_6F_2$:0.6Ce,0.4Tb.

### *3.3 Energy Dispersive X-ray spectroscopy (EDX)*

EDX spectrums of the prepared phosphors were obtained and the results confirms the present of all initial reactants. Figure 4 shows the EDX spectrum for $Y_6Ba_4(SiO_4)_6F_2$:0.6Ce,0.4Tb.

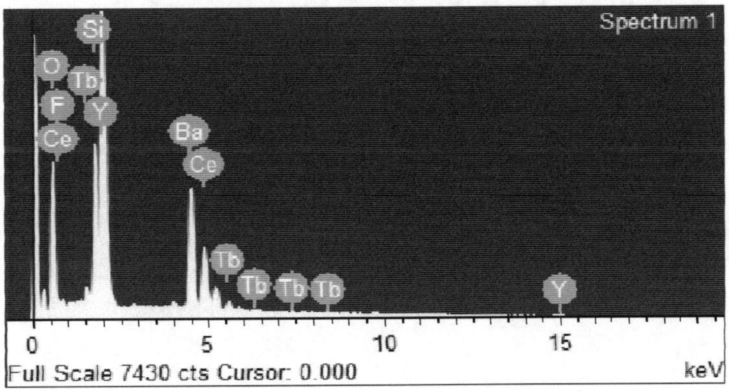


**Figure 4.** EDX spectrum.

*3.4 Thermoluminescence (TL) study*

TL study of the prepared phosphors YBSF:0.6Ce,xTb(x = 0.2,0.4,0.6,0.8 & 1mol%) was carried out by using Thermoluminescence Reader TL 1008 manufactured by Nucleonix system private limited. Samples were exposed to UV light of wavelength 254nm in UV chamber with exposure time of 2 min. Constant heating rate(β) of 3°C/s was kept and samples were heated from room temperature to 450°C with a holding time of 10 second at 450°C.Parameters like activation energy E(eV), order of kinetics(b) etc was calculated by using methods proposed by chen (Equation 1). Figure 5 represents a TL glow curve and definition of symbol used during the analysis of TL glow curve.

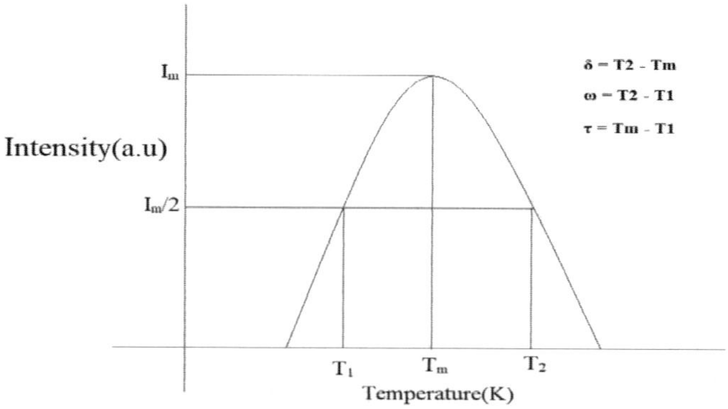


**Figure 5**. Schematic view of TL glow curve.

Chen's formula

$$E = C_\eta \frac{KT_m^2}{\eta} - b_\eta(2KT_m) \qquad (\eta=\delta, \tau \text{ and } \omega) \tag{1}$$

Values of $C_\delta$, $C_\tau$, $C_\omega$, $b_\delta$, $b_\tau$ and $b_\omega$ are obtained by equation given below

$$C_\tau=1.51 + 3(\mu_g\text{-}0.42) \tag{2}$$
$$C_\delta=0.976 +7.3(\mu_g\text{-}0.42) \tag{3}$$
$$C_\omega=2.52 + 10.2(\mu_g\text{-}0.42) \tag{4}$$
$$b_\tau=1.58 + 4.2(\mu_g\text{-}0.42) \tag{5}$$
$$b_\delta = 0 \tag{6}$$
$$b_\omega = 1 \tag{7}$$

Glow Curve Deconvolution (GCD) method [1,17,18] was used to determine the various parameters of the TL glow curve.YBSF:0.6Ce,0.4Tb and YBSF:0.6,1Tb showed better TL glow curve in comparison to YBSF:0.6Ce,xTb(x0.2,0.6,0.8mol%).Figure 6 represents the deconvulated glow curve of $Y_6Ba_4(SiO_4)_6F_2$:0.6Ce,0.4Tb and table 1 shows its various calculated parameter from chens formula. Figure 7 represents the deconvulated glow curve of $Y_6Ba_4(SiO_4)_6F_2$:0.6Ce,1Tb and table 2 shows its various calculated parameter from chens formula. TL glow curve of $Y_6Ba_4(SiO_4)_6F_2$:0.6Ce,0.4Tb and $Y_6Ba_4(SiO_4)_6F_2$:0.6Ce,1Tb shows second order kinetics and the average activation energy calculated are 1.115ev and 0.99ev respectively.

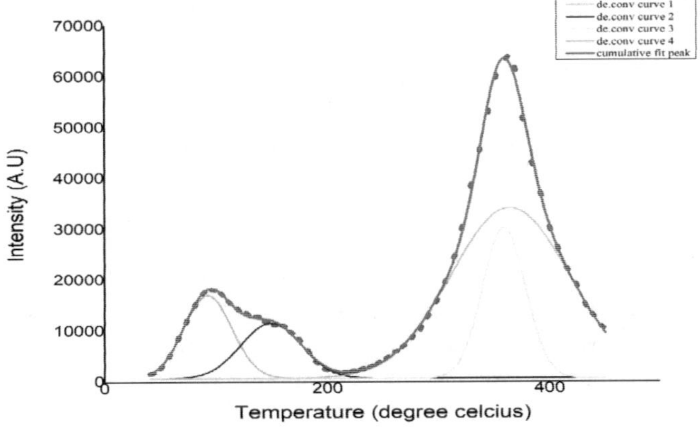


**Figure 6.** Deconvulated TL glow curve of YBSF:0.6Ce,0.4Tb

**Table 1.** Parameters of TL glow curve for YBSF:0.6Ce,0.4Tb.

| | Tm(K) | T1(K) | T2(K) | δ(K) | ω(K) | τ(K) | μg | γ | Activation Energy(ev) |
|---|---|---|---|---|---|---|---|---|---|
| **Peak 1** | 365.82 | 340.8 | 394.23 | 28.41 | 53.43 | 25.02 | 0.53 | 1.13 | 0.71 |
| **Peak 2** | 423.41 | 391.165 | 456.91 | 33.5 | 65.745 | 32.245 | 0.50 | 1.03 | 0.65 |
| **Peak 3** | 633.42 | 610.13 | 656.29 | 22.87 | 46.16 | 23.29 | 0.49 | 0.98 | 2.34 |
| **Peak 4** | 638.24 | 575.36 | 701.82 | 63.58 | 126.46 | 62.88 | 0.50 | 1.01 | 0.76 |

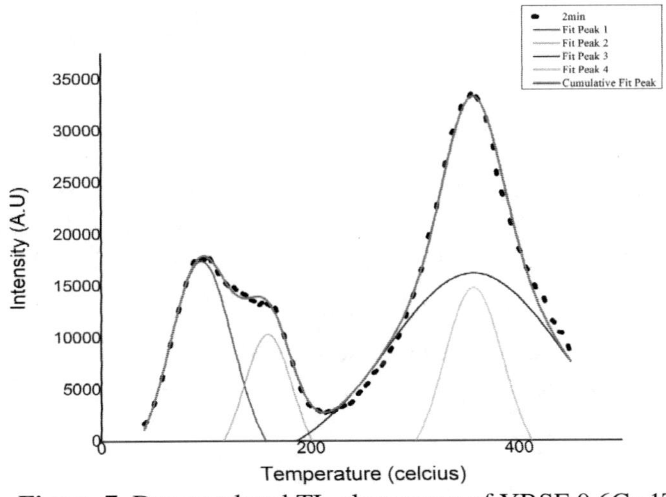


**Figure 7.** Deconvulated TL glow curve of YBSF:0.6Ce,1Tb.

**Table 2.** Parameters of TL glow curve for YBSF:0.6Ce,1Tb.

| | Tm(K) | T1(K) | T2(K) | δ(K) | ω(K) | τ(K) | μg | γ | Activation Energy(ev) |
|---|---|---|---|---|---|---|---|---|---|
| **Peak 1** | 369.462 | 339.13 | 401.23 | 31.77 | 62.1 | 30.33 | 0.51 | 1.04 | 0.565 |
| **Peak 2** | 433.01 | 409.91 | 456.57 | 23.56 | 46.66 | 23.1 | 0.50 | 1.01 | 1.08 |
| **Peak 3** | 630.79 | 540.47 | 721.11 | 90.32 | 180.64 | 90.32 | 0.5 | 1 | 0.45 |
| **Peak 4** | 630.32 | 601.33 | 660.57 | 30.25 | 59.24 | 28.99 | 0.51 | 1.04 | 1.88 |

## 4. Conclusion

$Y_6Ba_4(SiO_4)_6F_2$:0.6Ce,xTb apatite structured phosphors were successfully synthesized by solid state reaction method. Prepared phosphors has rod like structure. TL glow curve of the synthesized sample shows second order kinetics with average activation energy of 1.115ev and 0.99.The prepared phosphors can be used in the field of radiation dosimetry.

## References

[1] Salah N, Sahare P D and Kumar P 2006 *Phys. Status Solidi Appl. Mater. Sci.* **203** 898
[2] Dubey V, Suryanarayana N S and Kaur J 2010 Kinetics of TL Glow Peak of Limestone from Patharia of CG Basin ( India ) **9** 1101
[3] Bøtter-Jensen L and McKeever S W S 1996 and *Radiat. Prot. Dosimetry* **65** 273
[4] McKeever S W S, Bøtter-Jensen L, Larsen N A and Duller G A T 1997 *Radiat. Meas.* **27** 161
[5] Sio B 2014 *J. Lumin.* **145** 65
[6] Wang J, Lin H, Huang Q, Xiao G, Xu J, Wang B, Hu T and Wang Y 2017 *J. Mater. Chem. C* **5** 1789
[7] Zhu G, Shi Y, Mikami M, Shimomura Y and Wang Y 2013 *Opt. Mater. Express* **3** 229
[8] Ogundare F O, Folley D E, Chithambo M L and Arise T O 2020 *Nucl. Inst. Methods Phys. Res. B* **482** 53
[9] Sharma S, Brahme N and Dewangan D P B P 2018 *J. Mater. Sci. Mater. Electron.* **29** 1412
[10] Oglakci M, Topaksu M and Can N 2020 *Sensors Actuators A. Phys.* **315** 112299
[11] Mandlik N T, Sahare P D, Dhole S D and Balraj A 2020 *Nucl. Inst. Methods Phys. Res. B* **480** 105
[12] Faria L O, Lo D, Kui H W, Khaidukov N M and Nogueira M S 2004 *Radiation Protection Dosimetry* **112** 435
[13] Gong X, Lin Y, Chen Y, Huang Z, Huang Y and Luo Z 2005 *Chem. Mater.* **17** 1135
[14] Palan C B, Bajaj N S and Omanwar S K 2016 *Bull. Mater. Sci.* **39** 1619
[15] Ma X, Liao L, Guo Q, Liu H, Zhou T and Mei L 2018 *RSC Adv.* **8** 27332
[16] Qi S, Huang Y, Tsuboi T, Huang W and Seo H J 2014 *Opt. Mater. Express* **4** 5505
[17] January I- 2015 International Research Journal of Natural and Applied Sciences **1** 178
[18] Tang K, Cui H, Wang Y, Zhu H, Fan Q and Mao Y 2008 *Radiat. Meas.* **43** 1187

IOP Conf. Series: Materials Science and Engineering 1120 (2021) 012007 doi:10.1088/1757-899X/1120/1/012007

# Study of TL characteristics of $CaSO_4$: Dy phosphor with fly ash for thermoluminescence dosimeter application

**Umakant Shrivastava[1], Vikas Mishra[1], Narottam Purle[1], Mohammad Ziyauddin Khan[1], Rahul Pali[1], DVK Narasingham[1], Sukanto Sarkar[2] and Bhusan Chandel[2]**

[1] Department of Physics, Dr. C. V. Raman University, Kota Bilaspur (C.G.) India
[2] Ultratech Laboratories Pvt. Ltd., (*Accredited by BARC, Mumbai*) Kumhari, Durg, (C.G.), India

*Corresponding author's e-mail address: ukscvru@rediffmail.com

**Abstract.** In recent past fly ash is only being treated as waste and a source of air and water pollution but is in fact an eco - friendly and economically beneficial resource material for various applications and has also proven its worth over a period of time. The utilization of fly ash in several novel researches will also paves the path for systematic disposal of this waste material. The present paper involves the study of Thermoluminescence (TL) of different compositions of fly ash collected from National Thermal Power Corporation, Korba, Chhattisgarh, INDIA with ($CaSO_4$:Dy) Phosphor. Thermo luminescence Dosimeters are regarded as the badge which absorbs the radiation when persons having TLD badges are exposed to radiation. We have to investigate and compare the glow curve behavior of pure phosphor and 10 %, 20 % and 30 % fly ash doped phosphor when it is exposed to gamma irradiation $Cs^{137}$.

## 1. Introduction

The phenomenon of thermoluminescence (TL) is widely used for measurement of radiation doses from ionizing radiations viz. x-rays, $\gamma$ rays and $\beta$ particles. Dosimeters based on luminescence phenomenon are also used for measurement of radiation doses from neutrons. The application of the phenomena of TL for the dosimetry purpose was first proposed by an American scientist F. Daniels. Subsequently, various phosphors were developed by various laboratories and manufacturing of these phosphors was undertaken by various commercial companies. These are being used in personnel, environmental and medical dosimetry. TL is the emission of light from a wide band gap material called phosphor during heating, which has already been exposed to ionizing radiations (x, $\beta$, $\gamma$ rays and neutrons). During irradiation of phosphor, the electron-hole pairs are generated, which while mobile in the phosphor can be trapped at the metastable energy levels which exist in the forbidden gap due to vacancies, defects or by the addition of impurities /dopants to the host material. [1]

When ionizing radiation is incident on an insulating crystal some of the deposited energy is stored in the lattice at defect sites, colour centres, etc. Upon heating the crystal, this stored energy is released and a fraction of it may be emitted as visible light. This is called thermoluminescence. Within certain limitations, the amount of light emitted is proportional to the radiation dose previously absorbed by the TL material.

Various types of materials can be examined for both low dose dosimetry and high dose dosimetry. Low dose dosimeters can be used for medical applications while high dose dosimeters are useful in various fields such as medical & pharmaceutical industries, research laboratories based on high energy accelerators, etc. Hence, we are here focusing on the examination of low dose behavior of Gamma Irradiation source on the phosphor sample mixed with different percentages of fly ash.

## 2. Experimental

The fly ash used in this experiment was collected from the Super Thermal Power Station, Korba which is situated on the banks of the confluence of rivers Hasdeo and Ahiran. Korba is the power capital of Chhattisgarh State in India. The source of coal for the power plant is Kusmunda & Gevara Mines. There are total five samples prepared in this investigation. Out of five four samples are based on fly ash and one is based on pure phosphor. $CaSO_4$: Dy phosphor and Teflon mixture in the weighted ratio of 1:3 is prepared using Liquid Nitrogen by Solid State Sintering Method. $CaSO_4$: Dy Phosphor and $CaSO_4$: Dy Phosphor is doped with 10%, 20% and 30% of fly ash and pure fly ash sample are exposed to Gamma Irradiation source of ($Cs^{137}$) varies from low dose range of 1000-4000 mR or 8.77 mGy to 35.08 mGy. TL glow curves of $\gamma$-ray irradiated samples were obtained using a Nucleonix 1009I TL reader in the temperature range of 20 °C to 300 °C with a heating rate of 5 °C/sec.

## 3. Result analysis and discussion

### *3.1 XRD Analysis*

The XRD properties of the samples were obtained using a Bruker D2 Phaser diffractometer. The value for 2θ in a range from 10° to 80° is taken with an increase of 0.01° per sec. The XRD diffraction of fly ash doped $CaSO_4$: Dy phosphor is shown in figure 1(a) below. This XRD data is compared with standard JCPDS data of 100% fly ash in figure 1(b) and $CaSO_4$: Dy Phosphor in figure 1(c) and it is almost well matched with it. Due to some impurity or system's issue noise is created in sample [2]. The "d" value is calculated by the Bragg's formula by assuming λ value for X-rays.

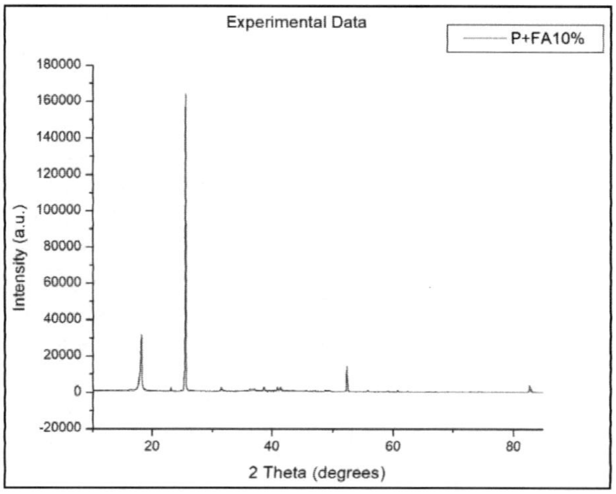


**Figure 1.** (a) Experimental XRD plot of P+FA10% sample.

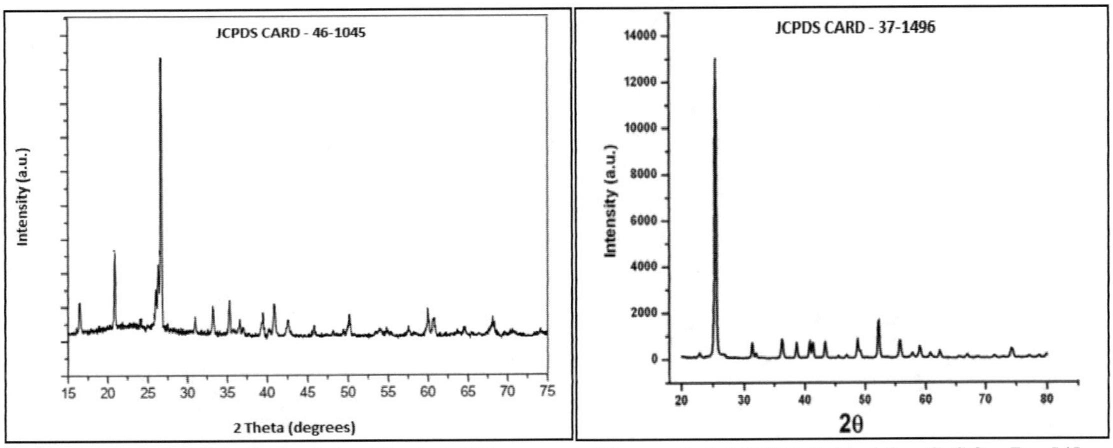


**Figure 1. (b)** JCPDS Card for Fly Ash [3] **Figure 1. (c)** JCPDS Card for $CaSO_4$: Dy [4]

**Figure 1.** XRD plot of experimental data and JCPDS cards of fly ash and $CaSO_4$:Dy phosphor.

The matching of JCPDS Cards shows orthorhombic crystalline phase is indicated by planes of $CaSO_4$:Dy. Only some major peaks are observed due to the mixing of fly ash in the $CaSO_4$: Dy Phosphor. Three major peaks are observed in the XRD pattern indicating the presence of main component of fly ash like $SiO_2$, $Al_2O_3$, and CaO phases which are dominant in the fly ash composition.

*3.2 TL Glow curve Analysis*
The TL techniques are used worldwide for the determination of radiation dose. The TL intensity should be linear with increasing the duration of radiation exposure in such a way that the value of doses absorbed by the material could be easily predicted. The below figures 2(a), 2(b), 2(c) & 2(d) shows the glow curves of 5 different samples given in the Table: 1 at different $\gamma$-ray irradiation doses i.e., 8.77mGy, 17.54mGy, 26.31mGy and 35.08mGy respectively of $Cs^{137}$.

In figure 2(a) all the samples of fly ash and phosphor are exposed to 8.77 mGy (1000mR) dose of $Cs^{137}$. Here we have recorded highest peak for P+FA20% sample and after that P+FA10% sample analogy with the peaks of traditional phosphor material. In Figure 2(b) all samples were exposed to 17.54 mGy (2000mR) dose. Here again P+FA20% sample shows highest intensity peak but not in analogy with phosphor material. While peak of sample P+FA30% possess nearby intensity with phosphor material and sample P+FA10% shows constant peak as it posses during the 8.77 mGy exposure.

Figure 2(c) shows that, with increment in the dose the intensity also increases with the samples with the increasing percentage of fly ash in it. All the samples were showing clear individual peaks at 26.31 mGy (3000mR) dose. Highest peak is recorded with the sample P+FA30%, then P+FA20% shows higher intensity while sample P+FA10% shows a little bit hump in comparison with the traditional phosphor. While in the figure 2(d) all the fly ash samples shows lower intensity in analogy with the phosphor when exposed to 35.08 mGy (4000mR). P+FA30% shows quite clear single peak with a little hump in the intensity.

**Table 1.** Composition of various samples.

| S.No. | Sample Name | Composition | |
|---|---|---|---|
| | | Phosphor | Fly Ash |
| 1 | PP | 100 % | 0 % |
| 2 | PFA | 0 % | 100 % |
| 3 | P+FA10% | 90 % | 10 % |
| 4 | P+FA20% | 80 % | 20 % |
| 5 | P+FA30% | 70 % | 30 % |

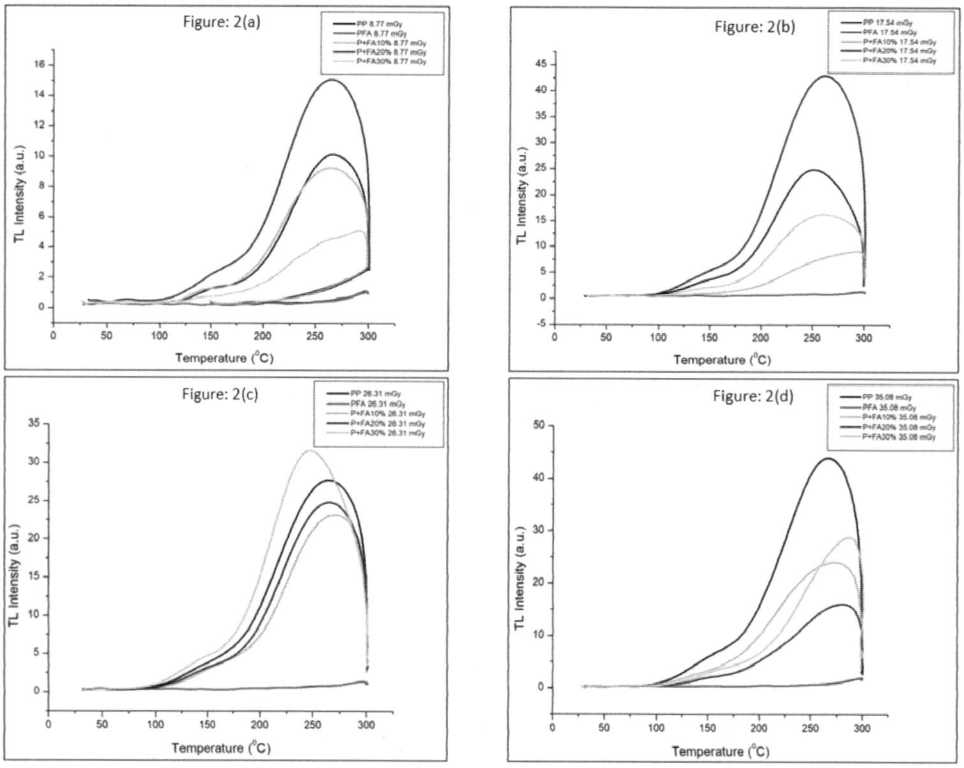


**Figure 2.** TL Glow Curves of Phosphor and Fly Ash samples at different γ-doses ranging from 8 - 35 mGy.

For sample P+FA10%, figures 3(a), 3(b) and 3(c) shows nice linear and individual peaks and figures 3(a) and 3(c) shows nearly the same intensity as it is desired intensity at a particular given dose. Apart from it The figures 3(d) and 3(f) of sample P+FA30% shows very nice single peaks but behaves considerably low with respect to the exposure of desired doses.

Figure 4 show the response of sample P+FA10% at all different doses. We can see in figure 4 that at low doses of 8.77 mGy and 17.54 mGy sample show intensity nearly equal to the desired value. While even at 26.31 mGy and 35.08 mGy we have got quite clear and individual peaks.

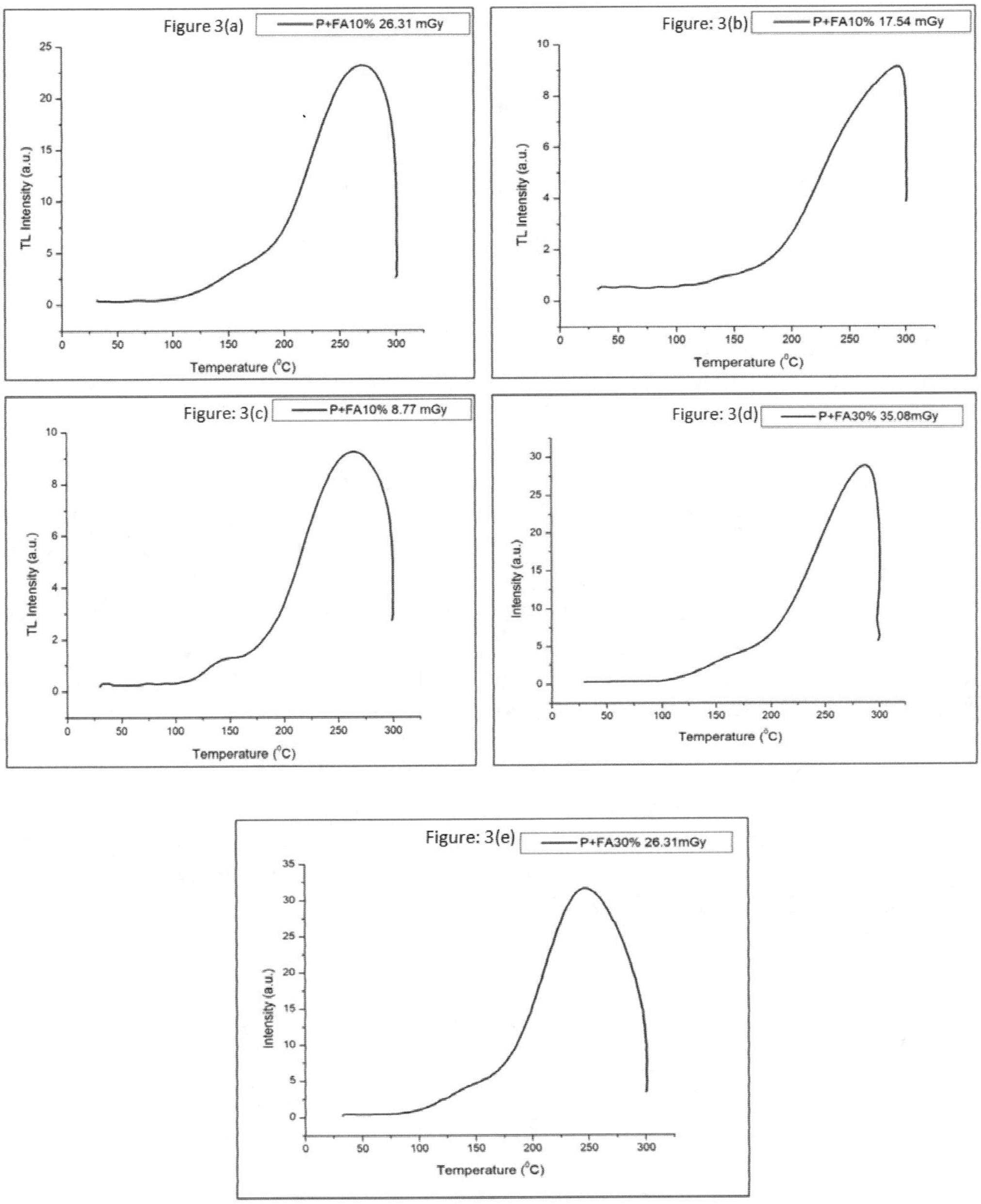


**Figure 3.** Individual Glow Curve peaks of sample P+FA10% and sample P+FA30% at different doses.

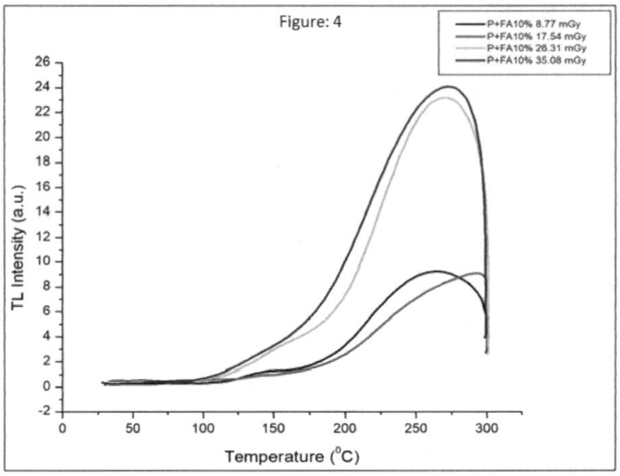


**Figure 4.** Glow curve of P+FA10% sample at different doses ranging from 8.77 mGy to 35.08 mGy.

### *3.3 TL Dose Response Curve Analysis*

The linear response arises from the recombination of localized entity which is an electron/hole pair and acts as a trapping center. The unoccupied electron/hole trap is suppressed during the linear response of the TL glow curve. Variation in the peak temperature position was observed with a change in radiation dose value. [5]

The linear fitting curve and TL dose response curve of fly ash, phosphor and fly ash composites exposed to low dose range of ~ (8 to 35) mGy respectively are shown in figures 5(a) and 5(b). Here PFA shows very nice linearity in initial and end points and also shows probability of showing positive response with respect to higher dose than 35.08 mGy as the intensity goes on increasing with increasing doses. Here in figures 5(a) and 5(b) sample P+FA10% shows linear patch near lower temperature zone and at high temperature zone as well but it started bending toward low intensity nearby 35.08 mGy. Sample P+FA20% initially loses its linearity and shows higher intensity than the phosphor material but in later part after 26.31 mGy it again get back into linear patch as sample PP in figure 5(a). Sample P+FA30% shows a small hump in between but apart from it shows linear curve with highest peak at higher dose value.

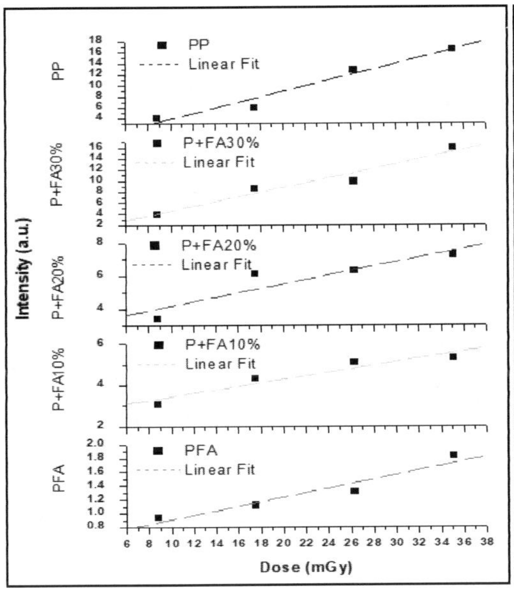


Figure 5 (a)

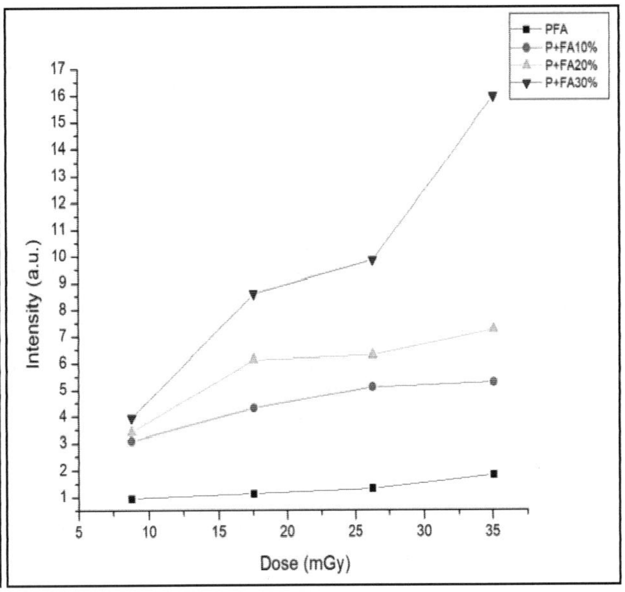


Figure 5 (b)

**Figure 5.** Linear fit and TL dose response curve of all samples to different γ-doses ranging from 8 - 35 mGy at 300 $^0$C temperature.

The number of luminescence and trapping centers has increased linearly inside sample P+FA10% and P+FA30% with increasing the dose values and has given rise to enhanced TL intensity. [5] Thus, a good fit linear curve of both samples from figure 5(a) and figure 5(b) suggests that P+FA10% and P+FA30% is likely to favorable material for used as an TL dosimeter at low dose range from ~ (8 to 35) mGy.

*3.4 Analysis of Trapping Parameters*

Several models have been applied to describe the TL process that arose in materials. The easiest one, named first order kinetics, assumes a thermal release of an electron from trap centers is followed by recombining with a hole in the recombination centers. In this model, the probability of the electron being retrapped is assumed to be impossible. While the second order model considers two possibilities are granted to the electrons upon the thermal release: recombine with holes or retrap again. An empirical model has been suggested to describe the situation when the TL peak of neither first order nor second order are satisfied, named general order kinetics. [6] Specific details of the trap and recombination centers provide valuable information about the material properties, defects, and history of radiation absorbed. This information is found in the trap parameters of the centers.

The parameters include the activation energy $E_A$ (eV), the kinetic order of TL processes b, the initial concentrations of trapped charges $n_0$ (cm$^{-3}$), the frequency factor S (s$^{-1}$), The TL glow curves are usually analyzed also by assuming different b values using many different methods. We can apply Glow Curve Deconvolution (GCD) method, Chen's Peak Shape method, Initial Rise (IR) method, Three Point Analysis (TPA) method, Whole Glow Curve method, Peak Position method, Isothermal Decay method, Curve Fitting method and many more but here we are focusing on initial four methods described above in this paragraph.

IOP Conf. Series: Materials Science and Engineering 1120 (2021) 012007 doi:10.1088/1757-899X/1120/1/012007

*3.4.1 Chen's Peak Shape Method*

Using this method we can calculate the activation energy ($E_A$), frequency factor (S), kinetic order (b), lifetimes of the peaks is calculated. Here we need to estimate three temperatures required for Chen's Peak Shape equations those are $T_M$, $T_1$, $T_2$ where $T_1$ and $T_2$ were temperatures at half intensity of glow curve in both the side of TM which is the maximum peak temperature. Using the value of $\tau$, $\delta$, and $\omega$ which can be evaluated as $\tau = T_M - T_1$, $\delta = T_2 - T_M$ and $\omega = T_2 - T_1$ respectively. Then we can find the geometrical shape factor $\mu = \delta/\omega$.

The value of $E_A$ can be calculated using the values of $E_\tau$, $E_\delta$ and $E_\omega$ where

$E_\tau = (1.51kT_M^2)/\tau - 1.58(2kT_M)$ (1)

$E_\delta = (0.976kT_M^2)/\delta$ (2)

$E_\omega = (2.52kT_M^2)/\omega - 2kT_M$ (3)

Where $T_M$ = peak temperature at the maximum TL intensity, $T_1$, $T_2$ = temperatures on either side of $T_M$, corresponding to the half-maximum intensity and k is the boltzman's Constant.

The calculation of trapping parameters using Chen's Peak Shape method is shown in the Table: 2 below. [7]

**Table 2.** Trapping parameter by Chen's peak shape method of the fly ash sample exposed to $Cs^{137}$ γ-irradiation.

| Sample | Peak Temperature ($^0C$) | Order of Kinetics (b) | Geometrical Form Factor (μ) | Activation Energy (eV) | Frequency Factor S ($s^{-1}$) | Lifetime of Peak (τ) |
|---|---|---|---|---|---|---|
| P+FA30% 35.08 mGy | 287.15 | 1.47 | 0.19 | 1.16 | $3.85x10^{11}$ | $8.02\ x10^{-2}$ |
| P+FA30% 26.31 mGy | 249.87 | 2.74 | 0.48 | 0.77 | $1.08x10^6$ | $12.3\ x10^{-2}$ |
| P+FA20% 17.54 mGy | 261.19 | 2.2 | 0.42 | 0.61 | $4.71x10^6$ | $12.3\ x10^{-2}$ |
| P+FA20% 8.77 mGy | 266.61 | 2.22 | 0.38 | 0.78 | $1.75x10^8$ | $10.1\ x10^{-2}$ |
| P+FA10% 26.31 mGy | 269.22 | 1.73 | 0.37 | 0.68 | $1.84x10^7$ | $12.1x10^{-2}$ |
| P+FA10% 17.54 mGy | 292.69 | 0.8 | 0.1 | 1.53 | $6.21x10^{14}$ | $6.6\ x10^{-2}$ |
| P+FA10% 8.77 mGy | 265.8 | 1.43 | 0.38 | 0.63 | $5.22x10^6$ | $13.5\ x10^{-2}$ |

*3.4.2 Initial Rise Method*

This method is based on the approximation that total rate change of trapped charges in initial rise part for temperature less than maximum peak temperature $T_M$ of any glow curve is constant with respect to change in temperature. This method is only application on single glow peak of any curve. Here activation energy is independent of order of kinetics thus TL intensity I is proportional to exp (-E/kT) and can be written as

$I(T) = R^2 \exp(-E_A/kT)$ (4)

Where I(T) is the intensity with temperature T, k is the boltzman's Constant, E is the activation energy, T is the temperature of glow curve and $R^2$ is the regression constant having value of 0.9997.We can get the value of activation energy $E_A$ by getting a slope in the plot of ln I(T) vs (-E/kT). First 11 data sets were taken to plot the graph and the value of glow curve peak temperature having intensity I was chosen in such a way that it belongs to (10-15) % of maximum peak intensity $I_M$. [5]

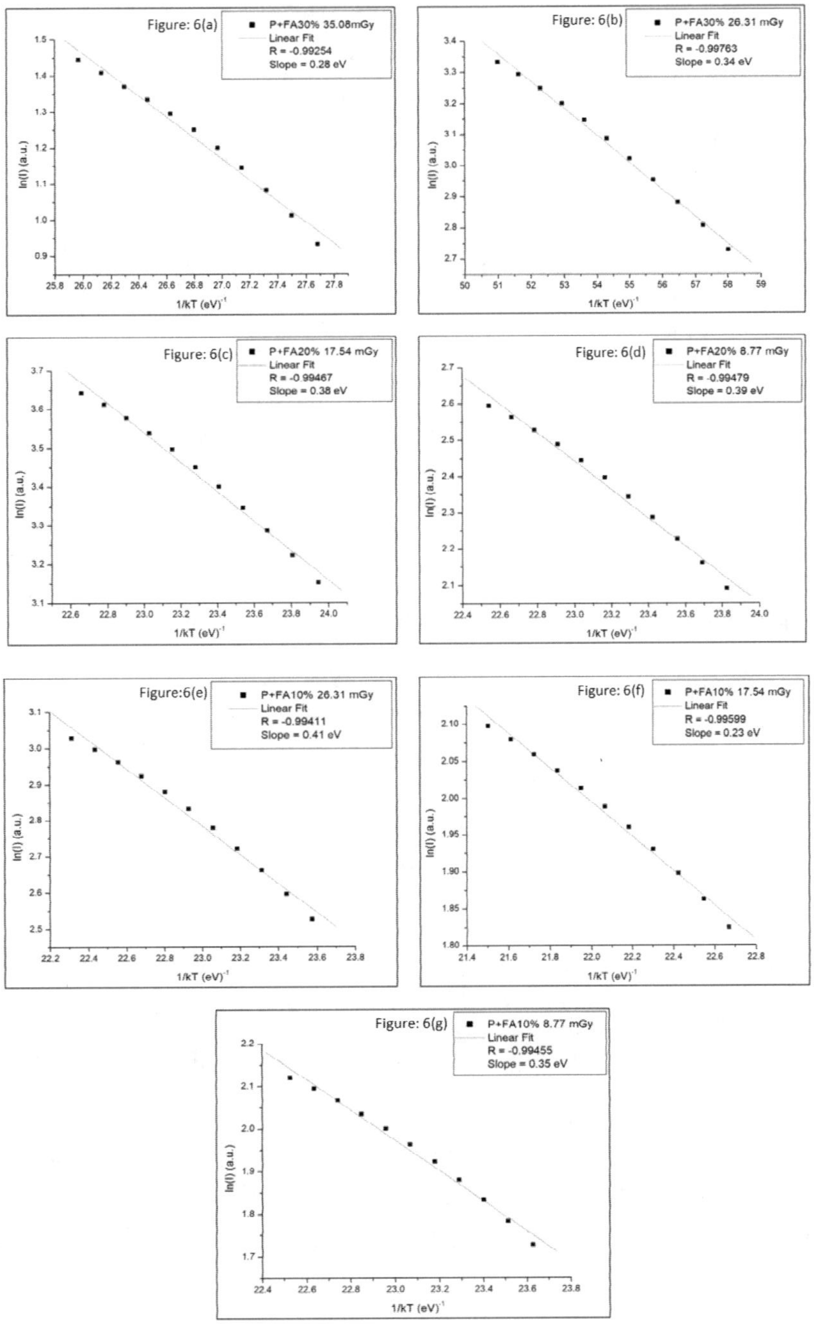


**Figure 6.** Plot of 1/kT Vs ln(I) for Initial Rise Method of different samples exposed to different γ-Irradiation $Cs^{137}$ source.

**Table 3.** Trapping parameter by IR method of the fly ash sample exposed to $Cs^{137}$ γ-irradiation.

| Sample | Peak Temperature (°C) | Activation Energy (eV) |
|---|---|---|
| P+FA30% 35.08 mGy | 287.15 | 0.28 |
| P+FA30% 26.31 mGy | 249.87 | 0.45 |
| P+FA20% 17.54 mGy | 261.19 | 0.38 |
| P+FA20% 8.77 mGy | 266.61 | 0.39 |
| P+FA10% 26.31 mGy | 269.22 | 0.41 |
| P+FA10% 17.54 mGy | 292.69 | 0.23 |
| P+FA10% 8.77 mGy | 265.8 | 0.36 |

The calculation of trapping parameters using Initial Rise method is shown above in the Table 3.

*3.4.3 Glow Curve Deconvolution (GCD) Method*

It is quite difficult to analyze a glow curve having more than one TL glow peak instead of single peak glow curve. In order to minimize and simplify the difficulty some mathematical expressions are defined by Vasilis Pagonis, George Kitis and Claudio Furetta of first, second and general order kinetics so that we can estimate theoretical values of activation anergy ($E_A$), order of kinetics (b) and frequency factor (S) those are:

For first order of kinetics

$$I(T) = I_M \exp\left[1 + \left(\frac{E}{kT} \cdot \frac{T-T_M}{T_M}\right) - \frac{T^2}{{T_M}^2}\left(1-\frac{2kT_M}{E}\right) \times \exp\left(\frac{E}{kT} \cdot \frac{T-T_M}{T_M}\right) - \frac{2kT_M}{E}\right] \quad (5)$$

For second order of kinetics

$$I(T) = 4I_M \exp\left(\frac{E}{kT} \cdot \frac{T-T_M}{T_M}\right) \times \left[\frac{T^2}{{T_M}^2}\left(1-\frac{2kT_M}{E}\right) \exp\left(\frac{E}{kT} \cdot \frac{T-T_M}{T_M}\right) + 1 + \frac{2kT_M}{E}\right]^{-2} \quad (6)$$

and for general order kinetics

$$I(T) = I_M\, b^{\frac{b}{b-1}} \left(\frac{E}{kT} \cdot \frac{T-T_M}{T_M}\right) \left[1 + (b-1)\frac{2kT_M}{E} + (b-1)\left(1-\frac{2kT_M}{E}\right) \left\{\frac{T^2}{{T_M}^2} \exp\left(\frac{E}{kT} \cdot \frac{T-T_M}{T_M}\right)\right\}\right]^{-\frac{b}{b-1}} \quad (7)$$

$$\mathrm{FOM} = \frac{\sum_p |y_{experimental} - y_{fit}|}{\sum_p y_{fit}} \quad (8)$$

Where I(T) is the TL peak intensity calculated from above equations at temperature T, $I_M$ is the highest TL peak intensity at a particular temperature $T_M$ and k is boltzman's constant. [8]

Equations (5), (6) and (7) are used to generate a theoretical glow curve on the basis of some parameters like highest peak temperature of single peak glow curve, order of kinetics and activation energy. TL parameters of theoretical glow curve can be estimated by using Chen's Peak Shape method and Initial Rise method.

The value of activation energy ($E_A$) calculated by Chen's Peak Shape method can be substituted in the equation (9) to find the value of frequency factor (S). The equation for S is

$$S = \frac{\beta E}{KT_M^2} \exp\left(\frac{E}{KT_M}\right) \left[1 + (b-1)\frac{2KT_M}{E}\right] \quad (9)$$

All the calculated trapping parameters using GCD method is given below in the Table 4.

**Table 4.** Trapping parameter by GCD Method of the fly ash samples exposed to $Cs^{137}$ $\gamma$-irradiation.

| Sample | Peak Temperature ($^0C$) | Order of Kinetics (b) | Activation Energy (eV) | Frequency Factor S ($s^{-1}$) |
|---|---|---|---|---|
| P+FA30% 35.08 mGy | 287.15 | 2.68 | 1.16 | $4.12x10^{11}$ |
| P+FA30% 26.31 mGy | 249.87 | 9.71 | 0.34 | $2.71x10^{4}$ |
| P+FA20% 17.54 mGy | 261.19 | 7.28 | 0.4 | $7.84x10^{4}$ |
| P+FA20% 8.77 mGy | 266.61 | 7.57 | 0.59 | $4.21x10^{6}$ |
| P+FA10% 26.31 mGy | 269.22 | 5.57 | 0.43 | $8.89x10^{4}$ |
| P+FA10% 17.54 mGy | 292.69 | 0.62 | 1.39 | $3.34x10^{13}$ |
| P+FA10% 8.77 mGy | 265.8 | 3 | 0.42 | $5.46x10^{4}$ |

## 4. Conclusion

Fly Ash mixed $CaSO_4$: Dy Phosphor has a orthorhombic crystalline structure confirmed by matching the JCPDS XRD data. TL response dose response shows the sample P+FA10% shows nearly equal intensity as the traditional phosphor materials. Linear fit and linearity curves indicate that the sample P+FA10% and P+FA30% shows much more linear pattern. The activation energy calculated by the Chen's Peak Shape method and GCD method were quite in analogy for P+FA10% sample. Here is very small gap between the $E_A$ value calculated from experimental and theoretical methods. The frequency factor calculated by different methods is very much of same order. Thus the study suggests that P+FA10% sample are expected to show promising behavior if employed in radiation dosimeter applications at low dose. Apart from that we also suggest P+FA10% sample along with PFA sample which shows most linear curve with respect to dose applied to be exposed to high dose value for the identification of its high dose application as well.

**Acknowledgement**

The author acknowledges the help and support provided by all faculty members of the Department of Physics, Dr. C. V. Raman University Bilaspur, Department of Physics, Department of Physics, Pt. Ravishankar Shukla University Raipur, and Ultratech Laboratories Pvt. Ltd. Kumhari Bhilai (*Accredited by BARC, Mumbai*) For their kind intellectual and technical support.

**References**

[1] Bhatt B Cand Kulkarni M S 2014 *Defect and Diffusion Forum* **347** 179
[2] Ovhal D A, Chopra V, Dhoble N S and Dhoble S J 2019 *International Journal of Photonics and Optical Technology* **5** 10
[3] Suriyanarayanan N, Kanan N K V and Bernardo E 2009 *Journal of Non-Oxide Glasses* **1** 247
[4] Mandlik N T, Bhoraskar V N, Sahare P D, Kumar B, Patil J, Kumar V, Kulkarni M S and Dhole S D 2011 *Radiation Protection and Environment (IARP)* **34** 185
[5] Joshi S, Gupta K K, Ubale S K and Dhoble S J 2017 *Radiation Measurements* **105** 47
[6] Bakr M and Omer M 2020 *Materials* **13** 1047
[7] Chen R 1969 *Journal of Electrochemical Society: Solid State Science* **16** 1254
[8] Pagonis V, Kitis G and Furetta C 2005 Numerical and Practical Exercises in Thermoluminescence, Springer Publication, ISBN : 0-387-26063-3 e-ISBN: 0-387-30090-2

IOP Conf. Series: Materials Science and Engineering 1120 (2021) 012008 doi:10.1088/1757-899X/1120/1/012008

# Biofabrication and structural characterization of cerium oxide nanoparticles

**Komal Kashyap[1], F Khan[1*], Dakeshwar Verma[2], Sonalika Agrawal[1] Ch Chandra[1], Pradeep Kumar Dewangan[1], Vinayak Sahu[1], Padma Rani Verma[1] and Vikas Kumar Jain[3]**

[1] Department of Chemistry, National Institute of Technology Raipur-492010, (C.G.) India
[2] Department of Chemistry, Government Digvijay Autonomous College, Rajnandgaon, Chhattisgarh 491441
[3] Government Engineering College, Raipur (C.G.) India

*Corresponding author's e-mail address: fkhan.chy@nitrr.ac.in, kkashyap.phd2018.chy@nitrr.ac.in

**Abstract.** Cerium oxide synthesis ($CeO_2$) nanoparticles were studied using a new and simple, eco-friendly process. The cerium oxide nanoparticles are synthesized with precursors of ammonium cerium nitrate and sodium hydroxide. Their chemical and physical properties are characterized by energy dispersive spectroscopy (EDS), X-ray diffraction (XRD), Fourier transforming infrared spectroscopy (FTIR), and UV-Vis spectrophotometer scanning electron microscopy (SEM), high-resolution electron transmission microscopy (HRTEM). XRD analysis indicated the cubic structure of the nanoparticles containing cerium oxide. The standard particle size of $CeO_2$ was determined at 30 nm using HR-TEM and XRD. SEM determine the surface morphology of biosynthesized cerium oxide nanoparticles. A Ce-O stretching mode was defined by the strong peaks in the FTIR range, and the peak of UV-Vis range absorbance revealed the 3.26 eV band energy difference.

## 1. Introduction

Nanoscience and nanotechnology in science and engineering are really a recent revolutionary development that is growing at a very rapid rate [1]. Nanoparticles (NPs) are defined as products with a minimum diameter between 1 and 100 nm and are of considerable interest to numerous sciences and industries because of their specific physicochemical properties [2]. Due to their larger surface area compared to bulk equivalents, nanosized particles have various physiochemical characteristics such as stronger optoelectronics, catalytic, magnetic, and thermal conductivity [3]. Cerium oxide nanoparticles (CNPs), due to their self-regenerating antioxidant effects, have gained considerable interest in nanotechnology [4]. Cerium is a rare earth metal lanthanide and is found in two oxide phases with crystalline fluorite lattice structure in combination with oxygen [5]. Cerium oxide has been reviewed for numerous applications including electrical, catalytic, mechanical, optical, adsorption, electrochemical, battery, energy storage, practical materials, sensing properties, and magnetic data storage [6]. Ceria has achieved substantial popularity as a redox and combustion catalyst because of all of its ability to change from a reduced as well as oxidized state due to large concentrations of oxygen in

IOP Conf. Series: Materials Science and Engineering 1120 (2021) 012008 doi:10.1088/1757-899X/1120/1/012008

the gas phase [7]. Several studies have been developed to aim to form $CeO_2$NPs using chemical reduction, electrochemical reduction, and photochemical reduction, such methods use extreme reducing and stabilizing agents in the synthesis process which in biomedical applications can cause some dicey effects [8]. Most of these methods, however, are complicated and/or involve either vacuum or harmful chemical compounds. In recent years, environmentally friendly and green nanoparticle synthesis has become famous for overcoming these problems. Numerous green synthetic procedures for silver and gold nanoparticles have been published [9]. The method of biological synthesis involves the use of plants (or microbes, fungi, bacteria, algae) for the protection of the synthesis of microbial crops, Plant extract is far more relevant than microbial application. Green approach with various extracts of plants, such as *Prosopis farcta, Hibiscus sabdariffa, Olea europaea, Acalypha indica, and Gloriosa superba* had been reported for different medical needs, for example, neurodegenerative diseases antiapoptotic, anti-inflammatory, cancer therapy, radiation safety, possible applications have been reported for biosynthesized nanoceria with different plant extracts [10]. The extracts can be derived from numerous plant parts, such as seeds, stem bark, nuts, flowers, and roots, which are also easy to collect and store and do not damage plants instead of the last and these plant extract has so many natural components, such as flavonoids, tannins, and terpenoids that are synthesized as capers and stabilizers. The use of plant extracts to develop NPs is dependent on the added benefit of the resource itself [11-13]. In the present research work, we have confirmed cerium oxide nanoparticles synthesis using *Trapa natans* plant extract using the green synthesis process. The synthesized nanoceria was characterized by UV-Vis spectroscopy, FTIR, XRD, EDX, and HR-TEM.

## 2. Experimental

### *2.1 Reagents and Chemicals*
The chemicals used in this study have been of good quality scientific value and have been used without further purification. Ammonium cerium nitrate $(NH_4)_2Ce(NO_3)_6$, 99.9 % Merck), and Sodium Hydroxide solution (NaOH) were used in this study. During the synthesis, all of the aqueous solutions were prepared using Milli-Q water.

### *2.2 Preparation of plant extract*
Before the synthesis process, the naturally accessible Trapa natans collected from the market were washed and thoroughly cleaned with deionized, double-distilled (DD) water. The fresh, clean, and dried pericarp of trapa was added to a round bottom flask consisting of an aliquot quantity of DD water and methanol at 100°C temperature to synthesize the cerium oxide nanoparticles. The wine-colored methanol solution was received after four hours of ideal condition. Then, the prepared solution was filtered. The resulting extract was stored at 4°C for the green synthesis of cerium oxide nanoparticles.

### *2.3 Synthesis of nanoparticles*
For the biosynthesis of cerium oxide nanoparticle, an extract aliquot (100 ml) containing 40 ml of 1 mM $CeO_2$ was processed and the solution mixture has been stirred under low temperature for 24 hours. A visual evaluation has observed a change in color in the reaction mixture. Calcinated, coated, and deposited the as-prepared nanoparticles in air-tight containers before application.

## 3. Characterization of Cerium oxide nanoparticles
CeNP was primarily characterized by such a UV spectrophotometer (SL-159, Elico,149 India) with wavelengths ranging between 200 and 600 nm. FTIR spectra of prepared cerium oxide nanoparticles had been obtained from FTIR Spectroscopy (Bruker Optics, GmBH, Germany). Surface morphology had been identified using HR-TEM (Jeol, Japan) and XRD patterns have been calculated using a diffractometer (Xpert-Pro, England) at room temperature with a current of 30mA, 40kV, and peaks of $2\theta$.

### *3.1 UV- Visible Spectroscopy*

The UV-Visible range has a broad absorption band at 373 nm (figure 1), possibly because the electrons are moved from the valence band (VB) to the conductive band (CB). For cerium oxide nanoparticles, the mean value was 3.324 eV reported. Besides, the bulk $CeO_2$ has an absorption peak at 357 nm (corresponding to 3.45148 eV energy) suggesting a clear blue change in the UV-Visible spectrum [13].

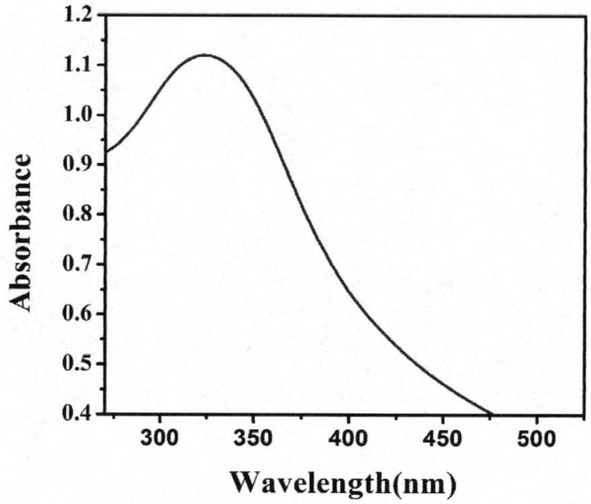


**Figure 1.** UV-Visible spectrum of cerium oxide nanoparticles

### *3.2 FTIR Analysis*

FTIR was a valuable method for understanding the importance of functional groups in the relation between metal particles and biomolecules, for analyzing the chemical structure of the surface of cerium nanoparticles, and for detecting and effectively stabilizing biomolecules to cap nanoparticles of metal [7]. There have been many specific groups that may have been significant in micro-reducing the $Ce^{+4}$ ions ) (figure 2).

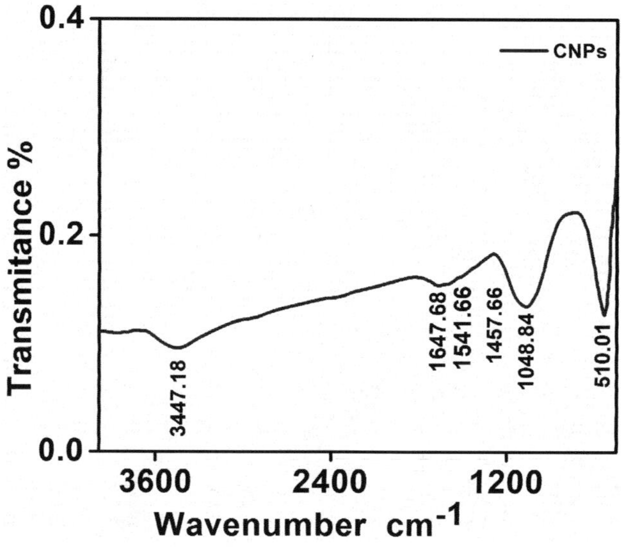


**Figure 2.** The FTIR transmittance spectrum of cerium oxide nanoparticles.

IOP Conf. Series: Materials Science and Engineering 1120 (2021) 012008 doi:10.1088/1757-899X/1120/1/012008

*3.3 XRD*

Figure 3 demonstrate the XRD pattern of the sample. All the diffraction peaks are identified and can be recorded as the cubic face-centered phase. The strong and sharp diffraction peaks indicate that the samples are finely crystallized. The high purity of the prepared ceria nanoparticles was shown by no additional peaks in the XRD. All diffractions concerning the CNPs are consistent with normal Cerianite crystallographic details (JCPDS No: 8 34-0394). This result verified the crystallization into a face-centered fluorite cubic structure of high purity single-phase $CeO_2$ [12-13]. In comparison, the diffraction peaks found at $2\theta = 28.66$, 33.03, 47.56, 56.39, 59.05, 69.34, 76.61 and 79.27 refer to planes (111), (200), (220), (311), (222), (400), (331), and (420) [14]. The relationship between the X-ray beam incident and the sample induces extreme X-ray reflections by constructive interference when Bragg's law is satisfied [1].

By using the Debye-Scherrer equation the average Crystallite size (D) of the synthesized $CeO_2$ nanoparticles at 500 ° C was determined.

$$D = 0.9\,\lambda / \beta \cos\theta \qquad (1)$$

Where $\lambda$ stands as the wavelength of the utilized X-ray (1.5406 Å), $\beta$ is the angular peak width at half maximum in radians and $\theta$ is Bragg's diffraction angle. The PXRD peaks further demonstrate that the crystallite sizes of the obtained NPs are smaller than 30 nm, as confirmed by the synthesized $CeO_2$ images from the SEM [15].

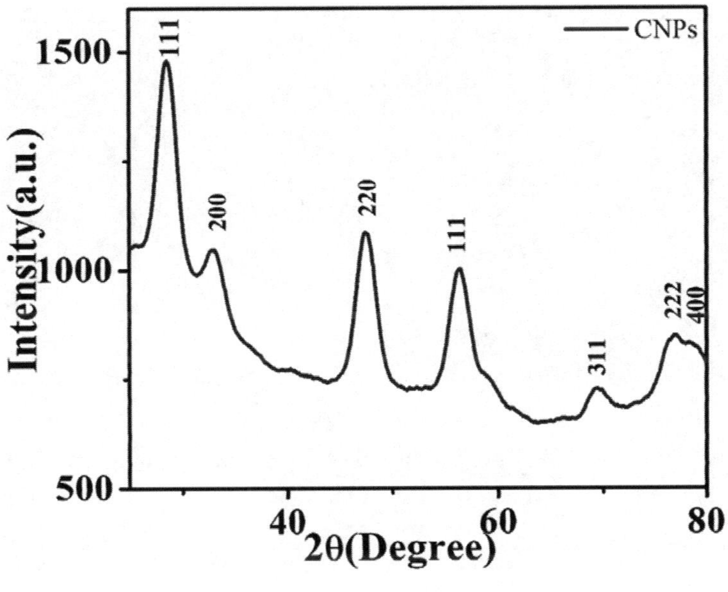


**Figure 3.** XRD of Cerium oxide nanoparticles using trapa pericarp extract.

*3.4 SEM and EDX*

Scanning Electron Microscopy is performed to show particle surface morphology. Usually, the electron beam is scanned in a raster scan pattern, and the beam direction is coupled with the observed signal to build an image. SEM images can achieve a resolution above 50 nanometres. SEM has examined the surface dynamics of nanoparticles packed with cerium oxide. The above figure 4 shows the SEM picture of nanoparticles with freshly synthesized cerium oxide. The

IOP Conf. Series: Materials Science and Engineering 1120 (2021) 012008 doi:10.1088/1757-899X/1120/1/012008

cerium oxide nanoparticles can be observed in a spherical shape with sizes varying from 10-26 nm. From the figure 4, it can be seen that the nanoparticles of cerium oxide aggregate to form a spherical-like structure.

The energy-dispersive x-ray spectroscopy (EDX) is an observational method designed for elemental detection or chemical sample analysis characterization [1]. A standard EDX distribution as seen in the figure 4 (d) shows that Ce and O are present in the product.

**Figure 4.** (a) SEM image showing the surface morphology of the cerium oxide nanoparticles (b) SEM image showing particle size of the cerium oxide nanoparticles (c) and (d) EDX pattern of cerium oxide nanoparticles.

### *3.5 Raman Spectroscopy*

Raman spectroscopy is a common spectroscopic vibration technique used to measure a substance's molecular motion and to classify and examine molecular species by providing proof of symmetry and molecular structure (figure 5) [16].

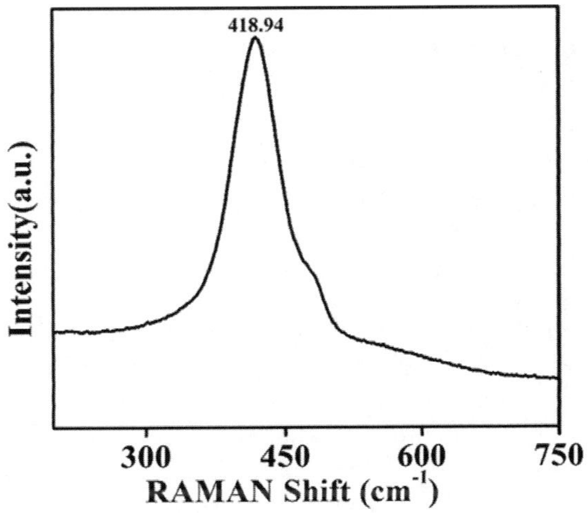


**Figure 5.** RAMAN spectrum of cerium oxide nanoparticles.

*3.6 HR-TEM*

High-Resolution Transmission Electron Microscopy (HRTEM) data, and unique field electron diffraction patterns (SAED) have been used to examine the crystalline morphology of the $CeO_2$ sample [17]. The $CeO_2$ NPs HR-TEM picture reveals the standard particles of spherical scale are agglomerated together (figure 6). In addition, all the particles are well crystallised and a normal particle size with less than 100 nm [14].

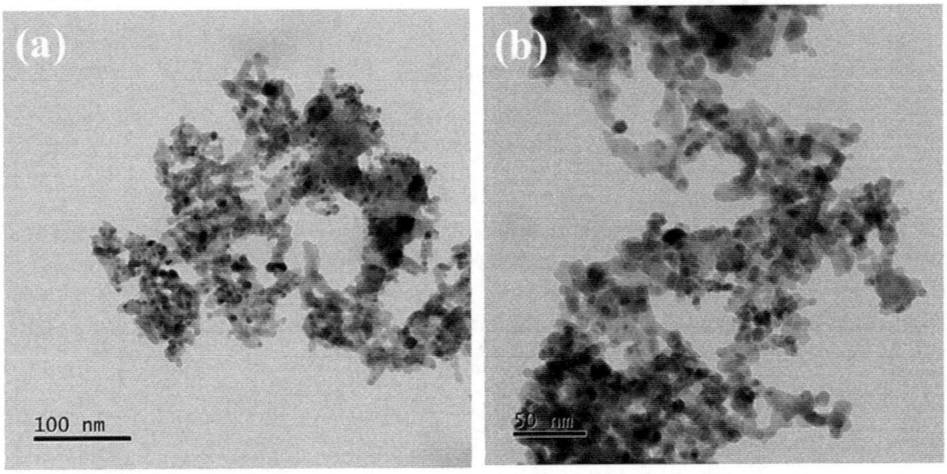


**Figure 6.** TEM images of cerium oxide nanoparticles.

## 4. Conclusions

Analyzing previous literature on ceria NPs showed that different methods of synthesis could provide different catalytic and physicochemical properties for cerium oxide NPs which could contribute to antioxidant or prooxidant properties [9]. The near-sphere-shaped bio-capped cerium oxide nanoparticles were successfully synthesized here via the green synthesis method using trapa pericarp extract as a reduction / stabilizing agent. The FTIR analysis confirmed that chemicals belong to the phytochemicals responsible for biodegrading behaviors. The average reduction in $CeO_2$NPs was 80 percent of $Ce^{3+}$ ions within 2 days. The structural and microstructural data of these $CeO_2$NPs was collected using the techniques XRD, EDX, FTIR, and SEM.

**Acknowledgment**
The authors are thankful to the Director, National Institute of Technology, Raipur for providing the laboratory and library. One of the authors, Komal Kashyap gratefully acknowledges the CSIR (file no 09/1116(0007)/2018 EMR-1) for fellowship.

**References**

[1] Pitchumani Krishnaveni, Priya L and Annadurai G 2019 *RJLBPCS* **5** 589
[2] Merrifield, Ruth C, Wang W Z, Palmer E R and Lead J R 2013 *Environmental science & technology* **47** 12426
[3] Sharmila, Govindasamy, Muthukumaran C, Saraswathi H, Sangeetha E, Soundarya S and Kumar N M 2019 *Ceramics International* **45** 12382
[4] Pavan Kumar, M A, et al. 2015 *Eur. Phys. J. Plus* **6**130
[5] Inbaraj B S, Chen and Huei B 2019 *Asian Journal of Pharmaceutical Sciences* **15** 558
[6] Farahmandjou M, Zarinkamar M and Firoozabadi TP 2016 *Revista mexicana de física* **62** 496
[7] Panahi-Kalamuei, Mokhtar, Alizadeh S, Mehdi Mousavi-Kamazani MM and Niasari M S 2015 *Journal of Industrial and Engineering Chemistry* **21** 1301
[8] Gavade S J, Mane S J, et al. 2015 *Journal of Nano Science and Nano Technology* **9** 89
[9] Charbgoo, Fahimeh, Ahmad M B and Darroudi M 2017 *International journal of nanomedicine* **12** 1401
[10] Yallappa, S, et al, 2013 *Spectrochimica Acta Part A: Molecular and Biomolecular Spectroscopy* **110** 108
[11] Elahi, Behrouz, Mirzaee M, Darroudi M, Oskuee RK, Sadri K and Amiri M S 2019 *Ceramics International* **45** 4790
[12] Hussain, Imtiyaz, Singh N B, Singh A, Singh H and Singh SC 2016 *Biotechnology letters* **38** 545
[13] Goharshadi, Elaheh K, Samiee S and Nancarrow P 2011 *Journal of colloid and interface science* **356** 473
[14] Tamizhdurai P, Sakthinathan S, Chen SM, Shanthi K, Sivasanker S and Sangeetha P 2017 *Scientific reports* **7** 46372
[15] Miri, Abdolhossein and Sarani M 2018 *Ceramics International* **44** 12642
[16] Yadav, Reddy L S, Manjunath K, Archana B, Madhu C, Naika H R, Nagabhushana H, Kavitha C and Nagaraju G 2016 *The European Physical Journal Plus* **131** 154
[17] Malleshappa, J, Nagabhushana H, Sharma S C, Vidya Y S, Anantharaju K S, Prashantha S C, Prasad B D, Naika H R, Lingaraju K and Surendra B S 2015 *Spectrochimica Acta Part A: Molecular and Biomolecular Spectroscopy* **149** 452

# Plasmonic nanostructures for sensor technology-a review

**Sudha Kumari*** **and Sapan Mohan Saini**

Department of Physics, National Institute of Technology, Raipur, Chhattisgarh - 492010, India

*Corresponding author's e-mail address: kumari.sudha93@gmail.com

**Abstract.** Presently, metal nanostructures have broadly been used as building blocks for sensing devices in various fields because of their remarkable optical, electrical, and mechanical features. Propagating Surface plasmon resonance (SPR) and localized SPR (LSPR) have been demonstrated as the principles of some highly sensitive sensors. LSPR-based nanostructures array has huge potential for high throughput screening platforms for sensing based on the kinetic measurement and change in refractive index. Highly sensitive plasmonic biosensors have been designed for biomedical fields, including point-of-care diagnosis and clinical purposes. Besides, the integration of plasmonic nano-arrays and connecting them with electrochemical sensing has noticeably blown up the application scenarios of the plasmonic nano-array sensors. In conclusion, fabrication methods for plasmonic well-designed nanostructures have been studied and the upcoming research trends of plasmonic sensing are highlighted.

## 1. Introduction

Plasmonic is a subfield of nanophotonics new ground of science and technology that exploits the distinctive optical properties of metallic nanostructures to drive light at nanometer length scales. This field of science and technology is emerging rapidly for challenging sensing applications. Plasmonic is the collective oscillation of the electron at the metal-dielectric plane. The mirror-like aspect of smooth noble - metal films changes when the metal is estranged into particles that are smaller than the wavelength of light [1]. Surface plasmon resonance (SPR) in the visible/near-infrared wavelength range is of meticulous significance because there is constructive optical confinement at these wavelengths. All conductive materials, such as metals (i.e. copper, silver, gold and aluminum), support plasmon because their plasmon resonance lies nearer to the visible region of the spectrum [2]. Gold and silver are generally appropriate as plasmonic materials but copper and platinum, palladium and aluminum are sometimes used [3, 4]. Silver has higher properties in the near-infrared region and has a sharper resonance peak but generally, gold is preferred due to its chemical stability resistance to oxidation. SPR based sensing is a real-time and label-free detection process to measure biomolecular interactions on the surface of the gold. Label-free detection techniques monitor biomolecular interactions by eradicating the need for secondary reactants. Furthermore, they give quantitative information for the binding kinetics [5, 6]. Surface Plasmon polarities (SPP) are traveling charge density waves at the surface of conducting materials present at the interface of two medium (metal-dielectric) with permittivities of opposed sign [7].

IOP Conf. Series: Materials Science and Engineering 1120 (2021) 012009 doi:10.1088/1757-899X/1120/1/012009

## 2. Design principle for sensing performance

In SPR sensors, the SP wave excited at the boundary between a metal layer and a dielectric medium is very sensitive to changes in bulk or local refractive index of the dielectric surrounding medium interfaced with the metal film. Assuming that there is a local change in the refractive index of $\Delta n$ upon binding of chemical or biological molecules in the surrounding dielectric medium (with bulk refractive index $n$). Then, a change in the refractive index $n + \Delta n$ in the local surrounding medium produces a change in the propagation constant $\beta + \Delta\beta$ of SP wave at metal-dielectric (liquid media) interface. Subsequently, this leads to the change in the phase-matching condition require exciting SPR at different resonance wavelengths or angles, which results in red-shifting of SPR reflection dip. Based on the distinctiveness of the beam interacting with the SP wave, the SPR sensor can be classified with different schemes such as angular, wavelength, intensity and phase modulation. In an angular interrogation-based SPR sensor, a monochromatic beam is used to excite SP waves at a metal-dielectric interface. The strength of coupling between the incident light and SP wave is recorded for multiple angles of incidence of light. The SPR resonance angle appears as a dip in the reflectance intensity in the angular spectra. By recording the shift in the angle or reflection intensity as a sensor signal, we can monitor the binding of chemical or biological molecules on the sensor surface. Similarly, wavelength interrogation-based SPR sensor, the SP wave is excited via a collimated broadband light source at a constant incident angle of the light. The strength of coupling between the incident wave and the SP wave is recorded for multiple wavelengths. The shifting wavelength that corresponds to the SPR peak is employed as a sensor signal to monitor various chemical or biological processes that occurred on the sensor surface. SPR sensors based on intensity modulation measures the coupling strength between the incident light and a SP wave at a fixed angle of incidence and wavelength. The change in the intensity of light at the SPR condition is employed as a sensor signal for the attachment of the biological or chemical molecules on the sensor surface. In phase modulation based SPR, the shift in phase or change in the polarization of light wave interacting with the SP is measured at both constant wavelength and incident angle with respect to bulk refractive index change or attachment of biological/chemical molecules on the sensor surface [8]. The SPs confined to the electromagnetic field carries the corresponding energy and are quickly transformed into high-energy electromagnetic waves, leading to a vast localized electromagnetic enhancement. Sensitivity and resolution can be calculated by studying the properties of the SPR sensor further real-time measurement can be performed. Although there are improvements in the present technology and instrumentation of SPR biosensor still needs major advances in miniaturization of SPR biosensing platforms and assimilation of SPR sensor set up with microfluidic devices. Furthermore, the bulky size and high cost of the instruments restrict their usage for portable, affordable and low-cost applications.

Plasmonic nanostructures or nanoarrays have motivated several colorimetric detections for chemical pollutants, environmental contaminants, biomolecules, and medicinal drugs. These colorimetric sensors have importance in that the interaction of analytes with sensor chip can alter the distance, orientation, size, morphology, or composition of plasmonic nanostructures (nanorods, nanoplates, nanoflowers, nanobipyramids and nanocubes), inducing apparent changes of optical properties like spectra and color [10]. Diverse nanostructures have been exploited for achieving optimal nanosensors for biosensing [11], chemical sensing [12] energy harvesting such as photocatalysis and photovoltaics [13, 14] and monitoring of environmental pollutants and metals ion detections [15,16].

Here, various nanostructures-based sensitivity based is listed from different papers:

**Table 1.** List of sensitivity adapted from different literatures based on plasmonic nanostructures.

| Nanostructures | Sensitivity(nm/RIU) | References |
|---|---|---|
| Hollow gold nanoshell | 408 | 17 |
| Gold nanorings | 880 | 18 |
| Gold nanoring trimers | 345 | 19 |
| Gold nanodisk trimers | 170 and 373.9 | 20 |
| Array of gold nanodisks | 167 and 327 | 21 |
| Nanocubes | 165 | 22 |
| Nanostars | 218 | 23 |
| Nanocrescents | 879 | 24 |
| Double Nanopillars with nanogaps | 652 and 1056 | 25 |
| Arrays of Plasmonic nanotubes | 250 | 26 |
| Nanopillars arrays | 675 | 27 |
| Plasmonic nanorods metamaterial | 30000 | 28 |
| EOT gold nanoholes arrays | 1580 | 29 |
| Gold grating on nitride substrate | 1140 | 30 |
| Rectangular plasmonic interferometer | 4923 | 31 |

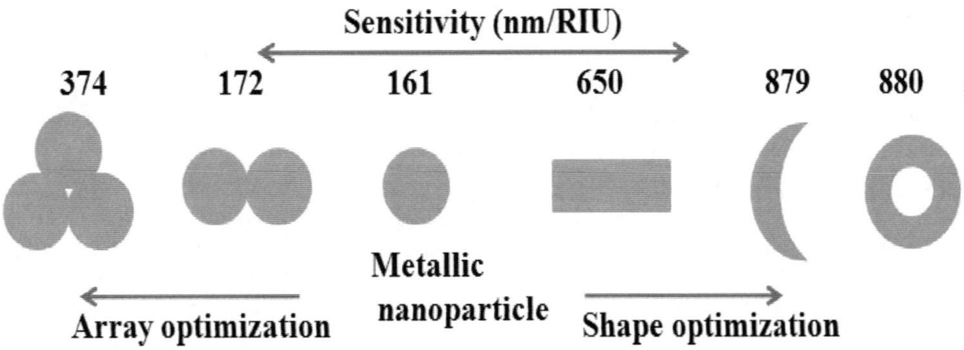


**Figure 1.** Optimized study of the different shapes of plasmonic sensors [32].

### 3. Outlook

In the research field of plasmonic waveguides and devices, there has been a lot of research going on for the betterment of society especially it is very useful for spectroscopy, chemical and biosensing applications. Up to now, many commercialized biosensors have been launched. Based on its sensing mechanism various biomedical problems can be solved. Through the studies of protein biomolecules on biological samples, disease diagnosis can be done at earlier stages. But there are many challenges to designing a sensor-based device. One of the most significant challenges is nanofabrication, especially the low-cost and high-throughput production of metal nanostructures with tailorable plasmonic properties. Hence, metal nanostructures can be used for sensing applications based on the

IOP Conf. Series: Materials Science and Engineering 1120 (2021) 012009 doi:10.1088/1757-899X/1120/1/012009

surface plasmon resonance or localized surface plasmon resonance. Appropriate manipulation of geometrical parameters is a prerequisite for explicit sensing purposes. We hope that this review on the sensors will help understand the basics and work principle of plasmonic sensing devices and provides a valuable perspective.

## References


[1] Anker J N, Hall W P, Lyndreas O, Shahn N C, Zhao J and Van Duyne R P 2008 *Nature materials* **7** 442

[2] Lal S, Link S and Halas N J *2007 Nature Photonics* **1** 641

[3] Debruijn H E, Kooyman R P and Greve H J 1992 *Appl Opt.* **32** 440

[4] Sambles J R, Bradbery G W and Yang F Z 1991 *Contemporary Phys.* **32** 173

[5] Helmerhorst E, Chandler D J, Nussio M and Mamotte C D 2012 *Clinical Biochemists Review* **33** 161

[6] Lee K L, Chih M J, Shi X, Ueno K, Misawa H and Wei P-K 2012 *Advanced. Materials.* **24** 253

[7] Abbas A, Linman M J, Cheng Q 2011 *Biosens and Bioelectron* **26** 1815

[8] Homola J 2003 *Anal Bioanal Chem* **377** 528

[9] Prabowo B A, Purwidyantri A and Liu K C 2018 *Biosensors* **8** 80

[10] Zeng J, Zhang Y, Zeng T, Aleisa R, Qiu Z, Chen Y, Huang J, Wang D, Yan Z and Yin Y 2020 *Nano Today* **32** 100855

[11] Chung T, Lee S Y, Song E Y, Chun H and Lee B 2011 *Sensors* **11**0907

[12] Colombelli A, Manera M G, Taurino A and Rella R 2016 18th Italian National Conference on Photonic Technologies Rome1-4

[13] Reddy I N, Reddy C V, Shridhar A, Cho M, Kim D and Shim J 2019 J Electro Analytical Chem. **842** 146

[14] Ye W, Long R, Huang H and Xiong Y 2017 *J. Mater. Chem. C* **5** 1008

[15] Choi I 2016 *Journal of Nanoscience and Nanotechnology* **16** 4274

[16] Amirjani A and Haghshenas D F 2018 *Sensors and Actuators B: Chemical* **273** 1768

[17] Sun Y and XiaY 2002 *Anal.Chem.***74** 5297

[18] Larsson E M Alegret J Kall M and Sutherland D S 2007 *Nano Lett.* **5** 1256

[19] Lin V K, Teo S L, Marty R Arbouet A, Girard C, Alarcon Llado E, Liu S H, Han M Y, Teo S L and Lin V K 2010 *Opt. Express* **18** 22271

[20] Tripathy S and Mlayah A 2010 *Nanotechnology* **21** 305501

[21] Lee S W, Lee K S, Ahn J, Lee J J, Kim M G and Shin Y B 2011 *ACS Nano* **5** 897

[22] Galush W J, Shelby S A, Mulvihill M J, Tao A, Yang P and Groves J T 2009 *Nano Lett.* **9** 2077

[23] Dondapati S K, Sau T K, Hrelescu C T A, Klar F D and Stefani J Feldmann 2010 *ACS Nano* **4** 6318

[24] Bukasov R, Ali T A, Nordlande P and Shumaker-Parry J S 2010 *ACS Nano* **4** 6639

[25] Kubo W Fujikawa S 2011 *NanoLett.* **11** 8

[26] McPhillips J, Murphy A, Jonsson M P, Hendren W R, Atkinson R, Hook F, Zayats AV and Pollard Robert J 2010 *ACS Nano* **4** 2210

[27] Cetin AE Yanik, AA Yilmaz, C Somu, S Busnaina, A Altug H 2011 *Appl. Phys. Lett.* **98** 111110

[28] Kabashin A V, Evans P, Pastkovsky S, Hendren W, Wurtz G A, Atkinson R, Pollard R, Podolskiy V A and Zayats AV 2009 *Nat. Mater* **8** 867

[29] Chen Z, Li P, Zhang S, Chen Y, Liu P and Duan H 2019 *Nanotechnology* **30** 335201

[30] Sharma A K and Pandey A K 2019 *Superlattices and Microstructures* **130** 369

[31] Khajemiri Z, Lee D, Hamidi S M and Dai-Sik Kim 2019 *Scientific Reports* **9** 13789

[32] Lee B, Chung T, Lee K, Lee Y and Cho J 2013 Proc. of SPIE89242Y

IOP Conf. Series: Materials Science and Engineering 1120 (2021) 012010 doi:10.1088/1757-899X/1120/1/012010

# Optical characterization of CTAB assisted Ni doped ZnS nanoparticles

**K S Ojha* and S M Saini**

Department of Physics, National Institute of Technology Raipur, Raipur (C. G.) India

*Corresponding author's e-mail address: kanhaiyaisojha@gmail.com, smsainin.phy@gmail.com

**Abstract.** Ni doped zinc sulphide nanoparticles at different concentration of CTAB have been prepared via solvothermal route. The structural characterization of the ZnS:Ni nanoparticles has been made using XRD technique and it reveals cubic zinc blende structure of ZnS. The crystallite size calculated from the XRD pattern is in close agreement to that estimated from the TEM image. The UV-Vis study of ZnS:Ni nanoparticles indicates the red shift in the absorption peak as compared to bulk ZnS. The emission spectra ZnS:Ni sample at 5 mmol CTAB concentration has been fitted into three Gaussian peaks.

## 1. Introduction:

The nano-crystalline form of Zinc sulphide, a II-VI semiconductor, is commercially important because it demonstrates outstanding structural, optical and opto-electronic properties compared to bulk ZnS [1-3]. It is being used as catalysts and phosphors [4] as well as also being used in electro- luminescence devices, solar cells, flat panel displays, light emitting diodes, laser diodes, optical sensors and biological labelling [5-11]. The doping of d-block elements similar to $Mn^{2+}$, $Co^{2+}$, $Cu^{2+}$ and $Ni^{2+}$ ions modify the electronic structure as well as transition probabilities due to change in optical properties of ZnS [12-16]. Therefore, the doping method is being used as a valuable tool for altering the absorption and photoluminescence properties of ZnS nanostructures for technological applications. The physical and chemical properties of Ni doped ZnS nanostructures have been investigated by several workers. Poornaprakash *et. al.* [17] studied the structural, optical along with other properties of Fe, Co and Ni doped ZnS nanostructure and reported that these divalent ions easily substitute the $Zn^{2+}$ ions due to their electron structures and less ionic radii. The luminescence properties of Ni, Cu doped ZnS nanoparticles has been studied by Murugadoss [14] and an increase in energy gap along with improvement in PL intensity with the increase of $Ni^{2+}$ concentration has been observed. Kumar et al. [18] synthesized the ZnS:Ni nanoparticles via chemical precipitation method and reported a blue shift in absorption edge as well as the decrease in PL intensity. Ni doped ZnS nanopowders have been prepared by Zhou et al. [19] using hydrothermal method and observed a significant decrease in PL intensity due to incorporation of $Ni^{2+}$ ions in ZnS. The photo-catalytic degradation efficiency of $Mn^{2+}$, $Co^{2+}$ and $Ni^{2+}$ incorporated ZnS nanoparticles has been studied by Rajabi *et. al.* [20] and Kaur *et. al.* [21], and a reduction in degradation efficiency has been reported. Wu *et. al.* [22] prepared Ni doped ZnS nanorods via hydrothermal method in presence of ethylenediamine and reported a blue shift in absorption band along with the improved intensity in emission bands. It has been also reported that the hexagonal wurtzite phase does not change with substitution of $Ni^{2+}$ on $Zn^{2+}$ sites. A decrease in energy gap along with the enhancement in PL intensity up to a certain concentration of Ni has been observed by Othman *et. al.* [23] in Ni doped ZnS nanocrystals prepared via sonochemical route. Numerous techniques have applied to synthesized the ZnS:Ni nanoparticles such as sol-gel, chemical precipitation, wet chemical method etc. Among these techniques, solvothermal route is simple, inexpensive and able to control the morphology. The above concerned literatures resemble that the described optical properties are uncertain. Therefore, more investigations on the optical properties of Ni doped ZnS nanoparticles are essential. In the present study, Ni doped ZnS nanoparticles have been synthesized via solvothermal route at different CTAB concentration.

## 2. Materials and Methods

Analytical grade raw materials Zinc acetate [Zn $(CH_3COO)_2.2H_2O$], nickel acetate [Ni $(CH_3COO)_2.4H_2O$, sodium sulphide ($Na_2S$) and CTAB are used to prepare the Ni doped ZnS nanoparticles. The aqueous solutions of zinc acetate and sodium sulphide were stirred separately for 90 minutes at 60°C and mixed together. After that the required aqueous solution of nickel acetate was stirred for 1h at 60°C and mixed with prepared solution of zinc acetate and sodium sulphide. Thus, prepared solution was again stirred for 1h at 70°C and cooled to room temperature and divided into four parts. In each solution, the required amount of CTAB was added and transferred into stainless steel Teflon lined autoclave for 24 h kept at 70°C. The solutions obtained from autoclave were filtered and the precipitates were cleaned several times with ethanol and doubly distilled water. The obtained precipitates of Ni doped ZnS were dried and used for further characterization.

## 3. Result

### *3.1 Structural and morphological study:*

The structural characterization of Ni doped ZnS nanoparticles has been studied using Cu-Kα line of Rigaku miniflex X-ray diffractometer and the observed pattern is represented in Fig 1. The observed peaks correspond to (111), (220) and (311) of ZnS demonstrating the cubic zinc-blende structure with lattice constant 5.406 A° and is in close agreement to that reported by earlier workers. In addition, diffraction pattern also exhibits that the intensity and the broadening of the observed peaks increases up to 5mmol of CTAB and then decreases which may be due to agglomeration of nanoparticles and doping of $Ni^{2+}$ ions.

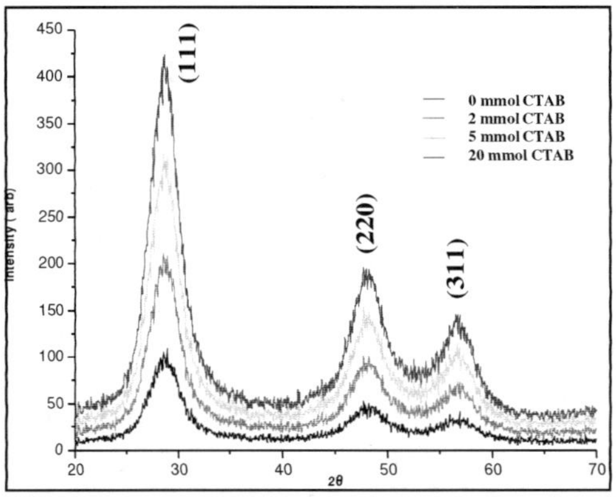


**Figure 1.** XRD pattern of Ni doped ZnS nanoparticles at different CTAB concentrations.

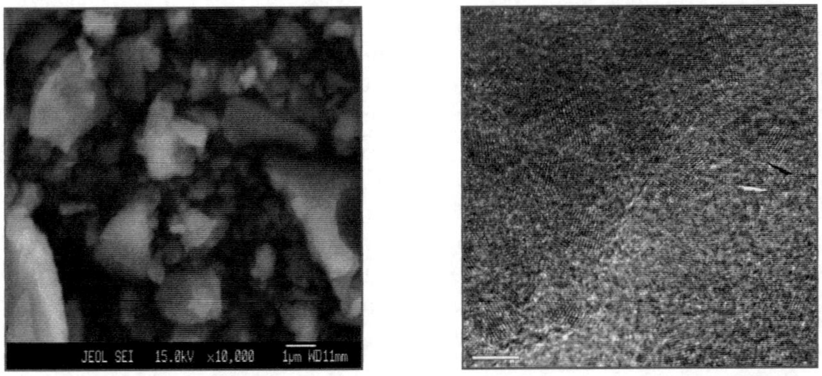


**Figure 2.** (a) SEM image and (b) TEM image of Ni doped ZnS nanoparticles at 5 mmol CTAB concentration.

The average crystallite size of Ni doped ZnS nanoparticle has been calculated using Scherrer's formula and it is found to be about 4.7 nm. No additional peaks have been observed in the diffraction pattern which indicated that there is no impurity in the synthesized nanoparticles. The surface morphology of synthesized sample has been studied using SEM micrograph at 5 mmol concentration of CTAB and represented in figure 2(a). The observed fringes in the TEM image of Ni doped ZnS nanoparticles, shown in figure 2 (b), indicate the good crystallinity in the synthesized sample.

*3.2 UV-Vis Study:*

The room temperature absorption spectra of Ni doped ZnS nanoparticles has been recorded using lambda 35 spectrophotometer (Perkin Elmer) in the wavelength range of 200-800 nm and depicted in figure 3. A blue shift in the absorption edge of Ni doped ZnS nanoparticles has been observed as compared to bulk ZnS which is due to the reduction in size at nanoscale. The absorption spectra also reveal that the absorption edge is red shifted with increasing concentration of CTAB and which may be due to the agglomeration of the particles. In addition, an absorption peak near at 212 nm has been also observed in each synthesized sample and the further studied is going on. The band gap calculated from the tauc plot is found to be varied from 4.9 eV to 4.1 eV with the increase of CTAB concentration.

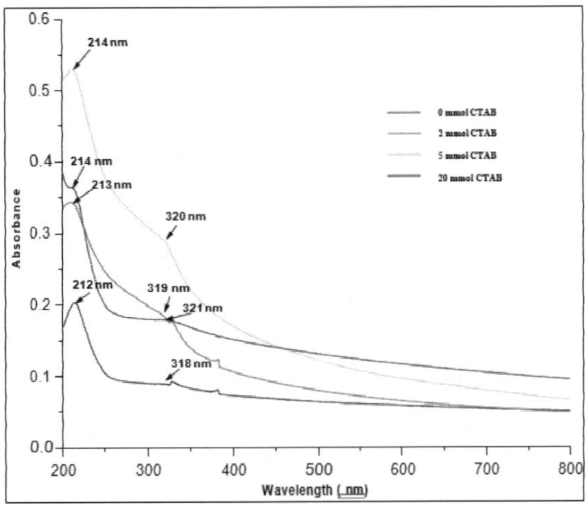


**Figure 3.** Absorption spectra of Ni doped ZnS nanoparticles at different CTAB concentrations.

*3.3 Photoluminescence Study:*

The nondestructive photoluminescence technique is applied to describe the impurity as well as defect energy states in semiconductor can be used even the impurity level is very minute [22]. The emission spectrum of Ni doped ZnS nanoparticles at 5 mmol CTAB concentration has been recorded in the wavelength range 450-700 nm using PerkinElmer Luminescent Spectrometer ((LS-55) with excitation source of 370 nm. The three Gaussian fitted emission bands centered at 492, 555 and 624 nm of observed PL spectrum are shown in figure 4. Emission peak centered at 492 nm assigned as blue emission and is due to electron-hole recombination between sulphur vacancy and sulphur interstitial while the green emission peak centered at 555 nm can be assigned to trapped electron-hole recombination between sulphur vacancy and deeper zinc vacancy [23]. A slight shift of 4 nm has been observed in the orange emission peak centered at 624 nm as reported by Wu *et. al.* [22] and is attribute to the electron-hole recombination between interstitial zinc states and zinc vacancy.

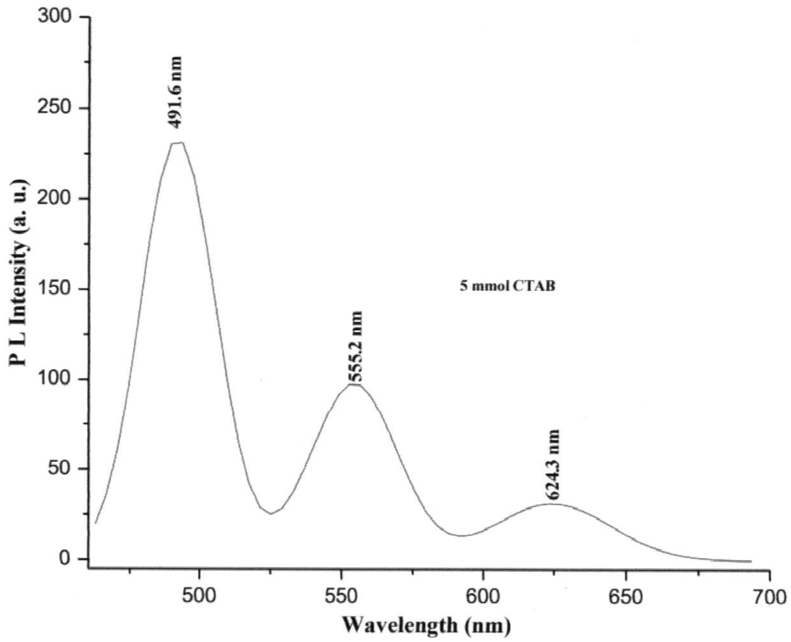


**Figure 4.** PL emission spectra of Ni doped ZnS nanoparticles at 5 mmol CTAB concentration.

## 4. Conclusion:

Ni doped zinc sulphide nanoparticles at different CTAB concentration have been synthesized via solvothermal technique. The XRD technique confirms the formation of nanoparticles having crystallite size 4.7 nm. The absorption study of the synthesized samples has been made using UV-Vis spectroscopic technique and it has been observed that the absorption edge is blue shifted in all samples as compared bulk ZnS. In addition, the absorption edge of Ni doped ZnS nanoparticles is red shifted up to 5 mmol CTAB concentration and then blue shifted. The observed broad and asymmetric PL spectra of synthesized sample at 5 mmol CTAB has been fitted into three Gaussian peaks lying in blue, green and orange regions are analysed.

## References

[1] Wang Y W, Zhang L D, Liang C H, Wang G Z and Peng X S 2002 *Chem. Phys. Lett.* **357** 314
[2] Dinsmore D, Hsu D S, Gray H F, Qadri S B and Tian Y 1999 *Appl. Phys. Lett.* **75** 802

IOP Conf. Series: Materials Science and Engineering 1120 (2021) 012010 doi:10.1088/1757-899X/1120/1/012010

[3] Denzier D, Olschcwski M and Sattler K 1998 *J. Appl. Phys.* **84** 2841
[4] Fukuoka A, Kimura J I, Oshio T, Sakamoto Y and Ichikawa M 2007 *J. Am. Chem. Soc.* **129** 10120
[5] Sang W, Qian Y, Min J, Li D and Wang L 2002 *Solid State Commun.* **121** 475
[6] Nada A A, Hamed H A, Barakat M H, Mohamed N R and Veziroglu T N 2008 *Int. J. Hydrogen Energy* **33** 3264
[7] Yamamoto T, Kishimoto S and Lida S 2001 *Physics B* **308** 916
[8] Hong K J, Jeong T S, Yoon C J and ShinY J 2000 *J. Cryst. Growth* **218** 19
[9] Zhang J, Yang X, Li S, Wang D and Qian Y 2000 *J. Cryst. Growth* **220** 231
[10] Sapsford K E, Pons T,Medntz I L and Mattoussi H 2006 *Sensors* **6** 925
[11] W. C. W. Chan and S. M. Nie 1998 Science **281** 2016
[12] Datta A, Biswa S, Kar S and Chaudhri S 2007 *Nanosci. Nanotechnol.* **7** 3670
[13] Hoa T T Q, The N D, McVite S, Nam N H, Canh T D and Long N N 2011 *Opt. Mater. (Amst)* **33** 308
[14] Murugadoss G 2012 *J. Lumin.* **132** 2043
[15] Liu L, Yang L, Pu Y, Xiao D and Zhu J 2012 *Mater. Lett.* **66** 121
[16] Khafri H Z, Ghaedi M, Asfaram A and Safarpoor M 2017 *Ultrason. Sonochem.* **38** 371
[17] Poornaprakash B, Chalapathi U and Vattikuti S V P 2017 *Appl. Phys. A* **123** 275
[18] Kumar R S, Veeravazhuthi V, Muthukumarasamy N and Thambidurai M 2015 *Superlattices Microstruct.* **86** 552
[19] Zhou X, Yang Q, Wang H, Haung F, Zhang J and Xu S 2018 *Adv. Powder Technol.* **29** 977
[20] Rajabi H R and Farsi M 2015 *J. Mol. Catal. A: Chem.* **399** 53
[21] Kaur J, Sharma M and Pandey O P 2015 *Opt. Mater. (Amst)* **47** 7
[22] Wu M, Wei Z, Zhao W, Wang X and Jiang J 2017 *J. Nanom.* **2017** Article ID 1603450 1-9
[23] Othman A A, Osman M A, Ali M A, Mohamed W S and Ibrahim E M M, 2020 *J. Mater. Sc.: Mater. Electron.* **31** 1752

IOP Conf. Series: Materials Science and Engineering 1120 (2021) 012011 doi:10.1088/1757-899X/1120/1/012011

# Adsorption isotherms and kinetics study for U(VI) removal by excellent sorbent materials graphene oxide and functionalised graphene oxide

**Khilawan Patel[1], Alekha Kumar Sutar[2] and Tungabidya Maharana[1*]**

[1] Department of Chemistry, National Institute of Technology, Raipur, 492010, India
[2] Department of Chemistry, Gangadhar Mehar University, Sambalpur, 768004, India

*Corresponding author's e-mail address: tmaharana.chy@nitrr.ac.in, khilawanpatel1209@gmail.com

**Abstract.** In this paper, the graphene oxide (GO) and carboxylic functionalized graphene oxide (COOH-GO) were synthesized for the removal of U(VI) ion in samples of water. The structures of prepared GO and COOH-GO were confirmed by X-ray diffraction (XRD), Scanning electron microscope (SEM), Fourier transformed infrared (FTIR) spectroscopy, Raman spectra, and high-resolution transmission electron spectroscopy (HR-TEM). An adsorption study was carried out for these and it has been found that COOH-GO has more excellent adsorption property at 4.5 pH. The batch adsorption was studied successfully by the effect of initial uranium concentration, pH of the medium, sorbent dose, temperature, and contact time efficiency. The adsorption isotherms were studied successfully for Langmuir and Freundlich. The kinetics were studied successfully and fitted for the pseudo-first-order and pseudo-second-order model. The maximum adsorption capacities at pH 4.5 were calculated for COOH-GO, GO, and graphite. The thermodynamic studies revealed that adsorption is endothermic and spontaneous.

## 1. Introduction

The recovery and removal of heavy radionuclide uranium metal are one of the most challenging problems in the current scenario. Uranium is radio-toxic, chemo-toxic, and a long half-life substance. With the fast growth of different activities, the discharge of uranium into water is a serious problem. Therefore, the removal of heavy toxic metals of uranium from water samples is a hot topic of research [1, 2]. The superior and most significant techniques sorption [3-5] has been broadly used for the purification of waste water (e.g., precipitation [6], bio reduction and reverse osmosis [7], and ion exchange [8, 9]) due to cost-effective, easy technique and no secondary toxic waste. However, most of the sorbents' materials have some disadvantages viz., less sorption ability, highly expensive, and less eco-friendly. Therefore, it is extremely important to develop cheaper & eco-friendly sorbent materials along with high sorption capacity.

In recent times, GO and COOH-GO has concerned with significant attention for their outstanding sorption capacity like cheapness, environmentally friendly, abundance, high stability, etc. All these properties have proved that GO and COOH-GO possess positive sorption ability for the removal of heavy metals like Pb [10], Hg [11], Co, and Sr [12]. Therefore, it is important to decorate GO and

COOH-GO with other large molecular compounds to find some better-quality properties in practical applications [13].

In GO and COOH-GO, the high number of oxygen-containing groups such as hydroxyl (-OH) and carboxyl (-COOH) are accountable to their high surface area, excellent sorption capacity, and specific electronic properties [14, 15]. However, it is commonly known that some cations like $Mg^{2+}$, $Ca^{2+}$, $Na^{+}$, and $K^{+}$, which are generally present in wastewater, most likely produce some interference for the adsorption capacity of sorbent [16].

The maximum sorption efficiency of GO for U (VI), as reported by Li*et al.* 2008, was 299 mg/g at pH 4 [17, 18]. Nowadays, heavy toxic metal pollution in water is a very serious problem for all kinds of living organisms. Thus, the sorption study of heavy metals in the water samples is currently a big area of research for researchers. Therefore, in the current perspective, low-cost materials such as GO and its derivatives are more useful in U(VI) sorption study.

Thus, in the present study, the surface of GO was modified to COOH-GO by using a monochloroacetic acid and formed a novel carboxylic functionalized graphene oxide ( COOH-GO) composite. The structure and morphology of GO and COOH-GO were characterized successfully by UV-VIS, FT-IR, XRD, HR-TEM, and Raman shift [19]. In this modified COOH-GO, different functional groups such as carboxylic group (-COOH), alcoholic group (-OH), and epoxy (-O-) groups are effectively decorated on the layer of the graphene sheet. The effect of initial concentration, pH, amount of sorbent dose, temperature, and contact time on the sorption behaviour of U(VI) onto GO and COOH-GO was studied. Further, the adsorption isotherms were developed for Langmuir and Freundlich isotherm models. The kinetic studies were also fitted for the pseudo-first-order and pseudo-second-order model. The thermodynamic parameters werealso studied for checking the sorption as endothermic or exothermic and spontaneous or nonspontaneous [20-22].

## 2. Experimental Part

### *2.1 Reagents, Materials, and Characterization*

The chemicals were used as received such as graphite fine powder (particle size 250 µm) and $NaNO_3$ (99%Sodium nitrate) of Loba Chemie Pvt. Ltd., $ClCH_2COOH$ (>99.0% Monochloroacetic acid), potassium permanganate (>98.5%$KMnO_4$), sulphuric acid (98%$H_2SO_4$), sodium hydroxide (>97%, NaOH,), hydrochloric acid (35%HCl), and hydrogen peroxide (30%$H_2O_2$,) from Sigma Aldrich (Mumbai, India). Uranium standard solution, Accustandard, USA was used as received. The pH meter of Hanna Instruments, USA; LED fluorimeter, model name LF-2a from Quantalase Inc. Indore, India; Merck millipore deionized (DI) water; Fourier Transform Infrared (FT-IR) Spectroscopy from Agilent Technologies, UK; Scanning Electron Microscopy (SEM) model name SEM Zeiss EVO 18, USA; Raman Spectra from Research India (model name RIAR-0403, India); High-resolution transmission electron microscope (HR-TEM) instrument (JEM-2100, Make- JEOL, Japan) were used.

### *2.2 Synthesis of Graphene Oxide (GO) and Carboxylic Functionalized Graphene oxide (COOH-GO)*

The GO was synthesized by using a modified Hummer's method [23]. The brief procedure has been discussed in our previous paper [19]. Further, the functionalization of GO has been carried out by using monochloroacetic acid. In this method, the synthesized graphene oxide powder (0.10 g) was taken in a 500 ml beaker and then added 100 ml millipore water and sonicated for 30 minutes. Then, 2.4 g monochloroacetic acid and 2g of sodium hydroxide were added to the graphene oxide solution and was sonicated for 3 hours. The above mixture was purified by repeated washing and was neutralized by adding 0.25M of hydrochloric acid dropwise until the pH of the resulting mixture becomes 7 [23]. Finally, the obtained residue of carboxylic functionalized graphene oxide (COOH-GO) was dried by using a hot air oven for 24 hours at 45°C [19, 24, 25].

*2.3Uranium (VI)Batch Sorption Study*

Uranium batch sorption study was carried out by using LED fluorimeter for COOH-GO, GO, and graphite at the desired pH. All the sorption procedure occurred under room temperature and atmospheric air unless the condition was specified. The sorption % of U(VI) and sorption equilibrium capacity (mg/g) of U(VI) were observed by using equations 1 and 2, respectively.

$$\text{U(VI) Sorption \%} = \frac{C_i - C_e}{C_i} \times 100\,\% \tag{1}$$

$$\text{U(VI) Sorption capacity (mg/g)} = \frac{(C_i - C_e) \times V}{m} \tag{2}$$

Where, $C_i$ is the initial concentration of U(VI) before sorption, $C_e$ is the equilibrium concentration or final concentration after U(VI) sorption, V = the volume of solution in mL, and m is the amount of adsorbent in mg.

*2.4 The Langmuir Adsorption Isotherm*

The Langmuir adsorption isotherm equation was given as follows (Eq. 3) [26]:

$$q_e = \left(\frac{q_{max} K_L C_e}{1 + K_L C_e}\right) \tag{3}$$

The Langmuir's adsorption isotherm can also be written as linear form (Eq. 4 and Eq.5), where, $q_e$ (mg/g) = Equilibrium sorption capacity of U(VI) adsorbed on adsorbent surfaces, $q_{max}$ = Maximum sorption capacity of U(VI) adsorbed to form a complete monolayer coverage on adsorbent surfaces, $K_L$ = Langmuir's isotherm sorption constant related to the binding affinity between U(VI) and sorbent materials. The most important features of the Langmuir isotherm equation was the enthalpy of sorption constant or Langmuir's isotherm sorption constant ($K_L$). The $q_{max}$ would be calculated by a linear Langmuir sorption isotherm equation given as follows:

$$\frac{1}{q_e} = \frac{1}{q_{max} K_L} \times \frac{1}{C_e} + \frac{1}{q_{max}} \tag{4}$$

$$\frac{C_e}{q_e} = \frac{C_e}{q_{max} K_L} + \frac{C_e}{q_{max}} \tag{5}$$

The separation factor (RL) was calculated by using Eq. (6).

$$R_L = \frac{1}{1 + C_i \times K_L} \tag{6}$$

Where, $R_L$= the dimensionless Langmuir constant which indicate the adsorption possibility either favourable (0 < RL > 1), unfavourable (RL > 1), linear (RL = 1) or irreversible (RL =0).

*2.5The Freundlich Adsorption Isotherm*

This was based on adsorption on a heterogeneous surface Eq. 7a [27]:

$$q_e = K_f Ce^{1/n} \tag{7a}$$

Where, $K_f$ = Freundlich constant and1/n = Adsorption intensity which calculated the adsorption process is either unfavourable (1/n > 2) or favourable (0.1 <1/n < 0.5)

Taking the logarithm in both sides, the linear form of equation (Eq. 7b) is given below:

$$\log q_e = \log K_f + \frac{1}{n}\log C_e \tag{7b}$$

### *2.6 Kinetic Study*

Adsorption rate of U(VI) on COOH-GO and GO were calculated for pseudo-first-order and pseudo-second-order kinetic models.
For pseudo 1st order model, the Eq. 8 is followed [28].

$$\ln(q_e - q_t) = \ln q_e - k_1 t \quad (8)$$

Where, $q_e$ = sorption capacity (mg/g) at equilibrium, $q_t$ = sorption capacity (mg/g) at time t and $k_1$ ($min^{-1}$) = rate constant at equilibrium
For pseudo 2nd order model, the Eq. 9 is used [29].

$$\frac{t}{q_e} = \frac{1}{k_2 q_e^2} + \frac{1}{q_e} \quad (9)$$

Where, $k_2$ ($g\ mg^{-1}min^{-1}$) = rate constant at equilibrium sorption.
For both the models the linear coefficient regression ($R^2$) values had been used forthe prediction of the best-suited sorption isotherm.

## 3.Results and Discussion

### *3.1FT-IR Spectra*

The FT-IR spectra of GO show peak values at 3418.9 $cm^{-1}$for hydroxyl (O-H) stretching vibration, 1725.2 $cm^{-1}$ for ketonic (>C=O) stretching vibration, and 1120.5 $cm^{-1}$ for epoxide (-C-O-C-) stretching vibration. The two doublet peaks were observed due to tautomerism of keto and enol form at 1627.5 $cm^{-1}$& 1725.2 $cm^{-1}$[30]. However, upon comparing FTIR of GO to COOH-GO, a slight shift is observed because of the increase of the -COOH domain group. The COOH-GO peak values at 3408.6 $cm^{-1}$ for hydroxyl (O-H) stretching vibration, 1731.2 $cm^{-1}$ for ketonic (>C=O) stretching vibration and 1113.8 $cm^{-1}$ for epoxide (-C-O-C-) stretching vibration were observed [19, 31, 32].

### *3.2Raman Spectra*

The Raman spectrum of GO and COOH-GO shows two peaks that are the G band and the D band. The D band value at 1332.4 $cm^{-1}$and G band value at 1579.5 $cm^{-1}$because of $sp^2$ and $sp^3$hybridised carbon atoms for GO were observed (Figure 1). The D band value ~1345.5 $cm^{-1}$andG band value ~1595.6 $cm^{-1}$because of $sp^2$ and $sp^3$ hybridised carbon atoms for COOH-GO were observed (Figure 1). The intensity ratio $I_D/I_G$ was calculated to be 1.1824 and 1.1952 for GO and COOH-GO, respectively [19,33-35].

IOP Conf. Series: Materials Science and Engineering 1120 (2021) 012011 doi:10.1088/1757-899X/1120/1/012011

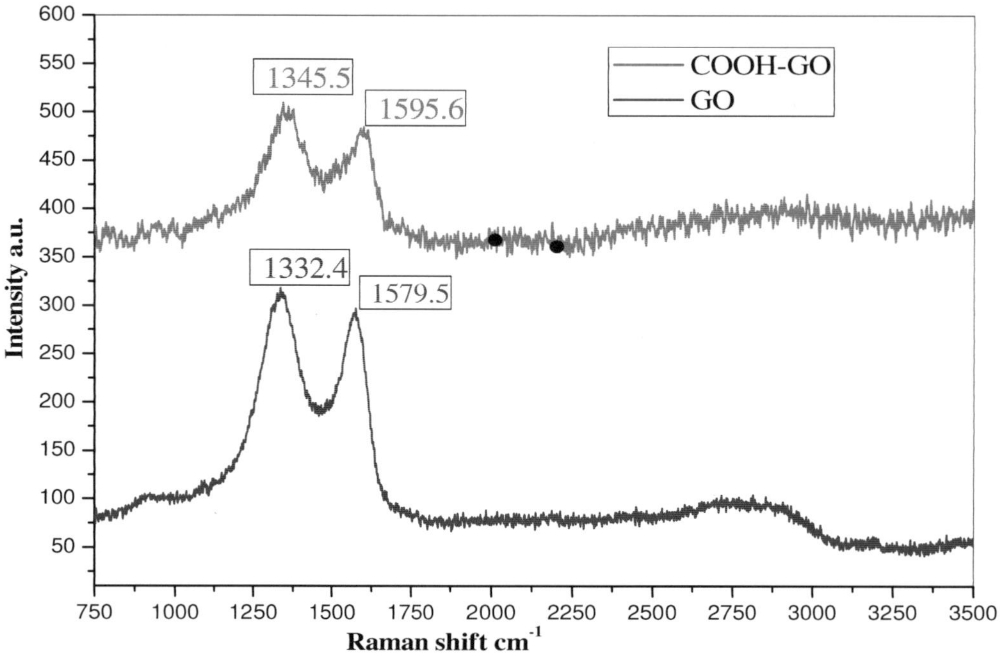


**Figure 1.** Raman Spectra of COOH-GO and GO.

*3.3HR-TEM for Carboxylic Graphene Oxide and Graphene Oxide*

The direct images of GO and COOH-GO were observed by using high-resolution transmission electron microscopy (HR-TEM). This technique is a very powerful and helpful tool for the atomic scale level of $sp^3$ hybridized carbon [36, 37]. The exfoliated sheet and larger surface area were observed for different nanometer level for GO and COOH-GO [38]. The more amorphous nature was observed for COOH-GO as compared to GO due to more number of carboxylic groups (figure 2a and 2b).

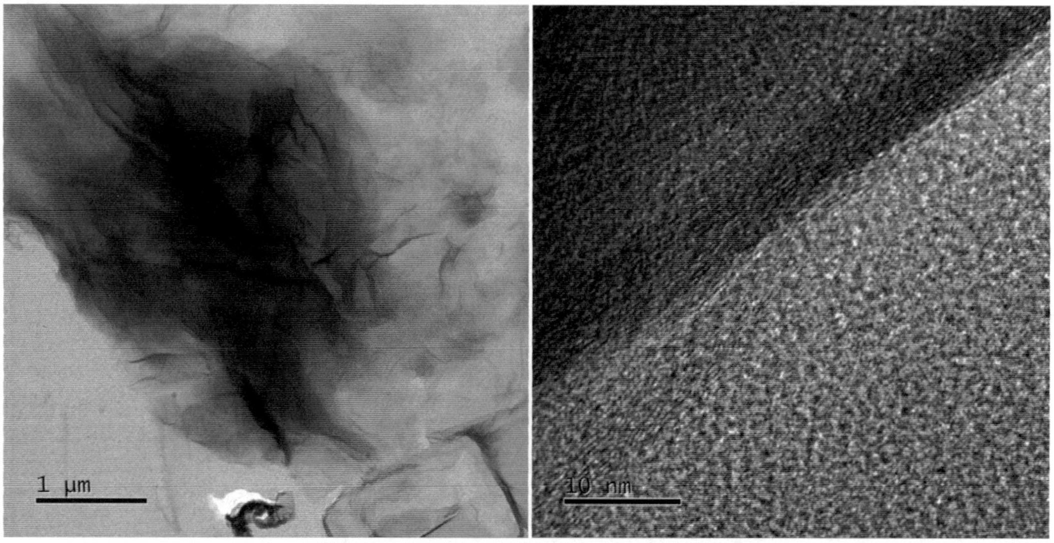


**Figure 2.** (a) HR-TEM image of GO.

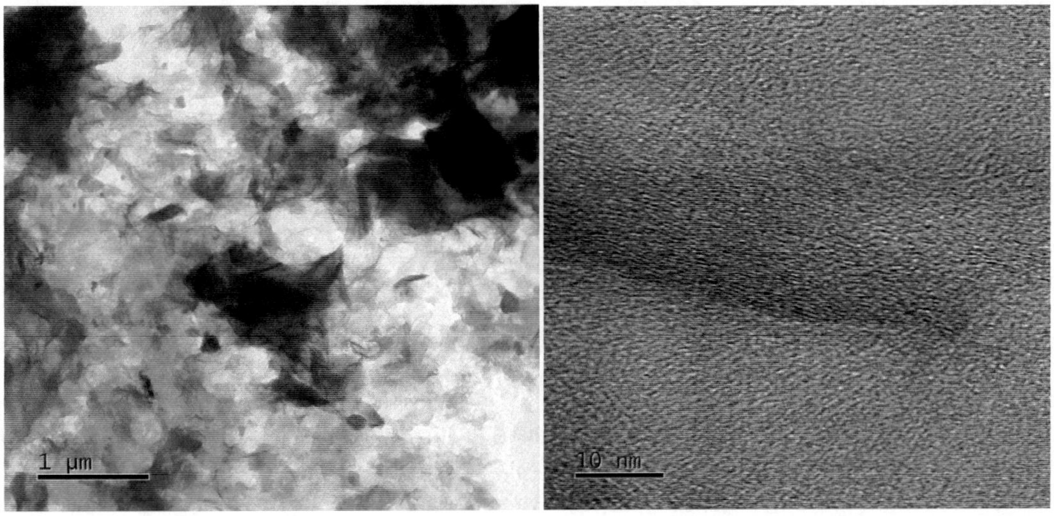


**Figure 2.** (b) HR-TEM image of COOH-GO.

*3.4Effect of pH on Uranium Sorption*

The solution pH affects strongly the sorption rate of the radionuclide. Therefore, the sorption of U(VI) was studied in pH range of 2-12 which is similar tothe previously reported paper [33, 39], and pH maintained by using 0.01M HCl and 0.01M NaOH (figure 3). The maximum sorption order was found to be COOH-GO > GO > graphite at 4.5 pH. The study suggested that the uranium removal is more due to the presence of more carboxyl functional groups in COOH-GO at low pH level. The COOH-GO and GO have been most likely decorated with negative surface charge and this group has an electrostatic attraction with U(VI) of $UO_2^{+2}$ positive charge species at lower pH region [19]. However, at higher pH, the sorption was decreased and was believed to be due to the formation of the negatively charged stable uranium carbonate complexes like $[UO_2(CO_3)_3]^{4-}$in solution [40, 41]. Thus, the electrostatic repulsion was observed between the negative complex of uranium and the negative sorbent surface of materials at a higher pH range.

The conditions for the experiments are, pH = 2-12, U(VI) concentration = 10μg/mL, mass of sorbent = 10 mg, volume of sample = 10 mL [Figure 3].

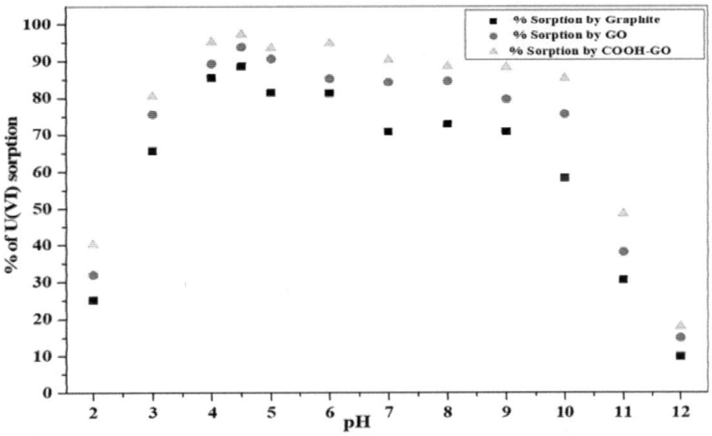


**Figure 3.** Effect of pH for sorbent materials.

### *3.5 Effect of Contact Time*

The effect of the contact time period was taken from 10 to 140 minutes with interval values 10, 20, 30, 40, 50, 60, 80, 100, 120, and 140 minutes for sorbent materials. This study shows 87.6%, 92.9 %, 95.9 % of sorption was occurred for graphite powder, GO, and COOH-GO, respectively [23]. These sorption results were well with those reported in the literature and the time required to attain equilibrium is 100 minutes [12, 19].

### *3.6 Effect of Sorbent Dose*

The sorption amount increases regularly from 5 to 70 mg with a very fast increase from 5 to 10 mg. The order of maximum sorption is in the order COOH-GO > GO >graphite [12, 23]. When the dose of sorbents varied from 5 to70 mg, the sorption percentage increased from 37% to 94% [19].

### *3.7 Sorption Isotherm*

The equilibrium study was developed for U(VI) removal in water samples. The initial concentrations were varied from 10 to 200 μg/mL and all other parameters were kept constant [23]. The sorption capacity of sorbent materials versus the equilibrium concentration of the U(VI) (Figure 4) shows that more adsorption capacity for COOH-GO was observed due to its more negative charge potential and porous nature than GO and graphite. The initial concentration of U(VI) affects the sorption of sorbent materials. The effect of initial concentration data ware used for two theoretical models Langmuir and Freundlich adsorption isotherm. Langmuir's isotherm shows monolayer adsorption of U(VI) onto the sorbent which has a finite number of sorption sites but Freundlich's isotherm supports the sorption occurs on the heterogeneous surface of the sorbent (Figure 5, 6) [42]. The plotted graph showed the Langmuir isotherm model best fits for the sorption of U(VI) due to the formation of a monolayer on COOH-GO,GO, and graphite. The sorption isotherm gives very important information about the surface nature, adsorption capacity, and binding affinity of the sorbent substances which help to know the binding mechanism of sorbate to the sorbent [43].

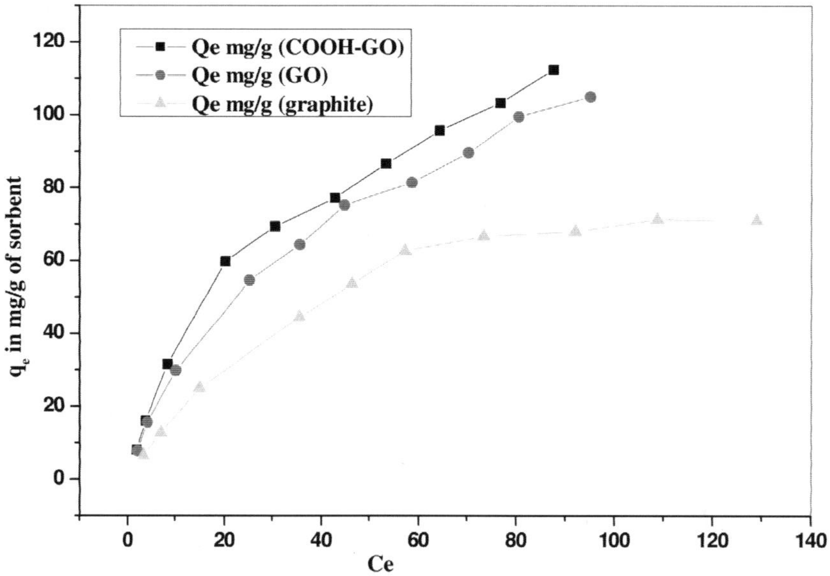


**Figure 4.** Sorption capacity graph for $C_e$ versus $q_e$

The sorption result was shown in Table 1 shows that the values of the linear regression coefficient ($R^2$> 0.999) was best fitted with Langmuir isotherm whereas Freundlich adsorption isotherm for COOH-GO, GO, and graphite (Table 2) has $R^2$< 0.99. So it was confirmed that the monolayer sorption of U(VI) on COOH-GO, GO, and graphite surface occurs. The calculated value of $q_{max}$ (maximum sorption capacity) was observed to be 148.36,128.53, and 116.14mg/g for COOH-GO, GO, and graphite, respectively (Table 1). The RL (separation factor) value is less than one and was observed that the adsorption process is favourable [23].

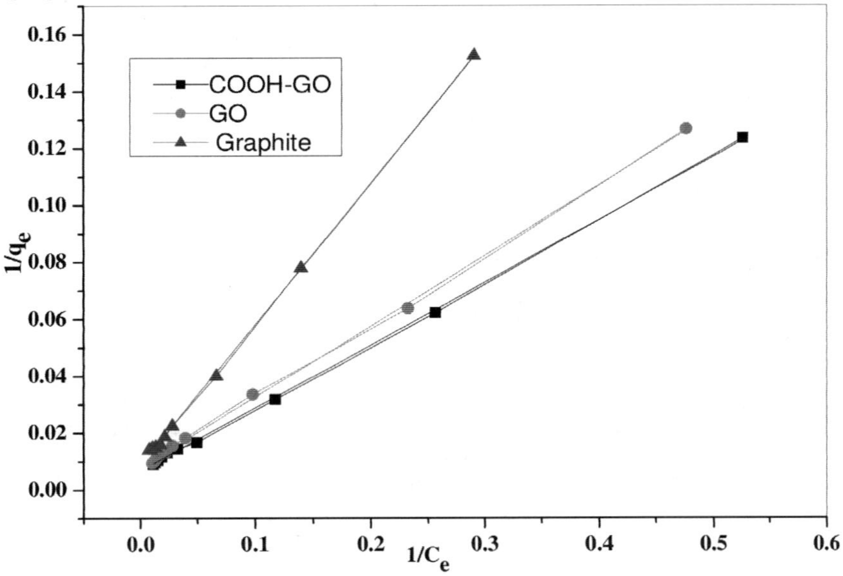


**Figure 5.** Langmuir adsorption isotherm between $1/C_e$ versus $1/q_e$

From above Eq. 4, Intercept = $1/q_{max}$, $q_{max}$ were calculated. The value of $K_L$ was calculated as 1/slope $\times q_{max}$, The value of $R_L$ were calculated as $1/(1+C_i \times K_L)$.

**Table 1.** Langmuir model for U(VI) sorption onto COOH-GO, GO, and graphite.

| Parameters | Sorbent materials | | |
|---|---|---|---|
| | COOH-GO | GO | Graphite |
| y = mx + c | y = 0.2203x + 0.00674 | y = 0.2483x + 0.00778 | y = 0.4939x + 0.00861 |
| **$R^2$** | 0.9997 | 0.9996 | 0.9997 |
| **$q_{max}$(mg/g)** | 148.36 | 128.53 | 116.14 |
| **$K_L$ (L/mg)** | 0.0305 | 0.0313 | 0.0174 |
| **$R_L$** | 0.2464 | 0.242 | 0.3645 |

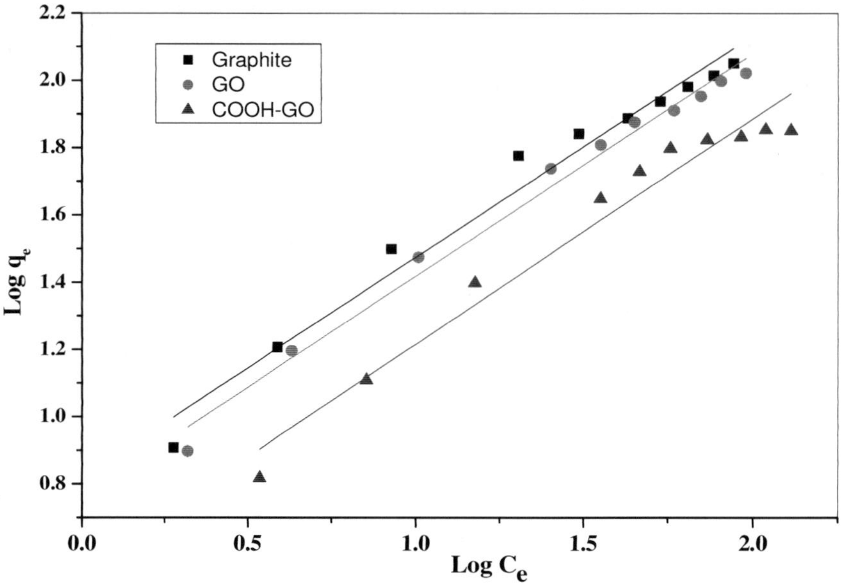


**Figure 6.** Freundlich isotherm between Log$q_e$ versus LogCe

From Eq. 8, slope, $K_f$ and Log $K_f$ were calculated.

**Table 2.** Freundlich model for U(VI) sorption onto COOH-GO, GO, and graphite.

| Parameters | Sorbent materials | | |
|---|---|---|---|
| | COOH-GO | GO | Graphite |
| **y = mx + c** | y = 0.6587x + 0.8153 | y = 0.6619x + 0.6619 | y = 0.6715x + 0.5437 |
| **$R^2$** | 0.9884 | 0.9839 | 0.9817 |
| **1/n** | 0.6587 | 0.6619 | 0.6715 |
| **$K_f$** | 6.5369 | 4.5911 | 3.4974 |

*3.8 Kinetic Studies for Sorption*

The kinetic study of both pseudo 1[st]order [28] and pseudo 2[nd]order rate equation [29] were applied for the sorption rate and its study of kinetic mechanism was done by using Eq. 8, and 9, respectively.

*3.8.1 Pseudo 1[st] order model*

The equation (Eq. 8) is:

$$\ln(q_e - q_t) = \ln q_e - k_1 t \qquad \text{Eq...(8)}$$

Where, $q_e$ = sorption capacity (mg/g) at equilibrium, $q_t$ = sorption capacity (mg/g) at time t and $k_1$ ($min^{-1}$) = rate constant at equilibrium.

The sorption kinetic curves were plotted for U(VI) sorption onto COOH-GO, GO, and graphite for both pseudo 1[st] order and pseudo 2[nd] order reaction and are shown in figure 7 and figure 8 (Table 3). In the initial 20 minute maximum and fast U(VI) sorption was observed than at 80 minutes [19, 23]. After 20 min, the sorbent become saturated with sorbate and eventually they attained the dynamic equilibrium [19, 43].

*3.8.2Pseudo 2nd order model*

The equation (Eq. 9) is: $\frac{t}{q_e} = \frac{1}{k_2 {q_e}^2} + \frac{1}{q_e}$

Where, $k_2$ (g $mg^{-1}min^{-1}$) = rate constants at equilibrium sorption.
For both the models the linear coefficient regression ($R^2$) values had been used for prediction of best suited sorption isotherm. The best-suited model was observed by using a linear regression coefficient ($R^2$) if the value is near 0.99 or one then this is best fitted kinetic model (Figure 7, 8). From table 3, it has been observed that for pseudo $2^{nd}$ order model, the linear regression coefficient $R^2$ > 0.99. The observed data shows that the pseudo $2^{nd}$ order kinetic model was followed. The rate-determining step depends on both concentration of U(VI) and sorbents [19, 43].

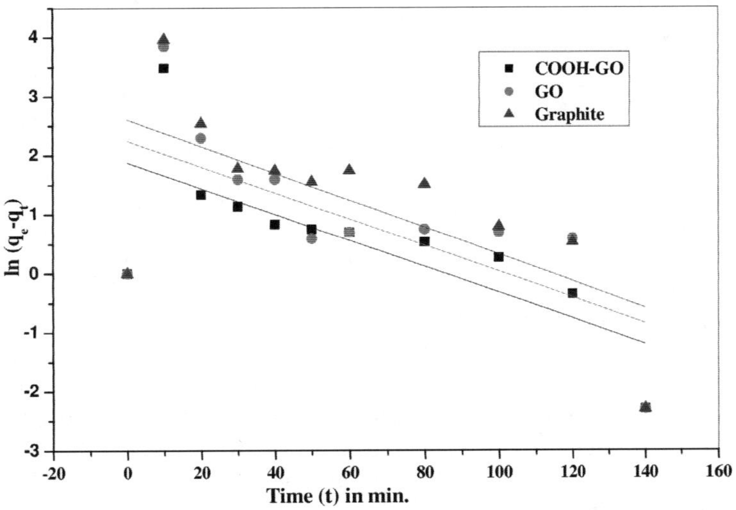


**Figure 7.** Pseudo $1^{st}$order kinetic plot between time and ln($q_e$-$q_t$).

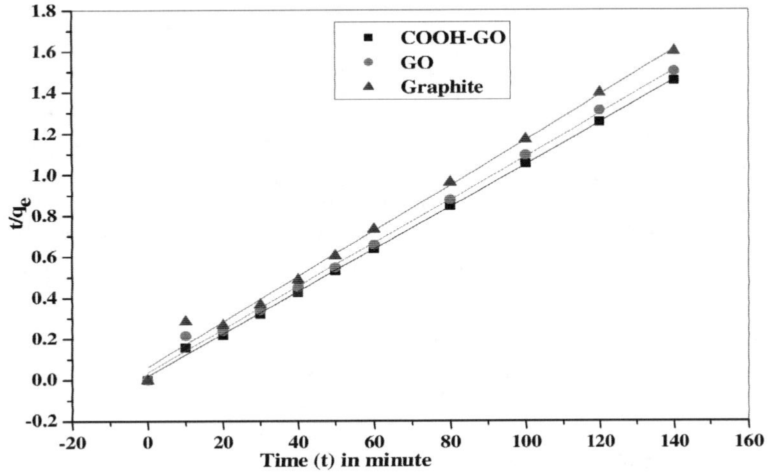


**Figure 8.** Pseudo $2^{nd}$order kinetic plot between time and t/$q_e$

**Table 3.** Parameters calculated for pseudo 1st order and pseudo 2nd order model.

| Observed Parameters | Sorbent materials | | | | | |
|---|---|---|---|---|---|---|
| | COOH-GO | | GO | | Graphite | |
| Order of Kinetics | Pseudo 1st order | Pseudo 2nd order | Pseudo 1st order | Pseudo 2nd order | Pseudo 1st order | Pseudo 2nd order |
| **Intercept** | 1.8778 | 0.0188 | 2.2446 | 0.0366 | 2.6066 | 0.0622 |
| **Slope** | -0.022 | 0.0103 | -0.0221 | 0.0105 | -0.0228 | 0.011 |
| **$q_e$ (mg/g)** | 6.5391 | 97.0873 | 9.4366 | 95.2381 | 13.5528 | 90.909 |
| **$q^2$ Square** | - | 9425.959 | - | 9070.295 | - | 8264.46 |
| **$k_1$** | -0.00016 | - | -0.00016 | - | -0.0002 | - |
| **$k_2$** | - | 0.005643 | - | 0.003012 | - | 0.00195 |
| **$R^2$** | 0.5372 | 0.999 | 0.4497 | 0.9997 | 0.4424 | 0.9929 |

*3.9 Thermodynamic Studies for Sorption*

The effect of temperature on U(VI) sorption was studied by using a water bath. The condition for sorption were $C_0$ = 10 μg/ml, mass of sorbent = 10 mg, pH = 4.5, and volume of sample =10 mL. The sorption capacity of sorbent materials was taken as a function of temperature in Kelvin. In this study, the three different temperatures were taken that is 298K, 308K, and 318K for all three sorbents (Figure 9). The sorption capacity of U(VI) was found to increase with the increasing temperature which indicates the sorption of U(VI) is endothermic (Table 4). This result is similar to previously obtained results of U(VI) sorption by activated carbon [44, 45]. The Thermodynamic parameters ΔG° ($kJmol^{-1}$), ΔH°($kJmol^{-1}$), and ΔS° ($Jmol^{-1}K^{-1}$) were also calculated for sorption. The value of ΔH°($kJmol^{-1}$) and ΔS° ($Jmol^{-1}K^{-1}$) were calculated by slope and intercept of linear line equation $\ln K_L$ versus 1/T by van't Hoff equation [46].

$$K_L = \frac{C_{ad}}{C_e} \tag{10}$$

$$\ln K_L = -\frac{\Delta H^o}{RT} + \frac{\Delta S^o}{R} \tag{11}$$

$$\Delta G^o = \Delta H^o - T\Delta S^o \tag{12}$$

Where, the value of $K_L$ (mL/g) = the distribution coefficient of U(VI), T = temperature in Kelvin, and R = gas constant (8.314 J/(mol K).

From Table 4, all these parameters were calculated from the linear fitting line Eq. 11. The value of ΔG°= -Ve (negative) indicates that the sorption is spontaneous in nature, ΔH°= +Ve (positive) means sorption is endothermic in nature, and ΔS° = +Ve means entropy of the system increases under these conditions [47].

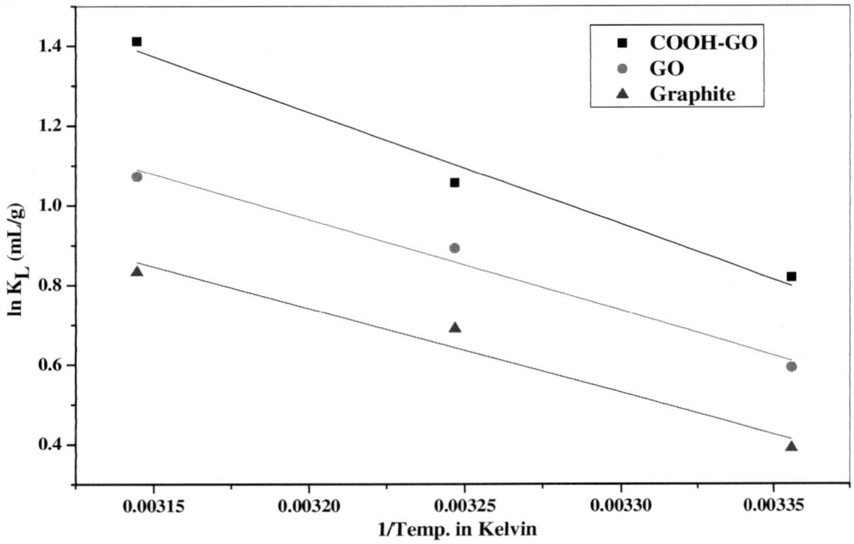


**Figure 9.** Thermodynamic graph between ln$K_L$ versus 1/T

**Table 4.** Thermodynamic parameters for U(VI) adsorption on sorbent materials.

| Parameters | Sorbent Materials | | | | | | | | |
|---|---|---|---|---|---|---|---|---|---|
| | COOH-GO | | | GO | | | Graphite | | |
| **Temp. in Kelvin** | 298 | 308 | 318 | 298 | 308 | 318 | 298 | 308 | 318 |
| **ΔG° ($kJmol^{-1}$)** | -2.02 | -2.70 | -3.73 | -1.46 | -2.28 | -2.83 | -0.96 | -1.76 | -2.20 |
| **R ($Jmol^{-1}K^{-1}$)** | | 8.314 | | | 8.314 | | | 8.314 | |
| **ΔH° ($kJmol^{-1}$)** | | 23.28 | | | 18.92 | | | 17.49 | |
| **ΔS° ($Jmol^{-1}K^{-1}$)** | | 84.76 | | | 68.58 | | | 62.14 | |

**Table 5.** Comparison of the U(VI) sorption capacities of GO and COOH-GO with other sorbent materials.

| Sorbent Materials | pH | Temp. (K) | Equilibrium time (min.) | $q_{max}$ (mg/g) | References |
|---|---|---|---|---|---|
| **GO** | 5 | 293 | 60 | 122.4 | [48] |
| **Reduced GO** | 4 | 293 | n/a | 74.1 | [33] |
| **Cyclodextrin-modified GO** | 5 | 288 | n/a | 97.3 | [49] |
| **Multi-walled carbon nanotubes (MWCNTs)** | 5 | 318 | 60 | 39.5 | [50] |
| **Activated carbon** | 3 | 293 | 180 | 28.3 | [51] |
| **Graphite** | 4.5 | 298 | 80 | 116.14 | This study |
| **GO** | 4.5 | 298 | 80 | 128.53 | This study |
| **COOH-GO** | 4.5 | 298 | 80 | 148.36 | This study |

## 4. Conclusion

The GO and COOH-GO were synthesized and characterized successfully by UV-VIS Spectroscopy, Fourier transformed infrared (FTIR) spectroscopy, X-ray diffraction (XRD), Scanning electron microscope (SEM), Raman Spectra, and high-resolution transmission electron spectroscopy (HR-TEM). The GO and COOH-GO synthesis was cost-effective, environment friendly, and have high efficiency for U(VI) removal. The batch sorption experiment was studied successfully by the effect of pH, contact time, amount of sorbent dose, initial U(VI) concentration, and temperature. The removal of uranium was found to be more for COOH-GO than GO and graphite due to the presence of more oxygen-containing domains in COOH-GO. The adsorption isotherms were studied successfully and the sorption was best fitted for the Langmuir adsorption model ($R^2 > 0.99$) as compared to the Freundlich adsorption model confirming monolayer adsorption. The kinetics was studied successfully and was best fitted ($R^2 >$ 0.99) for the pseudo-second-order model. The thermodynamic studies suggested that the adsorption was endothermic and spontaneous in nature.

## Acknowledgments

The research was supported by the Council of Scientific and Industrial Research (CSIR), Government of India, File No. 09/1116(0008)/2018-EMR-I, and the authors are thankful to the National Institute of Technology, Raipur, India for providing necessary research facilities.

## References

[1] Li Z J, Huang Z W, Guo W L, Wang L, Zheng L R, Chai Z F and Shi W Q 2017 *Environmental Science & Technology* **51** 5666

[2] Wang L, Tao W, Yuan L, Liu Z, Huang Q, Chai Z and Shi W 2017 *Chemical Communications* **53** 12084

[3] Tan L, Wang J, Liu Q, Sun Y, Zhang H, Wang Y and Song D 2015 *Colloids and Surfaces A: Physicochemical and Engineering Aspects* **466** 85

[4] Lingamdinne L P, Choi Y L, Kim I S, Yang J K, Koduru J R and Chang Y Y 2017 *Journal of hazardous materials* **326** 145

[5] De Decker J, Folens K, De Clercq J, Meledina M, Van Tendeloo G, Du Laing G and Van Der Voort P 2017 *Journal of hazardous materials* **335** 1

[6] Zhang C, Dodge C J, Malhotra S V and Francis A J 2013 *Bioresource technology* **136** 752

[7] Shen J and Schäfer A 2014 *Chemosphere* **117** 679
[8] Sreenivas T and Rajan K C 2013 *Separation and purification Technology* **112** 54
[9] Metwally S S, Ahmed I M and Rizk H E 2017 *Journal of Alloys and Compounds* **709** 438
[10] Zhang H, Gu L, Zhang L, Zheng S, Wan H, Sun J and Xu Z 2017 *Applied Surface Science* **406** 330
[11] Xu H, Qu Z, Zong C, Zhu W, Chen W and Yan N 2016 *Fuel* **182** 428
[12] Zhang L, Wei J, Zhao X, Li F, Jiang F, Zhang M and Cheng X 2016 *Chemical Engineering Journal* **302** 733
[13] Pan N, Li L, Ding J, Li S, Wang R, Jin Y and Xia C 2016 *Journal of hazardous materials* **309** 107
[14] Li Z, Chen F, Yuan L, Liu Y, Zhao Y, Chai Z and Shi W 2012 *Chemical engineering journal* **210** 539
[15] Cheng W, Ding C, Wu Q, Wang X, Sun Y, Shi W and Wang X 2020 *Environmental Science: Nano* **7** 1617
[16] Wan S, He F, Wu J, Wan W, Gu Y and Gao B 2016 *Journal of Hazardous Materials* **314** 32
[17] Li D, Müller M B, Gilje S, Kaner R B and Wallace G G 2008 *Nature nanotechnology* **3** 101
[18] Jasim D A, Lozano N and Kostarelos K 2016 *2D Materials* **3** 014006
[19] Patel K, Devi R, Dewangan P K, Sutar A K and Maharana T 2020 *Materials Science and Engineering Conference Series* **798** 012029
[20] Stopa L C B and Yamaura M 2010 *International Journal of Nuclear Energy Science and Technology* **5** 283
[21] Janu V C, Meena R K, Kumar N and Sharma R K 2020 *Journal of Environmental Chemical Engineering* **8** 104218
[22] Philippou K, Anastopoulos I, Dosche C and Pashalidis I 2019 *Journal of Environmental Management* **252** 109677
[23] Mohamud H, Ivanov P, Russell B C, Regan P H and Ward N I 2018 *Journal of radioanalytical and nuclear chemistry* **316** 839
[24] Paredes J I, Villar-Rodil S, Martínez-Alonso A and Tascon J M D 2008 *Langmuir* **24** 10560
[25] Cao H, Wu X, Yin G and Warner J H 2012 *Inorganic chemistry* **51** 2954
[26] Langmuir I 1916 *Journal of the American chemical society* **38** 2221
[27] Freundlich H M F 1906 *Z Phys Chem* **57A** 385
[28] Lagergren S *K 1898 SevenskaVetenskapsakad Handl* **24** 1
[29] Ho Y S and McKay G 1999 *Process biochemistry* **34** 451
[30] Ma J, Liu C, Li R and Wang J 2012 *Journal of applied polymer science* **123** 2933
[31] Pendolino F, Parisini E and Lo Russo S 2014 *The Journal of Physical Chemistry C* **118** 28162
[32] Bagri A, Mattevi C, Acik M, Chabal Y J, Chhowalla M and Shenoy V B 2010 *Nature chemistry* **2** 581
[33] Sun Y, Yang S, Chen Y, Ding C, Cheng W and Wang X 2015 *Environmental science & technology* **49** 4255
[34] Ding Z, Ma H and Chen Y 2014 *RSC advances* **4** 55290
[35] Xie Y, Helvenston E M, Shuller-Nickles L C and Powell B A 2016 *Environmental science & technology* **50** 1821
[36] Wilson N R, Pandey P A, Beanland R, Young R J, Kinloch I A, Gong L and Sloan J 2009 *ACS nano* **3** 2547
[37] Shen J, Hu Y, Shi M, Lu X, Qin C, Li C and Ye M 2009 *Chemistry of Materials* **21** 3514
[38] Shalaby A, Nihtianova D, Markov P, Staneva A D, Iordanova R S and Dimitriev Y B 2015 *Bulgarian Chemical Communications* **47** 291
[39] Ding Z, Ma H and Chen Y 2014 *RSC advances* **4** 55290
[40] Ivanov P, Griffiths T, Bryan N D, Bozhikov G and Dmitriev S 2012 *Journal of Environmental Monitoring* **14** 2968

[41] Wang Y, Liu X, Huang Y, Hayat T, Alsaedi A and Li J 2017 *Journal of Radioanalytical and Nuclear Chemistry* **311** 209
[42] Sierra-Trejo P V, Guibal E and Louvier-Hernández J F 2020 *Journal of Polymers and the Environment* **28** 934
[43] Yang A, Zhu Y and Huang C P 2018 *Scientific reports* **8** 1
[44] Kütahyalı C and Eral M 2004 *Separation and Purification Technology* **40** 109
[45] Kütahyalı C and Eral M 2010 *Journal of Nuclear Materials* **396** 251
[46] Xue G, Yurun F, Li M, Dezhi G, Jie J, Jincheng Y and Yujun Z 2017 *Applied Surface Science* **402** 53
[47] Fan L, Luo C, Sun M, Qiu H and Li X 2013 *Colloids and Surfaces B: Biointerfaces* **103** 601
[48] Liu S, Li S, Zhang H, Wu L, Sun L and Ma J 2016 *Journal of Radioanalytical and Nuclear Chemistry* **309** 607
[49] Song W, Shao D, Lu S and Wang X 2014 *Science China Chemistry* **57** 1291
[50] Fasfous I I and Dawoud J N 2012 *Applied Surface Science* **259** 433
[51] Mellah A, Chegrouche S and Barkat M 2006 *Journal of colloid and interface science* **296** 434

# In-vitro molecular interaction of boron doped carbon dots with human serum albumin

**Vinayak Sahu[1], Fahmida Khan[1*], Pradeep Kumar Dewangan[1], Komal Kashyap[1], Sonalika Agrawal[1] and Vikas Kumar Jain[2]**

[1] National institute of technology, Raipur (C.G.), India
[2] Government College of Engineering, Raipur (C.G.), India

*Corresponding author's e-mail address: fkhan.chy@nitrr.ac.in, vinayaksahu2208@gmail.com

**Abstract.** Carbon dots have drawn prodigious attention in biomedical and biological fields by virtue of its unique optical properties, but their effect on structure and behaviour of essential proteins have rarely been explored. In this study boron doped carbon dots (BCDs) which was synthesized by microwave treatment of citric acid, boric acid and urea, with maximum emission wavelength of 444 nm when excited upon long wavelength ultra-violet light. Synthesized BCDs were characterized by UV-Visible, FTIR, HRTEM and Fluorescence spectroscopy. Interaction between BCDs and human serum albumin (HSA) is carried out in physiological conditions. Stoichiometry of BCDs-HSA was found to be 1:1 and various thermodynamic parameters have also been calculated. The negative value of $\Delta G$ suggested that interaction between HSA and BCDs was spontaneous in nature.

## 1.Introduction

A novel class of carbon nanomaterial have been widely used in biomedical and biological field due their superior properties such as tunable fluorescence, photostability, excellent water solubility, resistance towards photo-bleaching, low toxicity and excellent biocompatibility. Adsorption of protein on the surface of NPs may changes the protein structure and may alter its biological activities. HSA is the major and most important protein of the blood plasma which is used as carrier in the circulatory system. It is produced in the liver and distributed in the blood plasma as a non-glycosylated protein. The binding interaction of serum albumin gives idea about clinical, bio-chemical and pharmaceutical applications [1-5]. HSA consists of three helical domains, and contain 585 amino acids each domain composed of two subdomains. Xu et al. studied the interaction between semi-conductor quantum dots i.e. CdTe and HSA, from this study it is revealed that CdTe binds to HSA and alters the structure of HSA [6]. Interaction between serum albumin and carbon based nanomaterials i.e. CDs and GQDs has also been reported but interaction between hetero atom doped carbon dots and HSA have rarely been explored.

In the present study synthesized BCDs have been used for in-vitro molecular interaction between BCDs and HSA has been done using multi spectroscopic techniques and various thermodynamic parameters were calculated.

## 2.Experimental

*2.1 Materials*

Boric acid, citric acid, urea was purchased from Sigma Aldrich Chemical Pvt. Ltd. Mumbai. HSA about 99% pure was procured from TCI chemical Pvt. Ltd. Warfarin and Ibuprofen were obtained from Pax Drug Company. All the chemicals used in the experiment were of analytical reagent grade.

*2.2 Preparation and purification of BCDs*

BCDs was prepared according to reported method [7] with slight modification, obtained raw solution of BCDs was dialyzed for 24 hours against distilled water to eliminate impurities. Purified BCDs was stored at 277 K for further use.

*2.3 Characterization*

In the UV-Visible spectrum of BCDs bands at 240 nm for π-π* and 340 nm for n- π* electronic transition obtained respectively. Fluorescence spectra of BCDs shows band at 444 nm. Colour of diluted solution of BCDs was light yellow and when it is exposed to UV light, it gives blue fluorescence as shown in inset of figure 1(a). When diluted solution of BCDs excited upon various excitation wavelengths it shows maximum fluorescence intensity at 444 nm when excitation wavelength 345 nm was applied. Figure1(b). FTIR spectra confirm the presence of –COOH and –OH functional groups in BCDs as shown in figure 1(c). Morphology and size of prepared BCDs were analysed by HRTEM which indicates the spherical and below 10 nm size of BCDs as shown in figure 1 (d).

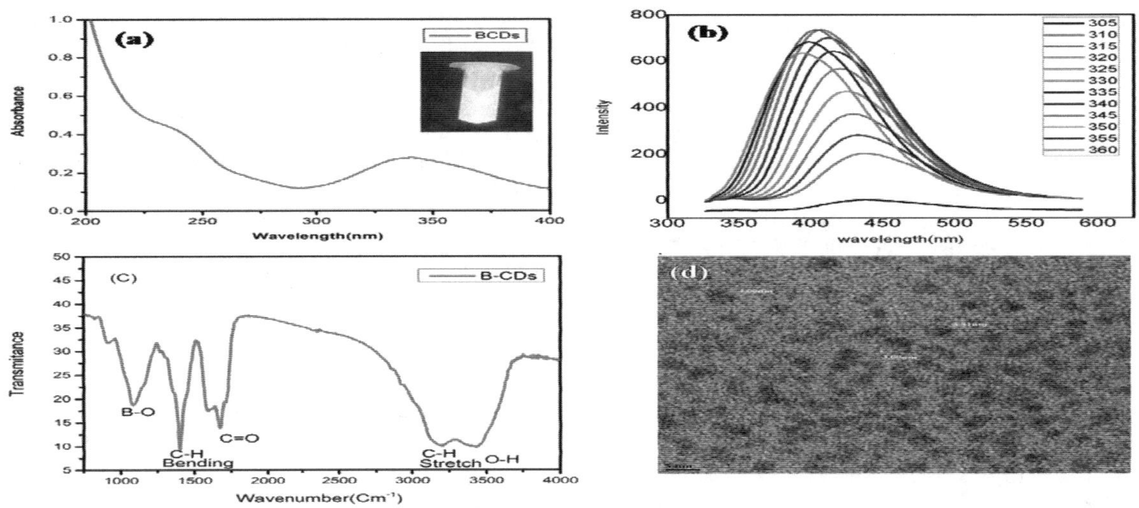


**Figure 1.** (a) UV-Visible spectra of BCDs, (b) FL excitation spectra, (c) FTIR spectra of BCDs (d) HRTEM image of BCDs

*2.4 Calculation of quantum yield of BCDs* [8]

$$Q = Q_R \frac{I}{I_R} \frac{OD_R}{OD} \frac{n^2}{n_R{}^2} \quad (1)$$

Where, $Q_R$ = quantum yield of reference, I = deliberate emission intensity is optical density (i.e. measured by UV-Vis spectrophotometer). The quantum yield of BCDs obtained is 4.05 using 350 nm excitation with reference quinine sulphate which has quantum yield 54% in 0.1 M $H_2SO_4$(n=1.33).

### *2.5 Interactions of HSA with BCDs*

A 2.0 x $10^{-6}$ M HSA solution was mixed with BCDs solutions ranging from 0.0 – 4.0 x $10^{-3}$ M. The mentioned solution was incubated for 10 minute for the completion of interaction. Then fluorescence spectra were taken at three different temperatures 295K, 300K and 305K respectively as shown in figure 2(a, b, c).

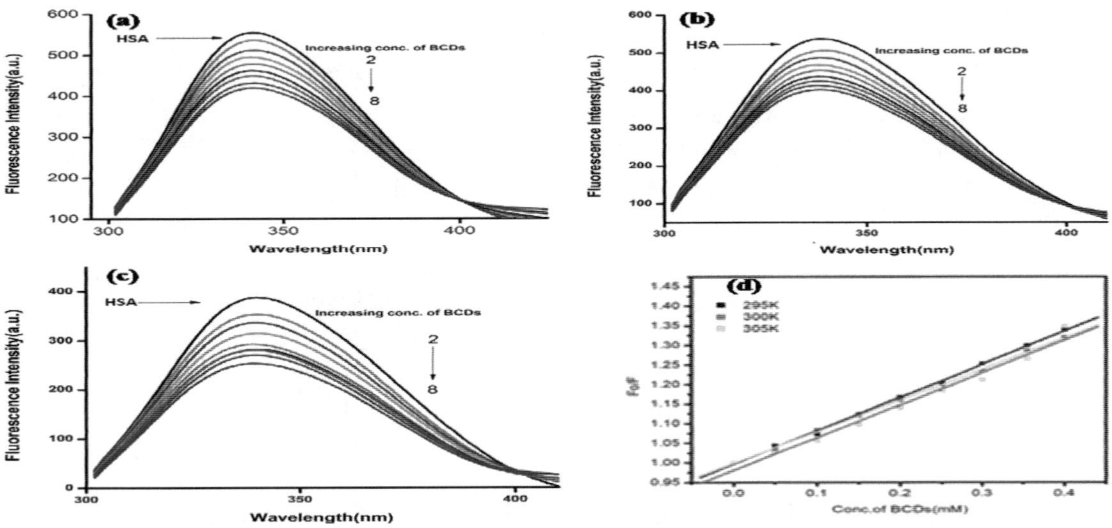


**Figure 2.** (a, b, c) FL quenching of HSA by BCDs at three different temperatures, (d) Linear calibration.

## 3.Results and discussion

### *3.1 Quenching of fluorescence of HSA by BCDs*

Fluorescence spectra of HSA is because of tryptophan (Trp) residue. It is fluorescence is depends upon its intrinsic environment and very sensitive towards different molecules[5]. HSA shows fluorescence band at 350 nm when excited upon 278 nm of wavelength. It is observed that fluorescence of HSA was quenched linearly when BCDs solution with increasing concentration was added.

The quenching of fluorescence of HSA by BCDs was analysed at three different temperatures using Stern-Volmer equation [9].

$$\frac{F_0}{F} = 1 + K_{SV}[Q] \qquad (2)$$

Where F (FL spectra without BCDs) and Fo (FL spectra with BCDs) [Q] is the concentration of BCDs. By plotting graph between [Q] and $F_O/F$ the value of was $K_{SV}$ determined (Table 1).

The apparent binding constant and binding sites can be calculated using following equation [10].

$$\log\frac{Fo/F}{F} = \log K_a + n\log[Q] \qquad (3)$$

By plotting graph between log ($F_o$/F) /F Vs log [Q], the value of Ka and n was calculated. There is 1:1 complex formation between HSA:BCDs. Thermodynamic parameters can be calculated using following formula (Table 2)

$$\log Ka = -\frac{\Delta H}{2.303RT} + \frac{\Delta S}{2.303R} \qquad (4)$$

$$\Delta G = \Delta H - T\Delta S \qquad (5)$$

3.2 Binding site

HSA consists of two domains (II & III) which are homologous and each domain consist of two sub-domain (A & B), to find where the BCDs is binding, site binding fluorescence probes i.e. warfarin for site I and ibuprofen for site II. For the investigation of binding site displacement reactions were performed on the basis of figure 3 (c) and 3(d), on addition of warfarin fluorescence intensity continuously increased and in case of ibuprofen addition fluorescence intensity increased and remains constant, which shows that ibuprofen is further available for binding. Results reveals that BCDs binds to Site I of HSA i.e. domain IIA.

*3.3 Conformational change investigation*

FTIR spectroscopic studies show that conformational changes of HSA in the presence of BCDs are shown in figure 3(b). HSA shows two amide bands, 1653 $cm^{-1}$ for C=O stretch, and 1543 $cm^{-1}$ for C-N stretching and N-H bending. After addition of BCDs there is shifting of spectra is observed up to 1656and 1545 $cm^{-1}$. These results shows that BCDs affects the C=O , C-N and N-H stretching frequencies [1].

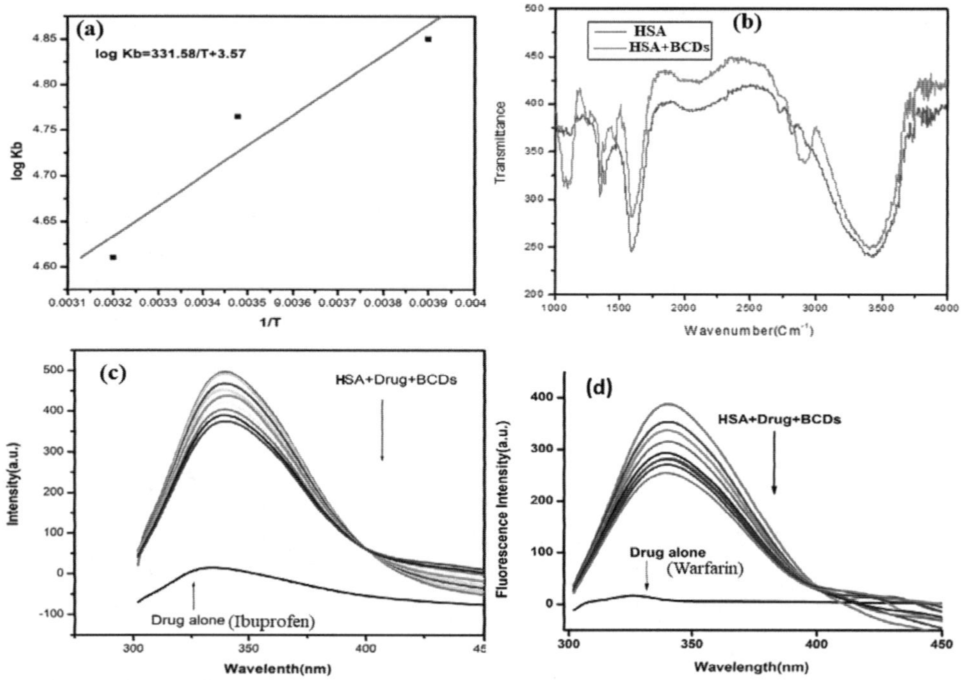


**Figure 3.** (a) Plot for calculation of binding constant, (b) FTIR spectra of pure HSA and with BCDs, (c, d) Displacement reaction using site probes ibuprofen and warfarin respectively.

**Table 1.** Stern-Volmer quenching constants $K_{sv}$ at three temperatures and pH.

| pH value | T(K) | Ksv ($10^3$ L $mol^{-1}$) | $R^2$ | S.D.[a] |
|---|---|---|---|---|
| 4.0 | 305 | 2.46 | 0.992 | 0.001 |
| 5.0 | 300 | 2.75 | 0.991 | 0.002 |
| 7.0 | 295 | 2.91 | 0.998 | 0.001 |

**Table 2.** Stern-Volmer quenching constants $K_{sv}$, associative binding constant and thermodynamic parameters of HSA-BCDs system.

| T(K) | $K_{sv}$ ($10^3$ L mol$^{-1}$) | Ka ($10^3$L mol$^{-1}$) | ΔH (KJ mol$^{-1}$) | $\Delta G$ (KJ mol$^{-1}$) | $\Delta S$ (J mol$^{-1}$) |
|---|---|---|---|---|---|
| 305 | 2.86 | 4.61 | -6.34 | -27.18 | 68.35 |
| 300 | 2.96 | 4.78 | -7.83 | -25.83 | 60.00 |
| 295 | 3.01 | 4.85 | -9.81 | -28.39 | 63.00 |

## 4.Conclusions

In this work, Boron doped carbon dots have been synthesized successfully in an aqueous medium using microwave treatment technique. Then molecular interaction of BCDs with HSA was investigated by multi-spectroscopic techniques. The results reveal that quenching of intrinsic fluorescence of HSA by BCDs occurred because of the formation of a 1:1 non-fluorescent ground state complex. Displacement experiments suggested that the BCDs bind to the sub domain IIA (sudlow site I) of HSA. Studies of various thermodynamic parameters show that the binding process of BCDs to HSA is spontaneous in nature. Results show that higher concentrations of BCDs can alter the confirmation of serum protein. Based on results obtained one can better understand the potential toxicity of BCDs at molecular level. BCDs-HSA complex may alter the function of the protein by some extent when introduced in the biological systems.

## Acknowledgment

The authors are thankful to the director of the National Institute of Technology Raipur for laboratory facilities. Financial support from CSIR, New Delhi (file no. 09/1116(0003)/2017-EMR-I) is gratefully acknowledged.

## References

[1] Vaishnav S K, Karbhal I, Satnami M L and Ghosh K K 2018 *J. Mol. Liq.* **255** 279
[2] Huang S, Qiu H, Lu S, Zhu F and Xiao Q 2015 *Journal of hazardous materials* **285** 18
[3] Treuel L and Nienhaus G U 2012 *Biophys. Rev.* **4** 137
[4] Hao A, Guo X, Wu Q, Sun Y, Cong C and Liu W 2016 *J. Lumin.* **170** 9096
[5] Huang S, Xie J, Cui J, Liu L, Liang Y, Liu Y and Xiao Q 2017 *Steroids.* 2017 **128** 136
[6] Huang S, Qui HN, Liu Y, Huang CS, Sheng JR, Su W and Xiao Q 2015 *Colloids Surf. B.* **136** 955
[7] Bourlinos A B, Trivizas G, Karakassides M A, Baikousi M, Kouloumpis, Gournis A D, Bakandritsos A, Hola K, Kozak O, Zboril R and I Papagiannouli I 2015 *Carbo* **83** 173.
[8] Guo L, Ge J, Liu W, Niu G, Jia, Q, Wang H and Weng P 2016 *Nanoscale.* **8** 729.
[9] Yin B T, Yan C Y, Peng X M,. Zhang S L, Rasheed S, Geng R X and Zhou CH 2014 *Eur. J. Med. Chem.* **71** 148
[10] Xiao C Q, Lai L, Zhang L, Wang S Y, Yuan S, Xu ZQ and Liu Y 2018 *J. Solution Chem.* **47** 1438.

# Structural and optical properties of Pr doped $CuInS_2$ thin films synthesized by chemical bath deposition

**S Sengupta[1*] and M A Pateria[2]**

[1] Chhatrapati Shivaji Institute of Technology, Dist. Durg, Chhattisgarh
[2] Department of Applied Physics, SSGI, Bhilai Chhattisgarh

*Corresponding author's e-mail address: sumita.2006@gmail.com

**Abstract.** The Pr doped $CuInS_2$ thin films have been prepared by chemically bath deposition technique at $80^0$Ctemperature. The semiconducting films are grown in bath containing aqueous solution of copper (II) chloride dihydrate, Indium (III) chloride, thiourea, TEA, ammonia (25%). Pr is used as the doping element. Chloride dehydrates is used for the copper ion source, Indium chloride for indium ion source and thiourea for sulphur ion source, TEA is a complexing agent and ammonia maintains the pH. The SEM Studies of Pr doped $CuInS_2$ films show that at lower volume polyhedral shaped particle appears on the surface. No voids are seen. But at higher volume of Pr; voids are seen on the surface. Particle looks like accumulated woolen balls or cubic like grains. Agglomeration of particle increases with volume and hence particle size also increases. In the XRD of prepared sample characteristic peaks of chalcopyrite phase appears between ($24^0$- $80^0$) in all the films. Hall measurement shows that resistivity of Pr (6ml) doped film is very high. The conductivity of all doped films is higher than undoped film. In general, increased doping will leads to increased conductivity. In the Photoluminescence study the emission peak appears at wavelength 575nm, 550nm and 525nm wavelength. High transmittance, low reflectance found in the visible region in the entire prepared sample. The absorption peaks fall in the visible region. High Photocurrent is found, and in dark current study the linear behavior between voltage and current are seen. Quite good photosensitivity is observed in Pr doped $CuInS_2$ films.

## 1. Introduction

In 1953 Hahn and co-workers amalgamated a bunch of substances which reveal chalcopyrite structure with a composition of I–III –$VI_2$. In 1954, Goodman and Douglass showed the possibility of semi conductivity in such materials. Henceforth these substances are recognized as chalcopyrite type semiconductors. Since last two decades' ternary chalcopyrite materials have drawn attention of the researchers worldwide due to their electro-optical properties which make them to be used to develop photovoltaic system as an alternative energy source. Out of different chalcopyrite semiconductors $CuInS_2$ and $CuInSe_2$ both presently probed in utilization in devices such as light emitting diodes, photovoltaic and solar cells etc. [1]. $CuInS_2$ thin films with chalcopyrite structure are having high scope as absorptive layers. Many exorbitant physical and chemical characteristics shown by $CuInS_2$ thin films like high value of coefficient of absorption $10^5$ $cm^{-1}$ within perceptible spectral range, more forbearance for occurrence of defects and a direct band gap of 1.5eV. Also in $CuInS_2$ the conversion efficiency of solar energy is most favorable to be used as photovoltaic devices, likelihood to avoid n and p-type

IOP Conf. Series: Materials Science and Engineering 1120 (2021) 012013 doi:10.1088/1757-899X/1120/1/012013

conductivity and high chemical stability [2]. Further $CuInS_2$ is more environmental friendly material than those containing Se or Cd [3]. For deposition of $CuInS_2$ thin film large number of techniques are there. For example; sol-gel [4], spray pyrolysis [5], d.c. magnetron sputtering [6], ILGAR [7], chemical bath deposition [8] etc. Various researchers have used chemical bath deposition (CBD) method for depositing different thin films. The reason is that it is the simplest method which requires only solution containers and mounting devices which make it economical. Large area processing is possible with this method. Also by varying the deposition time the thickness of the film can be controlled. With this method we can produce homogeneous, adherent and unbreakable films with good reproducibility. Therefore it is planned to synthesis films by CBD technique using appropriate fluxes, impurities and complexing agents to control reaction rate and hence to investigate the corresponding changes in structural and optical properties.

## 2. Experimental details

The solution of copper (II) chloride dihydrate [$CuCl_2.2H_2O$ (0.5M)] and Indium (III) chloride [$InCl_3$ (0.5M)] were taken in a beaker. Both the solutions were mixed uniformly by stirring. Then triethanolamine (TEA) and ammonia (25%) were added into the beaker. Then sulphur source thiourea $SC(NH_2)_2$ was added to the solution. All AR grade 99.9% pure chemicals were used for preparing the solution. Stirring was done for 10 minutes in order to obtain uniform and adherent film. The cleaned microscopic glass substrates were dipped vertically into the solution. The beaker is then kept inside the liquid bath at $80^0$C temperature for 1 hour. Definite volumes of 0.01M solutions of cadmium chloride ($CdCl_2$) and praseodymium nitrate [$Pr(NO_3)_3$] were added to the original solution to carry out sensitization and then the depositions were made. After 1 hour, the substrates were taken out and washed thoroughly with double distilled water and kept for drying at room temperature for 3-4 h. The synthesized films were homogeneous and well adhered to the substrate.

## 3. Results and discussion

The XRD of CIS:Pr films with varying volume of praseodymium were shown in figures 1, 2 and 3. Film with Pr(4ml) consist of very few peaks of chalcopyrite structure. Peaks corresponding to (004) and (116/312) were observed. No peak corresponding to wurtzite phase was observed. Two additional peaks for (022) plane and (331) plane was observed for $Cu_{1.8}S_1$ and $CuIn_3$ respectively. Less number of peaks reveals poor crystallinity. Also less number of peaks means that the crystal is oriented in preferred direction. Similarly in the XRD of CIS film doped with Pr (6ml) there were only two main characteristic peak (112) and (004) for chalcopyrite phase appeared with less intensity. In addition to this; one peak corresponding to wurtzite phase (010) also appeared. Thus this film shows polytypic behavior. X-ray diffractograms of CIS:Pr(8ml) doped film consist of very high intense characteristic peaks (112),(121),(123), (116/312) and (134) corresponding to chalcopyrite phase (JCPDS-98-062-8051). No peak corresponding to wurtzite was appeared. Thus the synthesized film is highly crystalline in nature as more characteristic peaks were appeared with chalcopyrite structure. Additional peak corresponding to (022), (004) and (165) plane was observed for $Cu_{1.8}S_1$, $Cu_2S_1$ and $In_2S_3$ respectively. Thus the synthesized film was multiphase. At last we can conclude that doping of Pr affects the crystallinity of the sample and crystallinity increases with the increase of volume of doping element Pr.

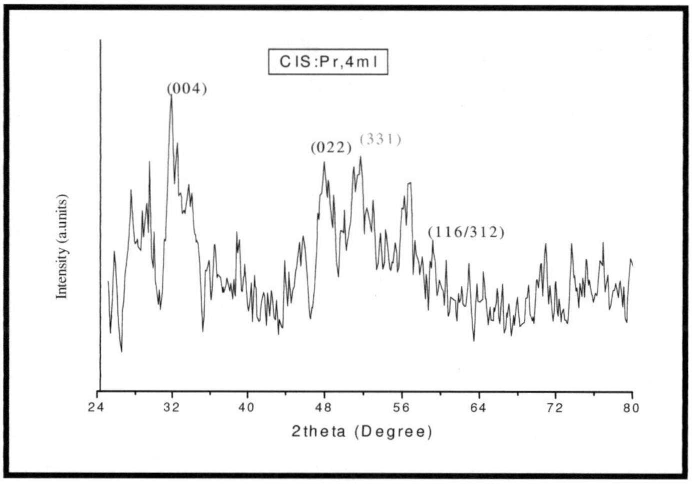


**Figure 1.** X-ray diffractogram of Praseodymium (4ml) doped $CuInS_2$ film.

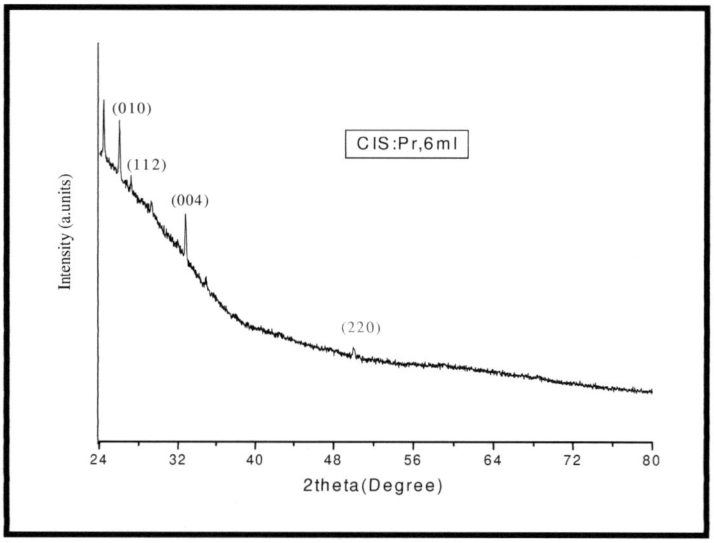


**Figure 2.** X-ray diffractogram of Praseodymium (4ml) doped $CuInS_2$ film.

In the dark current study the synthesized film is kept in dark region that means there is no light comes near to the prepared samples. The prepared thin films are good for conductivity. The concept of dark current in the semiconducting thin films is the movement of thermally excited charge carriers inside the material. Some of the charge carriers jump into the conduction band which is responsible for the conductivity. The dark current appears for limited voltage. We have observed the dark current appears in nano scale range.

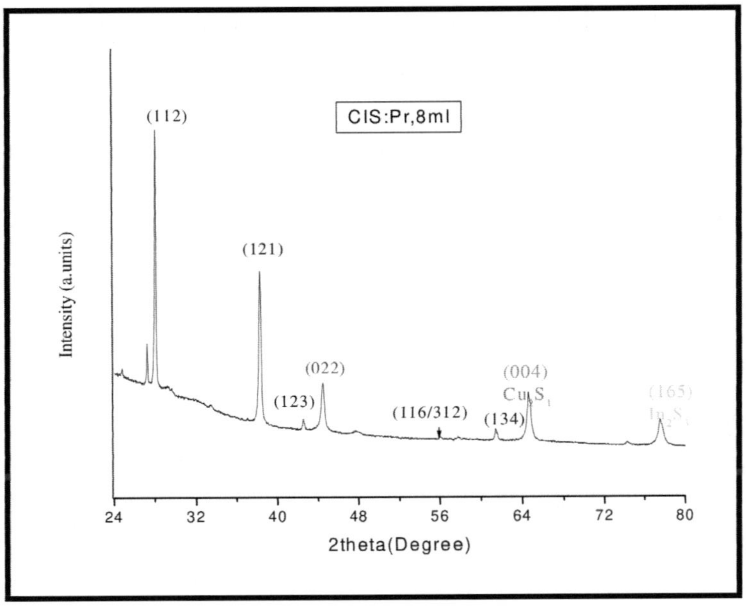


**Figure 3.** X-ray diffractogram of Praseodymium (8ml) doped $CuInS_2$ film.

It is clear from the figure that dark current varies linearly with applied voltage. This linear behavior shows that the electrode used is of ohmic type. In figure 4 rise and decay curves of Praseodymium (Pr) doped $CuInS_2$ films are shown. As volume of doping element increases the photo current also increases.

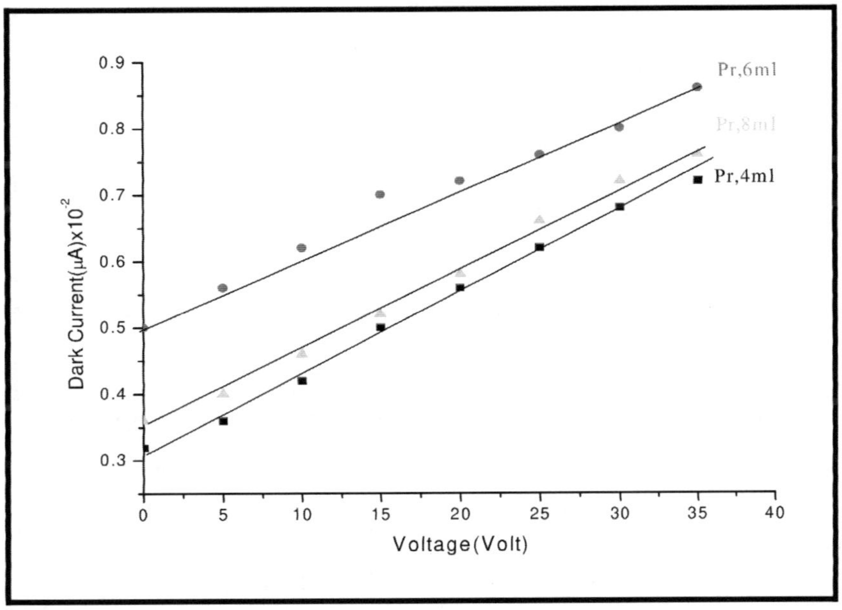


**Figure 4.** Graph between dark current and voltage of various Pr doped $CuInS_2$ film.

Figure 5 shows the rise and decay curves for CIS:Pr films with different concentration. The rise and decay curve looks identical in all the sample. When the radiation fall on the prepared thin film material the electron hole pair (photo carriers) becomes originate and those are responsible for the photocurrent

In the Present curves fast rise in the beginning followed by saturation. The increase in photocurrent is due to the generation of photo carries. The slowly increasing part is when recombination becomes dominant. The decay part is in the form of exponential equation. The ratio of photocurrent to dark current is found to be the order of $10^3$ which is quite good. The mobility of charge carrier increases with increasing the volume of doping element and it is due to the inter gains barrier height is lowered. The production of photocurrent increases when the volume of Pr increases. Thus more charge carriers are generated by the rare earth element. The role of rare earth element is to provide a platform to the charge carriers to come in the conduction band so as to give the conductivity. Decay rates appeared to be slower in the reported films.
The lifetime of the charge carriers are calculated by using the relation

$$\tau = \Delta\sigma\ st\ /\ tan\delta \quad (1)$$

where σ is the static photocurrent and tan$\delta$ is the slope of the initial state of the decay curve. The mobility is calculated by using the relation

$$G = \frac{\tau\mu V}{L^2} \quad (2)$$

Where G is photoconductivity gain, V is applied voltage between the electrodes and L is the distance between electrodes.

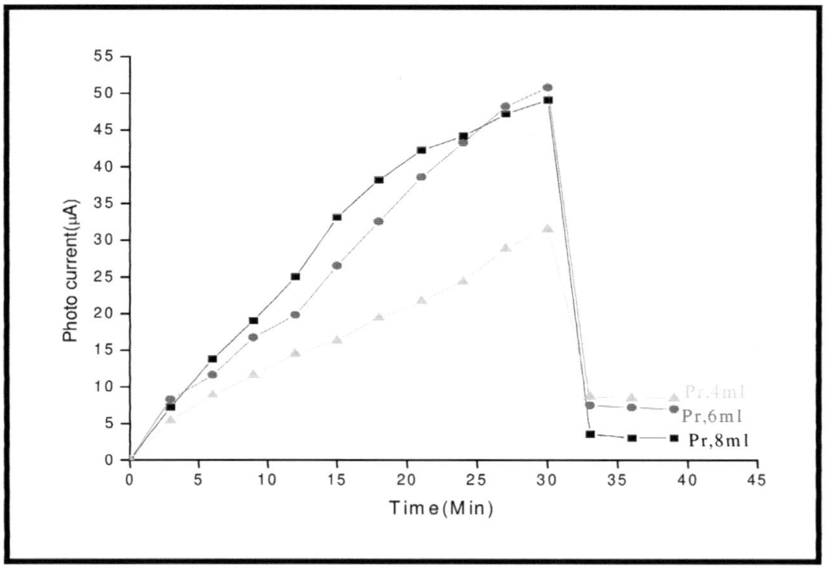


**Figure 5.** Rise and decay curve of various Praseodymium doped $CuInS_2$ film

**Table 1.** Value of Dark current, Photocurrent, Life time, Mobility and Current gain of undoped and various Pr doped $CuInS_2$ thin films at V=15 volt.

| Sample | $I_{dc}$(μA) | $I_{pc}$(μA) | Gain | τ (in sec) | μ($cm^2V^{-1}$ $s^{-1}$) |
|---|---|---|---|---|---|
| **Undoped CIS** | 0.007 | 22.5 | $3.2\times10^3$ | 59.1 | 1.44 |
| **CIS: Pr(4ml)** | 0.005 | 31.5 | $5.7\times10^3$ | 209.0 | 0.72 |
| **CIS: Pr(6ml)** | 0.008 | 52.8 | $6.6\times10^3$ | 171.8 | 1.02 |
| **CIS: Pr(8ml)** | 0.007 | 44.1 | $6.3\times10^3$ | 72.9 | 2.30 |

To show case the effect of doping element on morphology of the film; SEM images of the $CuInS_2$ films were taken. Figure 6 is the SEM micrograph of CIS: Pr (4ml) film at magnifications 2.5KX and 5KX. The surface of the film is non uniform with polyhedral shaped particles on the surface. The film is amorphous in nature. No voids are seen. But at higher volume (6ml) of Pr voids are seen on the surface (Figure7). More agglomeration of particles is seen. Particle looks like accumulated woolen balls or cubic like grains. Film morphology shows kind of Pin holes or cracks with 8ml volume of Pr (Figure 8). Agglomeration of particle increases with volume and hence particle size also increases. The grain size calculated for undoped film was 123nm and for doped film (Pr) was 100nm. Thus we can infer that grain size decreases with doping which confirms that films were nano sized.

The study of optical properties of $CuInS_2$ thin films is having special implication in the field of science, technology and industry for developing new optical devices. Optical absorption study provides useful information which is helpful to analyze band gap of thin film which are directly related to particle size. As we know that band gap of thin films is important for designing the photovoltaic cells. In figure 9 the absorption spectra of undoped film at various temperature is shown. In figure10 absorption spectra of Pr doped films are shown. When volume of doping element increases the spectra is red shifted towards higher wavelength side. The band gap energy ($E_g$) & absorption coefficient (α) of synthesized $CuInS_2$ film is calculated from the absorption statistics. The coefficient of absorption (α) can be computed by using relation [9].

$$\alpha = \frac{c(hv - Eg)^{1/2}}{hv} \quad (3)$$

**Figure 6.** SEM of CIS: Pr (4ml) at magnifications 2.5KX and 5KX.

IOP Conf. Series: Materials Science and Engineering 1120 (2021) 012013 doi:10.1088/1757-899X/1120/1/012013

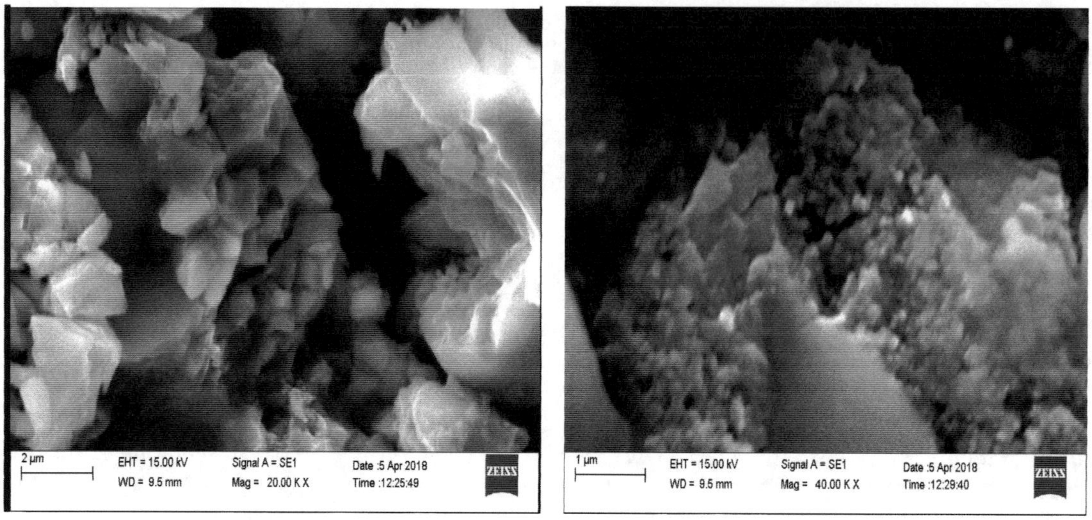


**Figure 7.** SEM of CIS: Pr (6ml) at magnifications 20KX and 40KX.

The value of absorption coefficient in all the films is found to be in the range of ten to the power five in the observable range (400-700nm). High coefficient of absorption clearly indicates that synthesized films are important for the fabrication of high absorptive layers of solar cell. For calculating band gap of films; a graph is plotted between $(\alpha h\nu)^2$ and $(h\nu)$ known as Tauc's plot. The extrapolation of nonlinear curve between $(\alpha h\nu)^2$ and $(h\nu)$ will lead to the value of $E_g$ of $CuInS_2$ thin films[10,11,12]. Using the hyperbolic band model the optical properties of $CuInS_2$ thin film was studied. Average particle size has been calculated by using equation (4).

$$E^2{}_{gn} = [E^2{}_{gb} + \frac{2\hbar^2}{m*R^2} E_{gb}\ \pi 2] \quad (4)$$

**Figure 8.** SEM of CIS: Pr (8ml) at magnifications 20KX and 30KX

where m* = effective mass of the electron, $E_{gn}$ & $E_{gb}$ are optical band gap of nano crystalline and bulk sample. The particle size is in nm range. Tauc's plot for Pr doped films is shown in figure 1.In all undoped and rare earth doped $CuInS_2$ thin films transmission spectra are recorded between 300-900nm ranges shown in figure 12. In Pr (6ml) $CuInS_2$ thin films very high value of transmittance ~90% is obtained in the visible range. Thus these materials are having good scope to be used in solar cells.

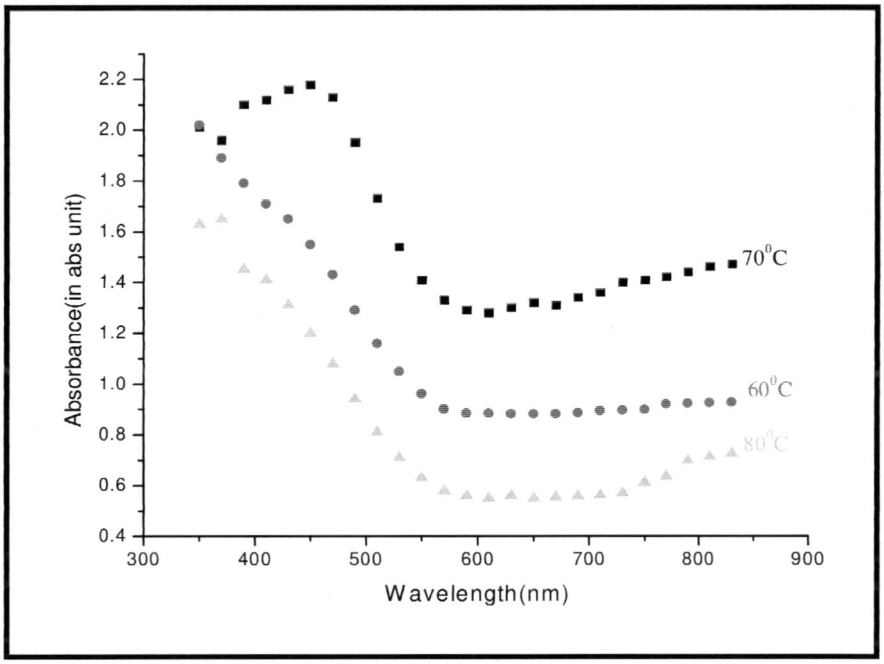


**Figure 9.** Absorption Spectra of Undoped $CuInS_2$ films at various temperatures.

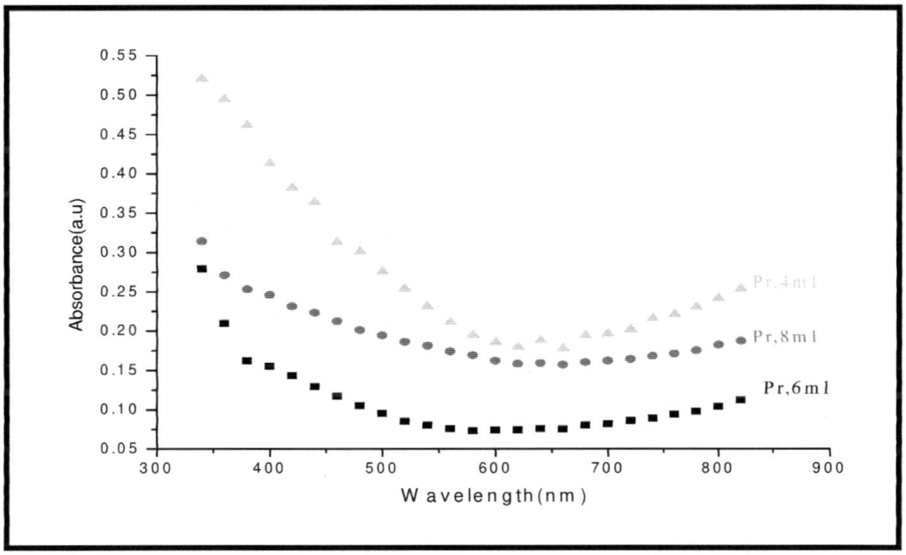


**Figure 10.** Absorption Spectra of various Praseodymium doped $CuInS_2$ film.

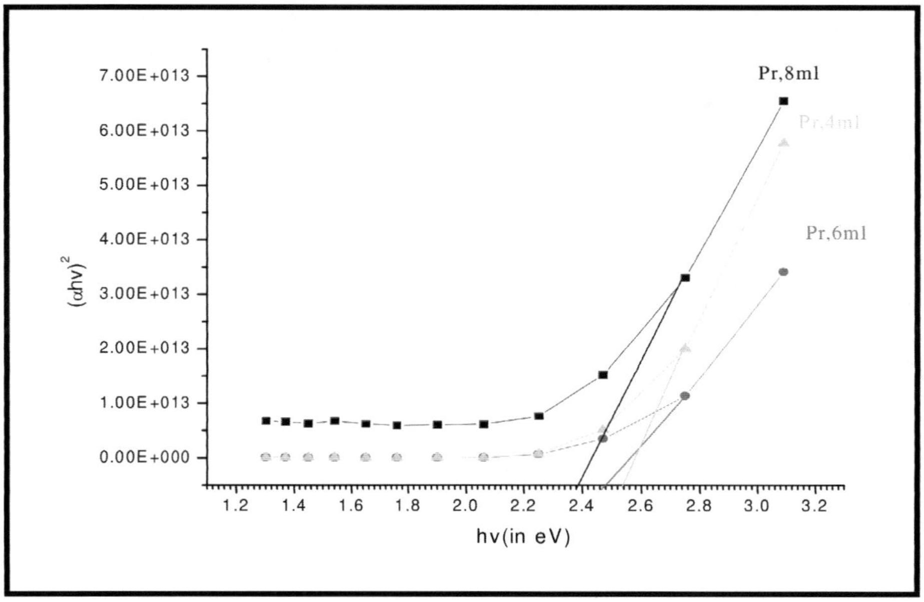


**Figure 11.** Graph between $(\alpha hv)^2$ and (hv) of various Praseodymium doped $CuInS_2$ films.

Hall measurement is the significant technique to get the concentration of charge carrier, mobility and resistivity of a sample which is conductive. In figure 13 and figure14 graph between Hall voltage and current for 6ml and 8ml doped films are shown. It is clear that as volume of Pr increases then resistivity decreases. It was found that the mobility reduces with the increasing volume of doping element. The reason may be the increased grain size. The resistivity of CIS: $CdCl_2$, Pr (6ml) is very high. The conductivity of all doped films is higher than undoped film. In general, increased doping will leads to increased conductivity. The reason is higher value of carrier concentration.

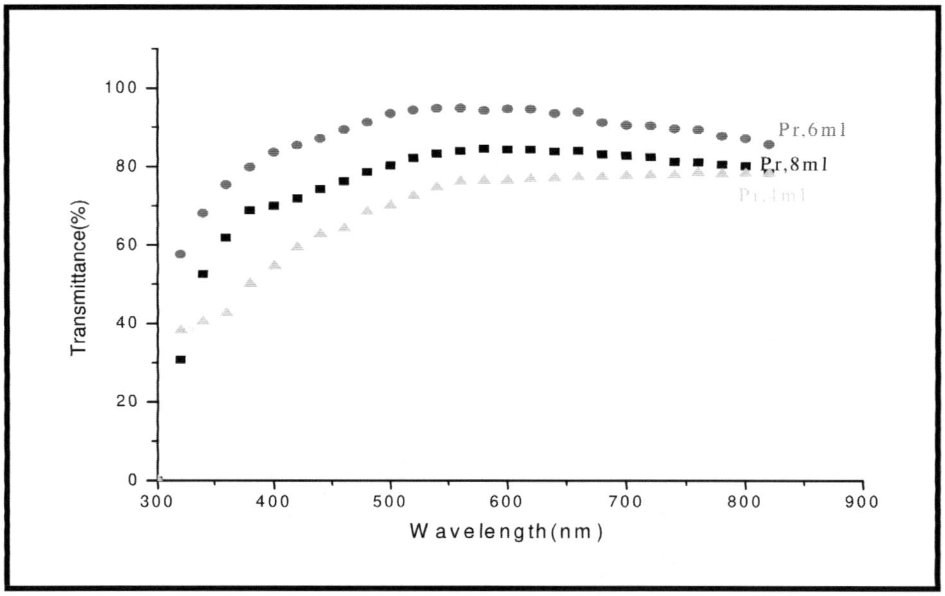


**Figure 12.** Transmittance spectrum of various Praseodymium doped $CuInS_2$ film.

IOP Conf. Series: Materials Science and Engineering 1120 (2021) 012013 doi:10.1088/1757-899X/1120/1/012013

**Table 2.** Peak position, band gap and particle size of undoped and Pr doped $CuInS_2$ thin films

| Sample | Peak Position | Optical band Gap (eV) | Emission Band energy (eV) | Particle Size (nm) |
|---|---|---|---|---|
| **Undoped CIS** | 580nm | 2.19eV | 2.12eV | 16.71nm |
| **CIS: Pr(4ml)** | 575nm | 2.54eV | 2.16eV | 13.51nm |
| **CIS: Pr(6ml)** | 550nm | 2.48eV | 2.25eV | 13.93nm |
| **CIS: Pr(8ml)** | 525nm | 2.38eV | 2.36eV | 14.72nm |

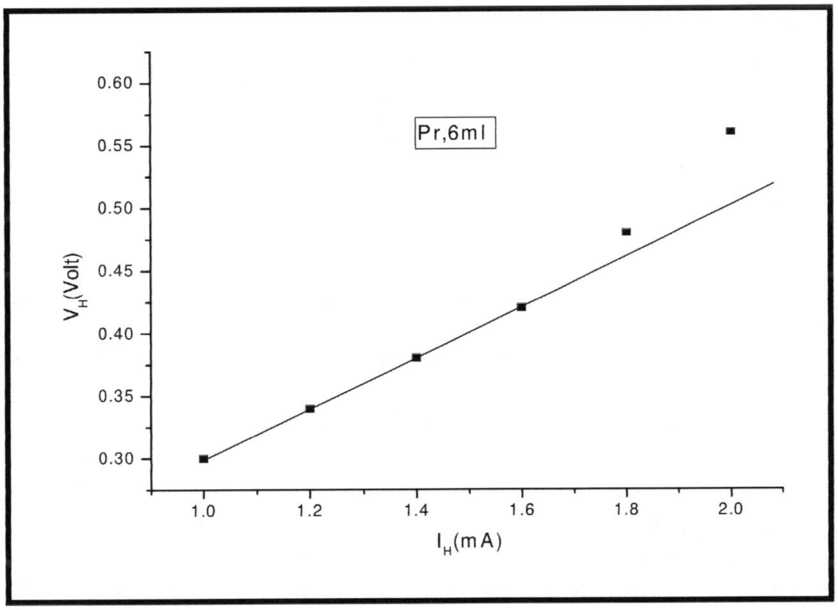


**Figure 13.** Graph between Hall voltage and current for CIS: Pr (6ml) doped film.

**Table 3.** Hall measurement results of various undoped and doped $CuInS_2$ thin films.

| Sample | $V_H/I_H$ (Ω) | $R_H$ ($cm^3/c$) | Type of carriers | Concentration of charge carrier [$cm^{-3}$] | σ at RT [Ω-cm]$^{-1}$ | Hall mobility [$cm^2/Vs$] | ρ[Ω-cm] |
|---|---|---|---|---|---|---|---|
| **CIS at 80⁰C** | 140 | 89.74 | Holes | $6.9\times10^{16}$ | 0.016 | 1.44 | 62.5 |
| **CIS:CdCl₂,Pr(6ml)** | 200 | 102 | Holes | $6.1\times10^{16}$ | 0.008 | 0.88 | 125 |
| **CIS:CdCl₂,Pr(8ml)** | 266 | 134.6 | Holes | $4.6\times10^{16}$ | 0.017 | 2.30 | 58.8 |

Room temperature (RT) photoluminescence emission spectrum of $CuInS_2$ films are taken by exciting the samples at a wavelength of 450 nm. PL spectra were recorded between 400-800 nm. In figure15 PL

IOP Conf. Series: Materials Science and Engineering 1120 (2021) 012013 doi:10.1088/1757-899X/1120/1/012013

emission spectrum of various Praseodymiums doped $CuInS_2$ film is shown. Blue shift occurs in the film which is may be due to quantum confinement appearing as a result of modified surface to volume ratio. Quantum size effect or confinement effects are followed by quantization of electronic energy levels that is molecule like energy levels even in the crystals and not exhibiting distinct band type structure. The emission in this case is evident of such structure in the prepared film. Such band edge luminescence can be caused by the recombination of excitons and or shallowly trapped electron–hole pairs.

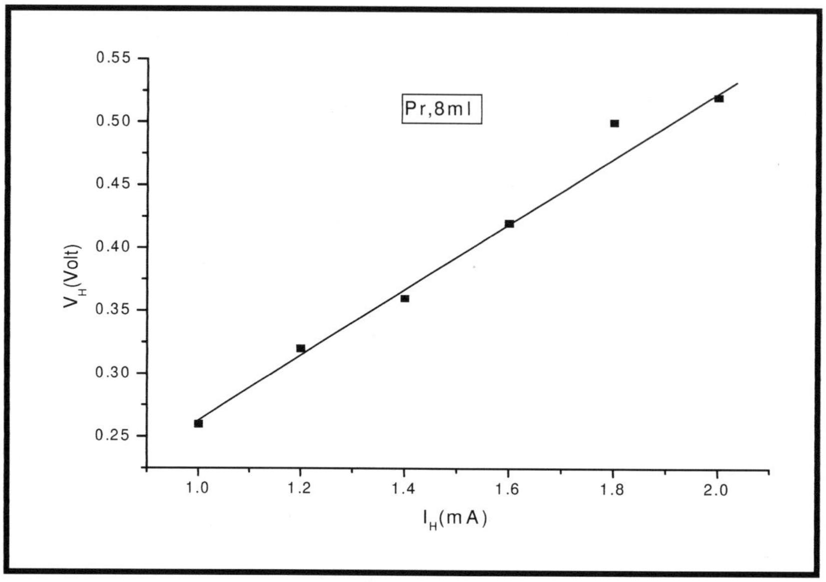


**Figure 14.** Graph between Hall voltage and current for CIS: Pr (8ml) doped film

The peak sharpness in terms of intensity increases by increasing amount of rare earth doping element. The emission peak appears at wavelength 575nm, 550nm and 525nm in 4ml, 6ml and 8ml doped Pr films. Band gap of doped films are higher than undoped film

The observation shows that when amount of doping element increases the band gap also increases. This result is expected at higher volume because coalescence of particle is there. The particle size falls in nm range. In Table 2 peak position, band gaps and particle size of undoped and Pr doped $CuInS_2$ films are given. Higher luminescent intensities were observed in nano range particles than bulk material. The reason may be due to more defect states and high value of surface to volume ratio. This will make availability of more carriers for photoluminescence.

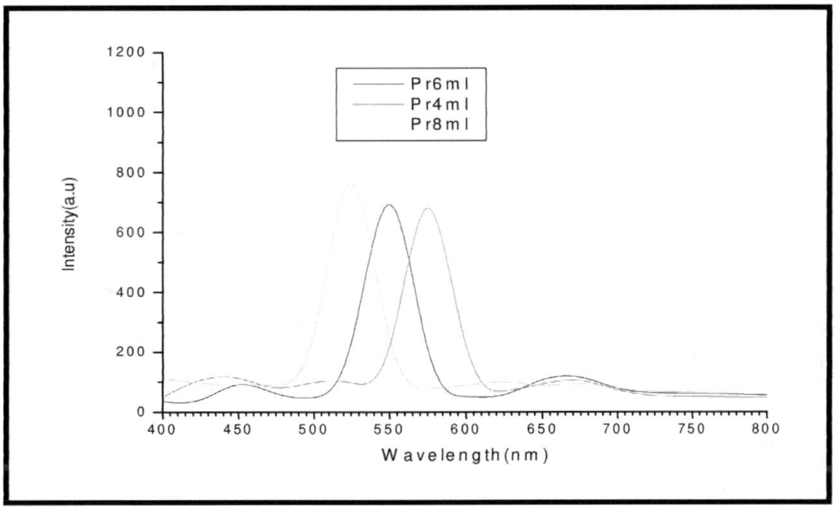


**Figure 15.** PL emission spectrum of various Praseodymium doped $CuInS_2$ film

## 4. Conclusions

The Pr doped $CuInS_2$ thin films have been prepared successfully by chemically bath deposition technique at $80^0$C temperature. The SEM Studies show that at lower volume polyhedral shaped particle appears on the surface. No voids are seen. But at higher volume of Pr; voids are seen on the surface. Agglomeration of particle increases with volume and hence particle size also increases. In the XRD of prepared sample characteristic peaks of chalcopyrite phase appears between ($24^0$- $80^0$) in all the films. High transmittance, low reflectance found in the visible region in the entire prepared sample. The absorption peaks fall in the visible region. High Photoconductive gain of $10^3$ is found. Quite good photosensitivity is observed in Pr doped $CuInS_2$ films We have observed that the dark current study show linear behavior and plot a curve between resistance and the voltage which is also found in linear manner in doped and undoped film. Hall measurement shows that all the films are of p-type. The conductivity of all doped films is higher than undoped film.

## References

[1] Mahendran C and Suriyanarayanan N 2010 *Physica B* **405** 2009
[2] Ben Rabeh M, Chaglabow N and Kanzari M *Chalcogenide Letters* **6** 155
[3] Krunks M, Bijakina O, Varema T, Mikli V, Mellikov E 1999 *Thin Solid Films* **338** 125
[4] Bourlier Y, Bernard R, Lethien, C, Roussel P, Zegaoui, M, Bouzaoui, M, Rolland N, Rolland P. A 2012 *Japanese journal of applied physics part-2* **5** 12580.
[5] Mohammed R, Hartiti B, Arba Y, Ridah A 2013 *Journal of Optoelectronics and Advanced materials* **15** 1328
[6] Guan R, Wang X and Sun Q 2015 *Journal of Nanomaterials* ID 579489
[7] Qiu J, Jin Z, Qian J, Shi Yong, Wu W 2005 *Journal of Crystal Growth* **282** 421
[8] Padam G K and Rao S U M 1986 *Solar Energy Material* **13** 297
[9] Deshmukh K, Bhushan S, Agrawal S 2010 *Recent Research in Science and Technology* **2** 2076
[10] Tell B, Shay J L and Kasper H M 1971 *Physical Review B* **4** 2463
[11] Onnagawa H and Miyashita K M 1984 *Jpn. J. Appl. Phys* **23** 965
[12] Nishikawa H, Shiroyama T, Nakamura R, Ohki Y, Nagasawa K, Hama Y 1992 *Physical Review B* **45** 586

IOP Conf. Series: Materials Science and Engineering 1120 (2021) 012014 doi:10.1088/1757-899X/1120/1/012014

# A first-principle density functional study of metal doped ZnO buckyball structure for electronic applications

**S Sharma[1*], A K Shrivastav[1], A Oudhia[2] and M L Verma[3]**

[1] Department of Physics, National Institute of Technology, Raipur, India.
[2] Department of Physics, Govt. Nagarjun P.G. Science College, Raipur, India.
[3] Department of Applied Physics, FET-SSGI Shri Shankaracharya Technical Campus, Junwani, Bhilai, India

*Corresponding author's email address: ssharma.phd2018.phy@nitrr.ac.in

**Abstract**. The nanostructured ZnO Buckyball (ZnOBB) and Al-doped ZnO Buckyball (ZnOBB: Al) has been optimized and their electronic properties were studied using the first-principle density functional theory (DFT) approach. The study includes the structural and electronic analysis of the system, done by observing the variation in the bond length, cell volume, Density of States (DOS) and Partial Density of States (PDOS). The DOS analysis provides information about Fermi level and bandgap while its further study with PDOS furnished an understanding of the orbitals which are contributing to the overall transition of electrons in the material. The optimized structure of ZnOBB: Al consists of dopant Al placed in the position of one of the Zn atoms in the cage of ZnOBB. Undoped ZnOBB showed a bandgap value ~1.5 eV, while ZnOBB: Al showed an increased bandgap value ~3.25 eV, which is attributed to the reduction in the size of ZnOBB after doping. This work proposes a doping mechanism, which may further explain the modifications of the bandgap. ZnOBB structure is itself a very contemporary topic of research and such kind of alteration in the bandgap with doping increases its applications in optoelectronic devices.

## 1. Introduction

There are diverse research works proceeding in the field of structural variation of the material and its impact on different properties, significantly enhancing their applicability in the field of nanotechnology. The structural variations from quantum dots to nanowires or nanoflowers, etc. Manifest several developments in the optical and electronic properties of that material [1, 2].

Among all different structures, Fullerenes have drawn the attention of researchers towards itself. After finding Fullerenes in 1985, they gathered a lot of appreciation and attention from researchers engaged in science. They have a unique cage-like structure that shows applications in the field of photovoltaics, sensors, portable power devices and biomedicines, etc. [3]. Some of the published works reported the buckyball structures of elements other than carbon. Here one or all atoms of carbon are replaced by any other element's atom, for example, the Buckyball structure of Boron (B80) [4]. The idea of replacing the carbon atoms in a buckyball structure with other element's atom to form a new buckyball structure had engaged our attention to form ZnOBB. We previously performed some simulations to understand ZnO and ZnS buckyball structures and looked for their structural and electronic properties [5]. In another study, we explored the optical properties after interaction between ZnOBB and Graphene sheets[6]. These theoretical studies enhanced our understanding of stable structures of ZnOBB, its electronic properties and the attainable variations in the properties of ZnOBB with new alterations. The possibilities of bandgap engineering in ZnOBB increased the scope for further study of the system.

ZnO is known for its wide bandgap, better chemical and thermal stability, high electron mobility, non-toxic nature, low cost and easy availability [7,8]. Group III elements are known to be an efficient dopant material with ZnO. Among all the elements in group III, Al is one of the most used materials. It is due to the low cost, non-toxic nature, as well as easy availability of Al. The Al doping in ZnO is known for the production of Transparent Conducting Oxides (TCO). TCOs are widely used in solar cells, LEDs, construction glasses, flat-panel displays, etc. [9]. Mostly Indium Tin Oxides (ITOs) and Fluorine doped Tin Oxides (FTOs) are known materials for making TCOs. But Al-doped ZnO has proven to be an adequate replacement for ITOs and FTOs, due to its low cost, non-toxic nature, high transparency and low resistivity [10]. Pristine ZnO shows unstable and low electrical conductivity, which might get overcome by using Al as a dopant while maintaining a high level of transparency [11]. Due to the possibility of numerous applications of ZnOBB: Al, we studied its structural and electronic behavior. We also gathered information on structural and electronic changes in ZnOBB after Al doping.

Here, in this work, material modeling is done by using a DFT based technique. Here we expanded our previous study of ZnOBB by replacing one of the Zinc atoms with an Aluminum atom in the cage of ZnOBB. To the best of our knowledge, ZnOBB: Al has been studied for the first time. In the future, we can further investigate the optical properties of the ZnOBB: Al, like conductivity, reflectivity and transparency, etc. And look for its applications in the Transparent Conducting Oxides.

## 2. Simulation and Theory

Spanish Initiative for Electronic Simulations on Thousands of Atoms (SIESTA) software is used for DFT based study. It is one of the most prominent codes used for interpreting the electronic and optical properties of materials [12]. Here we optimized ZnOBB and ZnOBB: Al individually and looked for the structural variation in ZnOBB: Al as compared to ZnOBB. Along with structural properties, we also studied the electronic properties of the system by obtaining its DOS and PDOS plots. The optimization of the system was carried out by systematically following steps including determination of Cut-off, K-Points and final optimizations. Two types of approximations are considered, Generalized Gradient Approximation (GGA) and local-density approximation (LDA). Here we employed GGA, for exchange-correlation functions [13]. For the modeling of core electrons, Troullier-Martins pseudopotentials were used [14]. The Double Zeta Polarization (DZP) basis set is used here. The system was optimized with sufficient CG steps, mesh cut off was 400 Ry and k-grid value 5x5x5 a.u.

## 3. Results and Discussion

### *3.1 Setup and Analysis of Structure*

For the structural analysis of ZnOBB and ZnOBB: Al, we optimized the structures first and looked for different parameters to study the system, such as bond length, Fermi energy, cell volume, etc. As shown in figure 1 (a) and (b), initially there are 12 Zinc atoms and 12 Oxygen atoms in the ZnOBB structure, which is fully optimized. Later we replaced one of the zinc atoms (Zn3) with one Aluminum atom (Al3) and optimized the structure again. ZnOBB and ZnOBB: Al, both have a ball-like structure, similar to a carbon buckyball. The changes that occurred in ZnOBB due to Al doping are discussed below.

IOP Conf. Series: Materials Science and Engineering 1120 (2021) 012014 doi:10.1088/1757-899X/1120/1/012014

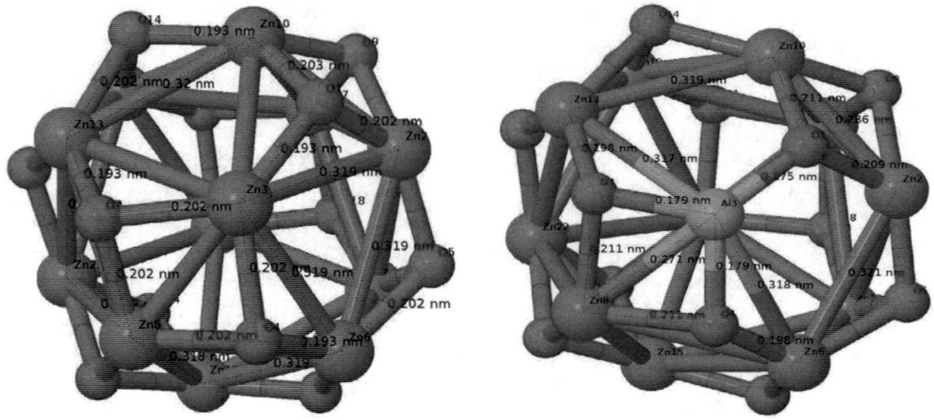


**Figure 1.** Optimized structures of a) ZnOBB b) ZnOBB: Al

**Table 1.** Obtained and Calculated parameters for the study of ZnOBB and ZnOBB: Al structures.

| SN | System used | Atomic volume ($A^{\circ 3}$) | Cell volume ($A^{\circ 3}$) | Packing Fraction (%) | Atomic Density ( $Kg/m^3$ ) | Fermi energy (eV) |
|---|---|---|---|---|---|---|
| 1. | ZnOBB | 149.53 | 395.78 | 37.78 | $4.13 \times 10^3$ | -5.1592 |
| 2. | ZnOBB: Al | 144.42 | 419.23 | 34.44 | $3.75 \times 10^3$ | -4.1803 |

**Table 2.** Bond lengths of ZnOBB and ZnOBB: Al structures.

| SN | Name of Compound | Bond Name | Bond Length(Å) |
|---|---|---|---|
| 1. | ZnOBB | Zn3-O7 | 2.02 |
| | | Zn3-Zn6 | 3.19 |
| | | Zn13-Zn10 | 3.20 |
| | | Zn13-O7 | 1.93 |
| 2. | ZnOBB: Al | Al3-O7 | 1.79 |
| | | Al3-Zn6 | 3.18 |
| | | Zn13-Zn10 | 3.19 |
| | | Zn13-O7 | 1.98 |

Bond length is one of the most important parameters to understand the system. Under different circumstances, these parameter changes, for instance, a change in bond length leads to a change in binding energy, which implies the change in stress or strain of the system [15]. Here we observed in table 1 and table 2 that, after doping ZnOBB with Al, when one of the Zn atoms is replaced with the Al atom in ZnOBB, a stable symmetric structure is formed. Here, we can see that after doping ZnOBB with Al, restructuring of

IOP Conf. Series: Materials Science and Engineering 1120 (2021) 012014 doi:10.1088/1757-899X/1120/1/012014

the system takes place. Such a kind of restructuring occurs so that the surface energy of the system is minimized and a stable structure is formed [16]. The reformed structure showed no significant changes in the Zn-Zn bonds. Mostly the bond formed by the replaced Al atom with the neighboring Zn atom showed some reduced value as compared to the same bond formed between Zn-Zn in the ZnOBB structure without any doping. This might be due to the smaller size of the Al atom as compared to the Zn atom. Now, in the case of bonds with oxygen, we observed quite visible changes. After replacing Zn with Al and comparing the bonds between Zn-O in ZnOBB with Al-O bonds in ZnOBB: Al, an increase in the bond length was observed. There is a scope to further study such behavior of bond length in detail.

*3.2 Electronic properties*
Under the study of Electronic properties of a material, we looked for the variations in the electron density and the bandgap of the material and also the participation of different orbitals in the Valence Band or Conduction Band. In the DOS study of material, several different peaks are observed in Conduction and valence bands. These peaks arise as a contribution of s,p,d and f orbitals.

Figure 2 and figure 3 shows the DOS of ZnOBB and ZnOBB: Al and PDOS of ZnOBB and ZnOBB: Al, respectively. Measurement of the bandgap is done by looking at the difference between the highest occupied molecular orbital (HOMO) and the lowest unoccupied molecular orbital (LUMO) [17]. With the bandgap, we can estimate various electronic and optical properties of the material [18].

*3.2.1. The Density of States*

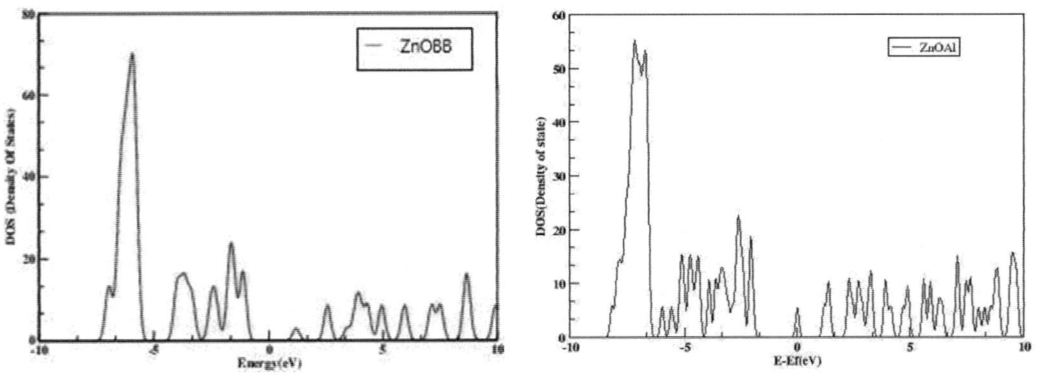


**Figure 2.** (a) DOS of ZnOBB (b)DOS of ZnOBB:Al

Figure 2 (a) and (b) shows the forbidden energy bandgap of ZnOBB and ZnOBB: Al, respectively. The Density of States of the material gives the details about the electron states of the system, knowing which we can determine the band structures and hence the bandgap of the material. Here as the structure of ZnO is buckyball, it showed variation in the bandgap as compared to other reported values for bulk structures. The bandgap observed for ZnOBB is around 1.5 eV, though with wurtzite structure in the bulk state it is reported around 3.37 eV [6, 19]. After doping ZnOBB with Al, we observed an increase in the value of the bandgap and DOS plots also showed some discrete peaks as compared to undoped ZnOBB. As shown in figure 2 (b), in the case of ZnOBB: Al band gap value is around 3.25 eV. Such an increase in the value of bandgap of ZnO after its doping with Al has been already reported in some of the experimental research works. In which the average crystallite size of ZnO nanoparticles decreases with increasing concentration of Al which might be due to the compressive strain produced in ZnO after doping it with Al [20]. Also, the bandgap might increase if the stress is increased in the system [15]. In some previous works, the contradictory fact that even though the bandgap is increased in Al-doped ZnO then also n-type conductivity rises has been reported. This was explained by stating that there might be no local energy levels generated due to Al but the extra electrons observed are transferred directly to the conduction band. The large radius electron polarons enhance n-type electrical conductivity, which is created by doping of Al in ZnO [11]. Such changes in the DOS after doping, where discrete peaks are observed and the bandgap is increased indicates the quantum confinement effect.

*3.2.2. The Projected Density of States*

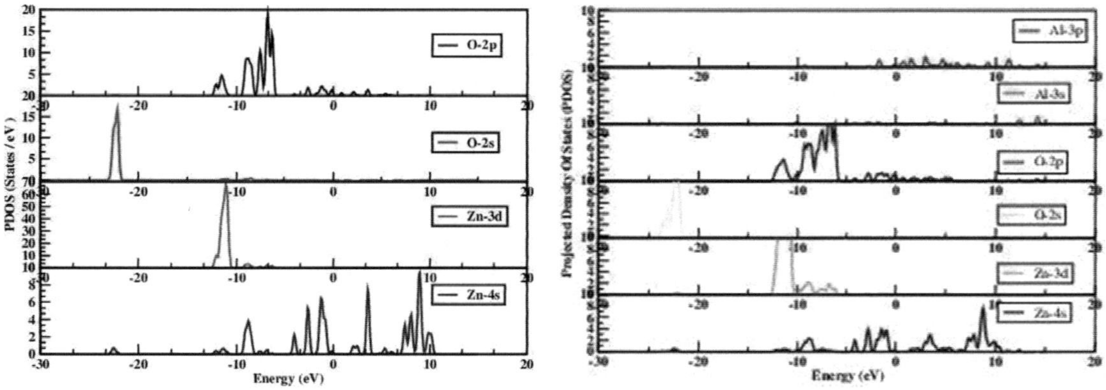


**Figure 3.** (a) PDOS of ZnOBB (b) PDOS of ZnOBB:Al

Plots of PDOS give the details about the relative contribution of different atoms and orbitals to the total DOS of the system. For understanding the electron states in detail, PDOS plots of the system are studied here. As shown in figure 3(a), The peak around -23 eV in valence Band, is originating due to the contribution of the O-2s orbital, which majorly showed its contribution in the valence band region. The higher energy region of the Valence band is mainly controlled by O-2p and Zn-3d states. It shows considerably strong interaction in the region from 0 to -13.3 eV, resulting in a strong ZnO bond. The 0-2p and Zn-3d orbitals also showed their maximum contribution in the valence band. The bottom of the CB is controlled by Zn-4s states, which probably had an almost equal contribution in both, conduction band and valence band. As compared to the previously reported results and graph obtained here, the bandgap value of ZnOBB is determined by both O-2p and Zn-4s orbitals [21].

After doping ZnOBB with Al atom, we obtained the PDOS plot for the same. In Al-doped ZnO, the valence band peak at around -23eV is obtained from O-2s states. The range in valence band from 0 to -15 eV is controlled by O-2p and Zn-3d. Zn-4s showed almost equal contribution in both the bands, just as explained for undoped ZnOBB. Here Al-3p showed its major contribution in the conduction band. As observed in figure 3 (b), the maximum contribution in the valence band comes from the Zn-4s, O-2p and Al-3p states. Similarly, in the conduction band, the major contribution comes from the Zn-4s and Al-3p orbitals [11]. It has been reported earlier that the donor levels appear in the conduction band because of the Al atom, which gets hybridized with O-2p orbitals and decreases the resistivity of Al-doped ZnO [22]. This statement holds a correct explanation for our results. The above-mentioned results for ZnOBB: Al matches with the previously reported experimental and theoretical works on Al-doped ZnO.

## 4. Conclusions

ZnOBB is a unique and contemporary system for research work. It showed some specific changes in the bandgap from existing ZnO structures. When one of the Zn atoms in ZnOBB was replaced by an Al atom, specific changes were observed. That replacement resulted in the restructuring of the system and hence changed the bond lengths. The bandgap of ZnOBB and ZnOBB: Al were calculated by using the DOS plots. Observed bandgap for ZnOBB and ZnOBB: Al was 1.5 eV and 3.25 eV, respectively. After doping ZnOBB with Al, the formation of discrete peaks in the DOS plot and an increased bandgap value indicated the quantum confinement effect in the system. The PDOS plots verified the contribution of different orbitals. These results justify the application of ZnOBB and ZnOBB: Al in the field of optoelectronic devices. In the future, there is scope for further studies of mechanical, optical and electronic properties of ZnOBB and its derivatives and we can further look for their various applications.

**Acknowledgment**

The authors want to acknowledge the management of Shri Shankaracharya Technical Campus-SSGI for giving access to their lab for working on SIESTA. They also want to thank the Department of Physics, NIT, Raipur for their kind support during this work.

## References

[1] Xu X, Zhuang J and Wang X 2008 *J. Am. Chem. Soc.* **130** 12527
[2] Oudhia A, Choudhary A, Sharma S, Aggrawal S and Dhoble S J 2014 *J. Lumin.* **154** 211
[3] Bhakta P and Barthunia B 2020 *Indian Acad Oral Med Radiol.* **32** 159
[4] Szwacki N G, Sadrzadeh A and Yakobson B I 2007 *Phys. Rev. Lett.* **98** 166804
[5] Sharma S, Shrivastav A K, Oudhia A and Verma M L 2020 *IOP conf. ser. Mater. sci. eng.* **798** 012032
[6] Sharma S, Shrivastav A K, Oudhia A and Verma M L 2021 *J. Inst. Eng. India Ser. E* (In Press) https://doi.org/10.1007/s40034-021-00204-6
[7] Lee J H, Ko K H, Park B O 2003 *J. Cryst. Growth* **247** 119
[8] Claeyssens F, Freeman C L, Allan N L, Sun Y, Ashfold M N R and Harding J H 2005 *J. Mater.Chem.* **15** 139
[9] Lewis B G and Paine D C 2000 *MRS Bull.* **25** 22
[10] Murugesan M, Arjunraj D, Mayandi J, Venkatachalapathy V and Pearce J M 2018 *Mater. Lett.* **222** 50
[11] Maldonado F and Stashans A 2010 *J. Phys. Chem. Solids* **71** 784
[12] Soler J M, Artacho E, Gale J D, Garcia A, Junquera J, Ordejon P and Sanchez D 2002 *J. Phys. Condens. Matter* **14** 2745
[13] Perdew J P, Burke K and Ernzerhof M 1996 *Phys. Rev. Lett.* **77** 277
[14] Troullier N and Martins J L 1991 *Phys. Rev. B* **43** 1993
[15] Hou X, Xie Z, Li C, Li G and Chen Z 2018 *Materials* **11** 188
[16] G. Cao 2003 Nanostructures & Nanomaterials (Imperial College 283 Press, London) p 19
[17] Djurovich P I, Mayo E I, Forest S R and Thompson M E 2009 *Org. Elect.* **10** 515
[18] Caglar M, Ilican S, Caglar Y and Yakuphanoglu F 2009 *Appl. Surf. Sci.* **255** 4491
[19] Chena H, Qua Y, Sunb L, Pengc J and Ding J 2019 *Physica E Low Dimens. Syst. Nanostruct.* **114** 113602
[20] Mallika A N, Reddy A R, SowriBabu K and Reddy K V 2014 *Ceram. Int.* **40** 8
[21] Upma and Verma M L 2019 *J. Electron. Mater.* **48** 4007
[22] Palacios P, Sanchez K and Wahnon P 2009 *Thin Solid Films* **517** 2448

IOP Conf. Series: Materials Science and Engineering 1120 (2021) 012015 doi:10.1088/1757-899X/1120/1/012015

# Synthesization and optical characterization of photovoltaic materials for perovskite solar cell application

**Y K Dongre[1], P Roy[2], A Khare[2] and S Tiwari[1]**

[1] Photonic Research Laboratory; SOS in Electronics and Photonics, PRSU, Raipur – 492 010, India
[2] Thin Film Research Laboratory, Department of Physics, National Institute of Technology, Raipur – 492 010, India
[1] Photonics Research Laboratory (PRL)

*Corresponding author's e-mail address: yogielectro@gmail.com

**Abstract.** The performance of Perovskite solar cells (PSCs) has rapidly increased over the last decade. The power conversion efficiency (PCE) of 25.2% has been reported. Drastic and rapid change in PCE of PSCs has attracted a lot of attention toward this alternative photovoltaic technology that can be manufactured using simple and low cost processing techniques, over the traditional silicon solar technology. Here in this work, we have synthesized Molybdenum trioxide ($MoO_3$) and Titanium dioxide ($TiO_2$) for charge transport materials application in PSC application. The analysis of materials has been carried out by photoluminescence and absorbance spectra. The results signify that $MoO_3$ has absorbed at the visible region around (600-800) nm and the absorbance of $TiO_2$ tuned at 400 nm. Due to high absorption range (600-800) at visible region $MoO_3$ is suitable for Hole Transport Material (HTM). The widely used Electron Transport Material (ETM) $TiO_2$ is also synthesized and characterized.

## 1. Introduction

In current era, most of the electricity is produced by using fossil fuels including coal, oil and Petroleum gases. The energy produced by fossil fuels is not green because it contains pollutant gases like ($CO_2$, S, C, N etc). The byproduct of fossil fuels is directly affected to our environment [1]. Clean, green and sustainable energy is best alternative for future energy, which is produced by solar technology. One other issue related to fossil fuels are its availability, it is limited in stocks and to be finished within some year and also it has unbalanced distribution in the earth [2]. Solar energy is a one of the best promising alternative energy source for today energy requirement. The solar cell poses the ability to act as a savior in the current scenario. On the basis of absorber materials there are variety of solar cells are available and classified from 1$^{st}$ to 4$^{th}$ generation.1st generation's cells are mainly based on Silicon technology, which have very high temperature processing and costly. The advancement of Si technology, the various cells such as single crystalline silicon (c-Si), multi-crystalline silicon and amorphous silicon (a-Si) were used with different cell architecture [3, 4]. Hydrogenated amorphous silicon (a-Si:H), Cadmium Telluride (CdTe), copper indium diselenide,

($CuInSe_2$) and its related alloys like Copper Indium Gallium diselenide, $CuInxGa_1$-$xSe_2$ (CIGS) are second generation solar cell, used in place of Si to reduce the cost of Si solar cell, with reported efficiencies of CIGS and CdTe 20% and 17%, respectively [5, 6]. Silicon (Si) processing cost is very high so the overall cost of device is also high; to reduce the cost and increases the efficiency of solar cells 3rd generation thin film solar technology was invented. Organic photovoltaic solar cells (OPVs) technology employs conducting polymers as low cost materials; ease of fabrication and tunable bandgap suitable alternatives to inorganic semiconductors (Si,CIGS and CdTe). The highest reported power conversion efficiencies for dye-sensitized and polymer single cells using chlorinated acceptor are relatively ~ 14% and 16% [7, 8]. The 4th generation solar cells are combination of organic and inorganic materials, which are basically perovskite materials. In 2009, perovskite material was first used in photovoltaic application by Miyasaka and his co-workers [9, 10]. The perovskite material has a highest reported efficiency of ~28.2% in tandem architecture [11]. The limiting factor of this technology is instability in environment and toxicity due to incorporation of lead (Pb).

### *1.1. Perovskite material structure*

Perovskite meanwhile, the mineral is composition of Calcium, Titanium and Oxygen in the form $CaTiO_3$. The generic form of perovskite structure is $ABX_3$ and the same crystallographic structure as the mineral [12]. The perovskite lattice structure and morphology arrangement is shown in figure 1.

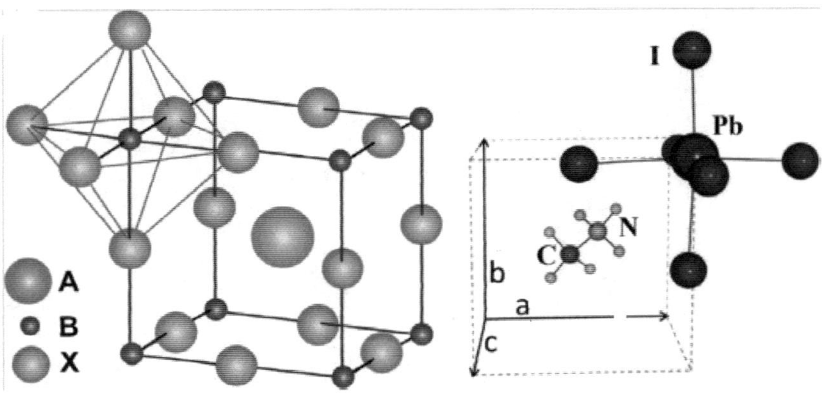


(a) (b)

**Figure 1.** (a) Perovskite structure and larger A-cation occupied in cubo-octahedral site, showing ($BX_6$) octahedral, (b) Cubic $CH_3NH_3PbI_3$perovskite Unit cell, Ref **[12].**

The generic form to think about a perovskite (positively-charged) is as a large atomic or molecular cation of type A in the centreof a cube. The positively-charged B cations are occupied at the corners the cube and the faces of the cube are occupied by a smaller atom X with negative charge (anion). The cation A can be an organic material such as Methylammonium $CH_3NH_3^+$ [MA], Formamidinium $NH_2CH{=}NH_2^+$ [FA] or inorganic material such as $Cs^+$ ion [13, 14].

### *1.2. Perovskite Solar Cell*

Organometal halides compound shortly named as perovskite represent an emerging active layer materials for photovoltaic technology. In recent years, perovskite shows capability of developing high performance photovoltaic devices with higher efficiency at a low cost. In 2009, first perovskite solar cell was discovered with reported PCE of around 3.81%. The current highest PCE was reported around 25.2% for single junction PSCs. Perovskite name comes from a Russian mineralogist L. A. Perovski and was discovered by Gus-tav Rose in 1839 in ural mountains of Russia [15]. Organo metal halide $CH_3NH_3BX_3$was discovered by Weber in 1978, the material that is responsible for the main part of perovskite solar cells. Here, B represents the metal elements and X substitutes for halide elements [16]. $CH_3NH_3BX_3$ has an oriented crystal structure arranged in $ABX_3$ formula, where (X can be oxygen or halogen, in PV application X is halogen). The cubo-octahedral site shared with twelve X anions, attached with larger Acation. B cation is stabilized in an octahedral site shared with six X anions. New and advanced thin film deposition techniques were adopted which led to the modification of the performance of the cell [16-18]. Methylammonium free stable and more efficient perovskite solar cell has been reported in 2020, with 20.5 % power conversion efficiency [19]. A new breakdown approach has been developed to replace the organic-inorganic composition to all inorganic materials, which shows 15.6% power conversion efficiency [20].

## 2. Experimental

### *2.1. Preparation of Titanium dioxide ($TiO_2$)*

Sol-Gel method is used for synthesis of Titanium dioxide ($TiO_2$); here Titanium tetra isopropoxide (TTIP) material is dissolve in the mixture of Ethanol and Acetic acid in proper molar concentration. Titanium tetra isopropoxide [$Ti(OCH(CH_3)_2)]_4$, from Sigma Aldrich, 97% pure], and iso-propanol [$(CH_3)_2CHOH$, from Sigma-Aldrich, 99.7% pure], is dissolved in mixture of ethanol ($C_2H_6O$) and acetic acid ($CH_3COOH$) to form $TiO_2$ liquid solution. The mixture of ethanol and an acetic acid undergoes constant steering for 50 minutes at 60ºC. After 50 minutes, titanium tetra isopropoxide is mixed in prepared solution and the resultant solution is stirred for 5 minutes at same temperature, leading the synthesis of $TiO_2$ solution. The formation process along with deposited film is shown in figure 2.

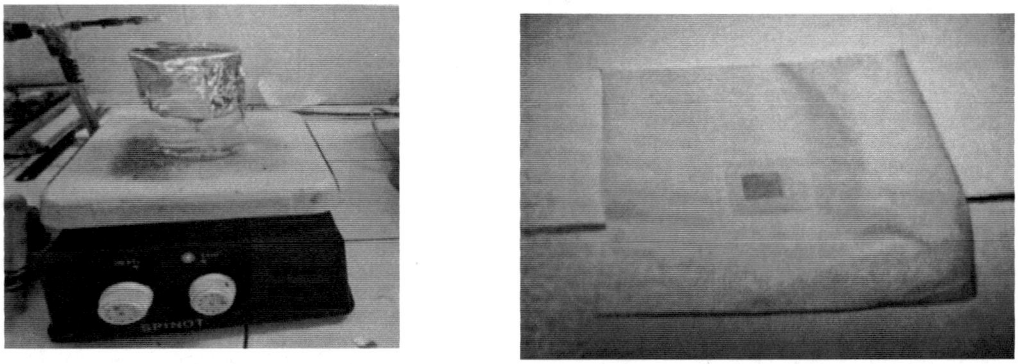

(a) (b)

**Figure 2.** Synthesis and deposition of ($TiO_2$), (a) Preparation of $TiO_2$ solution, (b) Deposition of ($TiO_2$), on ITO.

### 2.2. *Chemical Compounds for $MoO_3$ synthesis: Using Ethanediol, Isopropanol and Hydrogen Peroxide*

In this work, Isopropanol basically Isopropyl alcohol with the chemical formula $CH_3CHOHCH_3$ is used.

Isopropyl alcohol (IUPAC name propan-2-ol) flammable and colorless with chemical formula $C_3H_8O$is used. It poses strong odor. Isopropyl group linked to a hydroxyl group, to make the simplest example of a secondary alcohol, where the alcohol carbon atom is attached to two other carbon atoms. Basically it is a structural isomer of 1-propanol. Hydrogen peroxide is a chemical compound with the formula $H_2O_2$, in its pure form; it is a pale blue, clear liquid, and slightly more viscous than water. Hydrogen peroxide is unstable by chemical nature and slowly decomposes in the presence of base or a catalyst [21].

### *2.3. Synthesization of blue color ($MoO_3$) Molybdenum solution*

The synthesis of $MoO_3$ is based on the preparation of Molybdenum solution with the treatment of ultrasonic reaction. Firstly, the ethanediol and isopropanol were mixed in a proper volume (in ratio of 1:9; ethanediol to isopropanol), then $MoO_3$ powder is mixed in the complex solvent, 30% aqueous $H_2O_2$ was added to enhance the chemical reaction. The total solution was ultrasonicated for around 30 minutes. At last, the thermal treatment at around 200°C is provided with the help of hydrothermal method. Finally, a dark black colour $MoO_3$ solution is formed and proceeded for one day freezing, the colour of $MoO_3$ solution have been changed to dark blue as shown in figure 3.

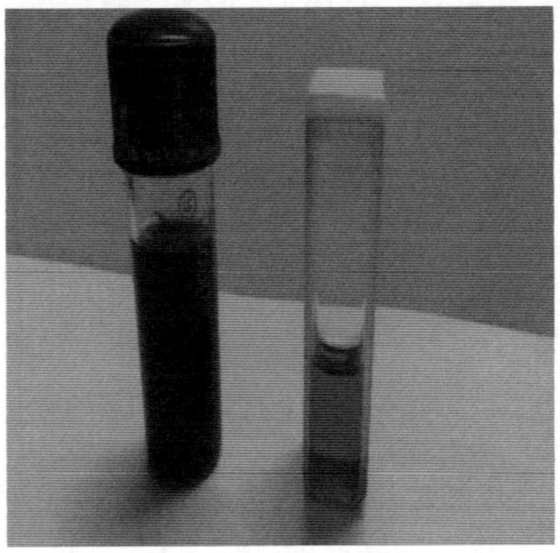

**Figure 3.** Dark blue color; Molybdenum oxide ($MoO_3$) synthesized solution.

## 3. Result and Discussion

### *3.1. $MoO_3$ Absorbance analysis*

High conductivity of $MoO_3$ materials varies from $10^{-9}$ to $10^{-5}$ $(\Omega\ cm)^{-1}$ [22] which raises the free charge carrier concentration and also increases the carrier mobility of the material. The high work function (6.7ev) [23] and energy levels of $MoO_3$ (5.3 to 2.3) matches with perovskite materials [24], making it suitable choice for carrier extraction (electron blocking as well as hole transporting) material to promote carrier transport and reduce the carrier recombination. Interface state of $MoO_3$ and perovskite is low, that helps reduce the recombination at the interface and it builds an ohmic contact with the metal electrode, and also provides the mechanical strength to devices. Apart from that the nontoxicity and environmental stability of $MoO_3$ is made suitable candidate for HTL in perovskite solar cell. Spectrophotometer model ELICO SL 159UV-VISIBLE is used for the measurement of the ultraviolet and visible spectrum of the prepared solutions. Firstly, the UV-VISIBLE measurement is performed on the molybdenum solution in the room temperature then at increased temperature up to $200^0$C. The obtained results are shown in figure 4. At the annealing temperature of $200^0$C, the small broad spectra ranging from 600nm to 800nm is absorbed, and it becomes more evident

IOP Conf. Series: Materials Science and Engineering 1120 (2021) 012015 doi:10.1088/1757-899X/1120/1/012015

and this fascinating phenomenon may be attributed to the free electron being trapped in the oxygen vacancies in $MoO_3$, which results $MoO_3$ to act as hole transport layer as well as Electron Blocking layer.

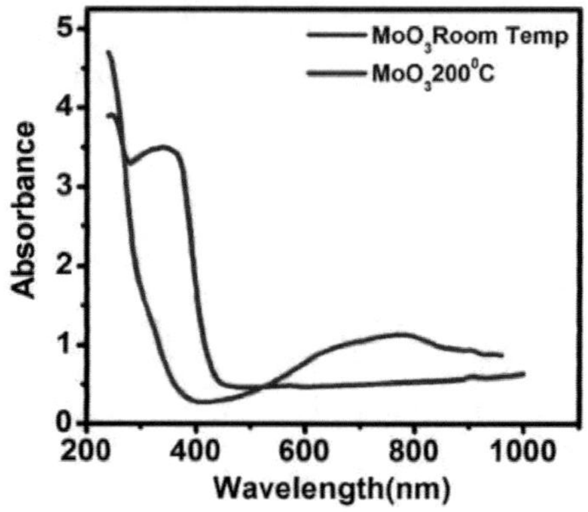


**Figure 4.** $MoO_3$ Absorbance analysis: Spectra of synthesized $MoO_3$ at room temperature and $200^0$c.

*3.2. Titanium dioxide ($TiO_2$) Absorbance and Thickness Spectrum analysis*

$TiO_2$ thin film prepared by Sol-Gel method and deposited by using spin coater over the surface of ITO substrate has been investigated by UV-VIS Spectrophotometer at 400nm. The absorbance peak is obtained in range (600-800) nm, the absorption spectra is evident that $TiO_2$ is suitable for charge transport application. The thickness measurement has been done by using Spectroscopic Ellipsometer, the measured thickness of deposited $TiO_2$ is around 53.2nm.

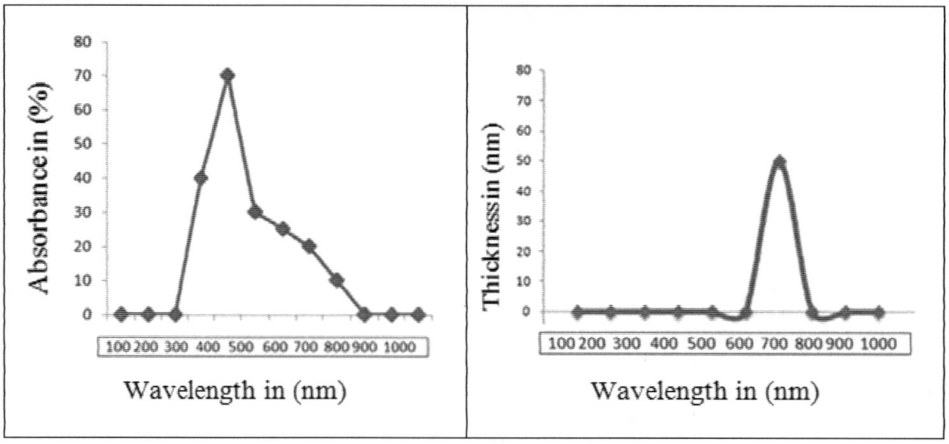


(a) (b)

**Figure 5.** $TiO_2$ spectrum analysis (a) Absorbance spectra of prepared $TiO_2$, (b) $TiO_2$ thickness spectra.

## 4. Conclusion

In this work, the synthesis and characterization of $MoO_3$ and $TiO_2$ film has been carried out. The $MoO_3$ film is synthesized by sol-gel method followed by ultrasonic reaction and deposited using spin coater on ITO substrate. Results suggest that the $MoO_3$ and $TiO_2$ affect the performance of the device, at the wavelength window from 600nm to 800nm a growing small broad spectrum is absorbed for the $MoO_3$ at annealing temperature $200^0$C. Due to high temperature annealing process, the small broad spectra becomes more evident and this fascinating phenomenon may be attributed to the free electron being trapped in the oxygen vacancies in $MoO_3$, which result $MoO_3$ serve as hole transport layer as well electron blocking layer. Absorbance spectra of $TiO_2$ thin film (53.2nm) prepared by Sol-Gel method deposited on ITO substrate has been investigated by UV-VIS Spectrophotometer. At 400 nm, the absorbance is high (~ 70%) and which tends to decrease from visible to IR region of solar spectrum. In this work the results suggest that $MoO_3$ can be used as eminent HTL material. This material is capable of enhancing the charge transportation of the device, prevents charge recombination, and also increases the carrier mobility of the cell, and prevents the active materials from the external environments (i.e. moisture, rain, air and temperature).

**Acknowledgment**
The author would like to thank my supervisor Prof. Sanjay Tiwari to provide me an opportunity to work in Photonics Research Laboratory, Raipur, and also their moral and technical support, for completion of work. I am thankful for their support in required materials and equipments.

**References**

[1] Perera F 2018 *Int. J. Environ Res.* **15** 16
[2] IEA 2017 World energy outlook, http://www.iea.org/weo
[3] Ranabhat K, Patrike L, Andrianov K, Sofronova E and Lapshinsky V 2016 *J. Appl. Engg. Science* **14** 481
[4] Green M A, Emery K, Hishikawa Y, Warta W and Dunlop E D 2014 *Progress In Photovoltaics Research and Applications, Appl.* **22** 701
[5] R G 2010 *Renewable Energy Data Book*, p1–132
[6] Tiwari S, Yakhmi J V, Carter S and Scott J C 2017 Optimization of Bulk Heterojunction Organic Photovoltaic Devices, Handbook of Ecomaterials, published by Nature Springer (2018-01-01) 1-36
[7] S Tiwari S, Yakhmi J V, Carter S and Scott J C 2017 Advances in polymer based photovoltaic cells Review of pioneering Materials, Design and Device Physics, Handbook of Ecomaterials, published by Nature Springer. DOI https://doi.org/10.1007/978-3-319-48281-1_59-1, Online ISBN 978-3-319-48281-1
[8] Bagher A M, Mahmoud M, Vahid A and Mohsen M 2015 Types of Solar Cells and Application, *American Journal of Optics and Photonics* **3** 94
[9] Cui Y 2019 *Nature Communications* **10** 2515
[10] Kojima 2006 Novel photoelectrochemical cell with mesoscopic electrodes sensitized by lead-halide compounds, *In Proc. 210th ECS Meeting*, DOI:10.1149/ma2007-02/8/352
[11] Kojima 2009 *J. Am. Chem. Soc.* **131** 6050
[12] Giorgi G, Fujisawa J, Segawa H and Small Y K 2013 *The Journal of Physical Chemistry Letters* **4** 4213
[13] Borriello I, Cantele G and Ninno D 2008 *Physical Review B* **77** 235
[14] Kagan C R, Mitzi D B and Dimitrakopoulos C D 1999 Science **286** 945
[15] Roy P, Sinha N K, S Tiwari and A Khare 2020 Solar Energy **198** 665
[16] Im J M, Lee C R, Lee J W, Park S W and Park N G 2011 *Nanoscale* **3** 4088
[17] Lee M M, Teuscher J, Miyasaka T, Murakami T N, Snaith H J 2012 *Science* **338** 643
[18] Hadlington S 2012 *RSC Chemistry world*, DOI: 10.1126/science.1228604
[19] Gao X X et al 2020 *Advanced Material*, online https://doi.org/10.1002/adma

[20] Mingyu J, Kim M and Choi H W 2020 *Science* **369** 1615
[21] Jhou Y, Hernandez C F and Shim J 2011 Nanoscale **336** 327
[22] López-Carreño L D, Pardo A, Zuluaga M, Cortés-Bracho O L, Torres J and Alfonso J E 2007 *Physica state solid*I, https://doi.org/10.1002/pssc.200775931
[23] Tang J X et al. 2011, Correlation between the electronic structures of transition metal oxide RSC publication, pubs.rsc.org
[24] Cao F, Wang M and L Li 2020 *Nano Select* **1** https://doi.org/10.1002/nano.202000005

# Impact of hole transport layer on the performance of all-inorganic tin (Sn) based perovskite solar cells

**P Roy[1*], Y K Dongre[2], S Tiwari[2], D Chaturvedi[3] and A Khare[1]**

[1] Thin Film Research Laboratory, Department of Physics, National Institute of Technology, Raipur – 492 010, India
[2] School of Studies in Electronics and Photonics, Pt. Ravishankar Shukla University, Raipur – 492 010, India
[3] Department of Chemistry, Govt. Kavyopadhyay Hiralal college Abhanpur-493661, India

*Corresponding author's e-mail address: proy.phd2018.phy@nitrr.ac.in

**Abstract.** The power conversion efficiency (PCE) of perovskite solar cells (PSCs) has shown vast increment from 3.81% to 25.5% within a decade. Despite the fast progress and many advantages, these cells are facing an issue of toxicity and poor device stability. In this work, we deal with the issue of toxicity by using Tin (Sn) instead of Lead (Pb). The PSCs are simulated using Solar Simulator Capacitor (SCAPS-1D version 3.3.07) software. We used all-inorganic, lead free material, Cesium Tin–Germanium Tri-iodide ($CsSn_{0.5}Ge_{0.5}I_3$) as an absorber material. We further investigate the influence of hole transport layer (HTL) on the performance of Sn based PSCs. Simulation and analysis of following $CsSn_{0.5}Ge_{0.5}I_3$ based PSC devices have been done: using (i) HTL free configuration (ii) inorganic HTL and (iii) organic HTL. The simulated cells are in planar n-i-p architecture. This work depicts the fact that Sn based PSCs have a potential to serve as high performance solar cells. The results signifies the impact of HTL on the performance of PSCs, in this study, we also suggest alternatives to the widely used HTM (Spiro-OMeTAD) that not only produce comparable photovoltaic performance but also are cost effective. We also analyse the influence of using various metal cathodes on the simulated cell. The result suggests that presence and type of HTM is having an impact on the contact selection criteria. The influence of HTL thickness variation on the performance of the PSCs are also studied.

## 1. Introduction

The Perovskite Solar Cells (PSCs) have revealed their potential by showing rapid growth in performance (3.81%-25.5%) within a shorter span of time [1]. This has made PSCs the fastest growing PSCs to date. However, the incorporation of toxic Lead (Pb) is still the matter of concern for researchers [2]. It is a well-known fact that toxicity due to lead causes harmful impact on our ecosystem and also causes severe health risks. This factor is creating hindrance in the market acceptance of the final product [3]. Generally, $MAPbX_3$ (where X is Cl, Br or I) is widely adopted absorber layer in the PSCs. In order to replace Pb, Tin (Sn) have been used by several researchers which is nontoxic. Unfortunately, Sn based PSCs suffers from drawbacks: (i) instability in ambient conditions, (ii) degrades easily due to oxidation of $Sn^{2+}$ into $Sn^{4+}$, (iii) efficiency as high as Pb based PSCs are not yet achieved. All such issues are restricting the development of Sn based PSCs. To date the highest efficient Sn based PSC has attained PCE of 9.6%[4]. In this work, we demonstrate an all-inorganic type of PSC using $CsSn_{0.5}Ge_{0.5}I_3$ (Caesium germanium tin triiodide) as light harvester layer. Min cheng et al. [5] demonstrated $CsSn_{0.5}Ge_{0.5}I_3$ based PSC using Spiro OmeTAD and PCBM as hole transport and electron transport material respectively. The device fabricated by them attained PCE of 7.11%. Moreover, the device exhibited excellent stability by working efficiently with decrement in performance by less than 10% for more than 500 hours under one sun illumination condition and nitrogenous atmosphere. Here, we present a $CsSn_{0.5}Ge_{0.5}I_3$ based PSC in (a) HTL free configuration (b) using inorganic HTL and (iii) using organic HTL. The motive of using different aspect of hole transport layer (HTL) is to investigate its impact on the performance of PSCs. We also discuss the influence of HTM on the performance of PSC and suggest alternatives to the conventional HTM Spiro-OMeTAD. The existing research suggests that Spiro-OMeTAD participates and accelerates the process of degradation rendering poor lifetime of the cell. It has been observed that Sn based perovskite while deposited over $TiO_2$ exhibits non uniform surface coverage [6]. The problem of ion immigration occurs when Titania ($TiO_2$) is used as ETL. In order to deal with it, here we use Zno-nanoparticles (NP) as ETL. An efficiency of 24.5% has been attained in this work.

## 2. Simulation Setup

We use SCAPS-1D (version-3.3.07) to simulate and analyze the PSCs [7]. This software in a open source solar cell simulating software which is highly user friendly. It is basically build to study inorganic solar cells comprising heterogeneous structure like CdTe, CIGs, CZTs etc. PSCs incorporate the material perovskite as an absorber layer which is hybrid combination of both organic and inorganic solar cells. The fact that this material has wannier type exciton, and the cell has planar heterogeneous structure similar to inorganic cells enables the analysis of this cell using SCAPS-1D [7].

### *2.1 Device Structure*

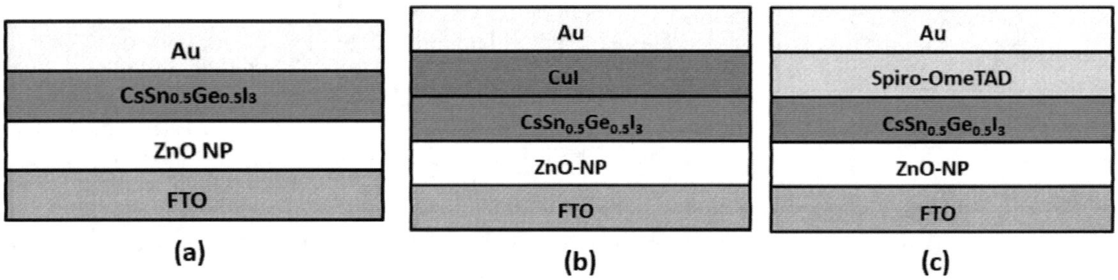


**Figure 1.** Schematic of device architecture of simulated Sn based PSCs.

The simulated cells have planar architecture structure, the schematic shown in Figure 1. The energy band diagram of the simulated cells is shown in Figure 2. The simulated PSCs have following layers: **(i)** Metal cathode (Au) **(ii)** Hole transport layer (HTL) [p-type 2,2,7,7'-tetrakis-(N,N-dimethoxyphenyl-amine)-9,9'-spirobifluorene (Spiro-OMeTAD) and copper Iodide (CuI) **(iii)** absorber layer as Cesium Tin–Germanium Tri-iodide ($CsSn_{0.5}Ge_{0.5}I_3$) **(iv)** Electron transport layer (ETL) as Zinc oxide nanorod (Zno-NR) **(v)** Flourine doped Tin Oxide (FTO) as Transparent Conductive Oxide (TCO). We simulate three different PSCs, in (a) HTL free configuration (b) using inorganic HTM (CuI) (c) using organic HTM (Spiro-OMeTAD) shown in Fig. 1.

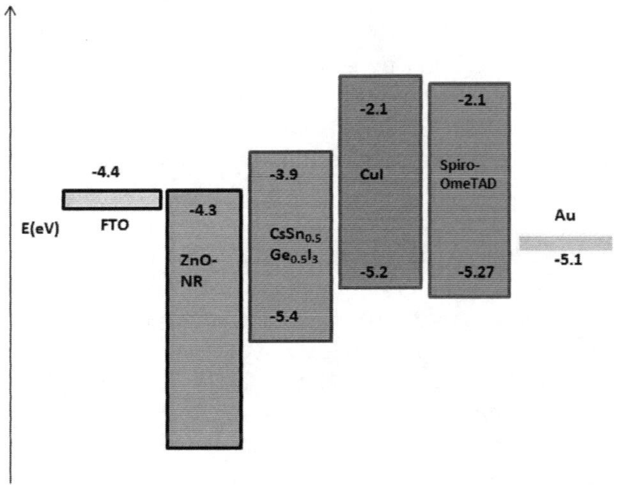


**Figure 2.** Energy-band diagram of the simulated cell.

*2.2 Parameters*

The electrical parameters adapted for simulation are selected from existing literature. Table-1 summarizes all the parameters used during simulation. The based solar cell is simulated at AM1.5G at 300K. The ETL, HTL and absorber layer defect density was taken as $10^{15}$ $cm^{-3}$, $10^{15}$ $cm^{-3}$ and $10^{14}$ $cm^{-3}$ respectively [8–12].

**Table 1.** The parameters used for simulation.

| Parameters | Spiro-OMeTAD | CuI | $CsSn_{0.5}Ge_{0.5}I_3$ [13] | ZnO- NR | $SnO_2$: F |
|---|---|---|---|---|---|
| **Thickness (µm)** | 0.5 | 0.5 | 0.4 | 0.5 | 0.25 |
| **Band gap(eV)** | 3.17 | 3.1 | 1.5 | 3.27 | 3.5 |
| **Electron affinity(eV)** | 2.1 | 2.1 | 3.9 | 4.3 | 4.4 |
| **Dielectric Permittivity** | 3 | 6.5 | 28 | 9 | 9 |
| **CB effective density of states (1/cm3)** | 2.5E+18 | 2.2E+19 | 3.1E+18 | 2E+18 | 2.2E+18 |
| **VB effective density of states (1/cm3)** | 1.8E+19 | 1.8E+19 | 3.1E+18 | 1.8E+20 | 1.8E+19 |
| **Electron thermal velocity(cm/S)** | 1E+7 | 1E+7 | 9.74E+2 | 1E+7 | 1E+7 |
| **Hole thermal velocity(cm/S)** | 1E+7 | 1E+7 | 2.13E+2 | 1E+7 | 1E+7 |
| **Electron mobility(cm2/VS)** | 2E-4 | 4.39E+1 | 0 | 1E+2 | 2E+3 |
| **Hole mobility(cm2/VS)** | 2E-4 | 2E+2 | 1E+18 | 2.5E+1 | 1E+2 |
| **Shallow uniform donor density ND (1/cm3)** | 0 | 0 | | 1E+18 | 2E+19 |
| **Shallow uniform acceptor density NA(1/cm3)** | 1E+18 | 1E+18 | | 0 | 0 |

## 3. Result and Discussion

*3.1 Performance*

Apart from the carrier generating absorber layer, the other two transport layers, electron transport layer (ETL) and hole transport layer (HTL) renders a performance deciding factor as they are responsible for the charge extraction of the cell [14]. The current research is going on to find out an efficient perovskite

IOP Conf. Series: Materials Science and Engineering 1120 (2021) 012016 doi:10.1088/1757-899X/1120/1/012016

absorber layer exhibiting good photovoltaic performance along with stable behavior. An effort is also being made to come up with a HTM being capable of improving the photovoltaic performance and also should be capable of maintaining stability by getting rid of the hygroscopic behavior. The role to hole transport material (HTM) is to extract the holes from the absorber layer and efficiently transfer it towards cathode. To date, the most widely used HTM is Spiro-OMeTAD, PTAA, PEDOT:PSS and P3HT [15–18]. These organic HTM based cells gave high efficiency and $V_{oc}$ but still its commercialization is not yet developed due to the its unstable nature in presence of moisture, temperature, light and also its high cost. Therefore, in this work we simulate three configurations of non-toxic PSCs, (i) Cell-1, HTL free PSC, a PSC in which HTL is omitted is termed as HTL free PSC. The removal of HTL from PSC fabrication not only simplifies the fabrication process but also it reduces the cost to a great extent. The simulated cell has configuration of Au/ $CsSn_{0.5}Ge_{0.5}I_3$/ZnO-Np/FTO. (ii) Cell-2 using inorganic HTL (CuI), the inorganic materials tend to have higher conductivity and carrier mobility and also exhibit stability. The device configuration is Au/ CuI/$CsSn_{0.5}Ge_{0.5}I_3$/ZnO-Np/FTO. (iii) Cell-3 using widely used Spiro-OMeTAD. The device configuration is Au/ Spiro-OMeTAD/$CsSn_{0.5}Ge_{0.5}I_3$/ZnO-Np/FTO. The role of electron transport material (ETM) is to extract electrons from the absorber layer and transfer it towards anode for collection [19,20]. We further use ZnO NP as ETM, which has high electron mobility , carrier concentration also requires low temperature processing condition. The simulated cell attains performance enclosed in Table-2. The JV curve of the simulated cell is shown in Figure 3. The results shows the fact that HTL free PSCs will poses lower efficiency as compared to PSC with HTL. The theoretically simulated cells showed the difference of 10% in PCE. The conventionally used HTM (Spiro-OMeTAD) is widely used on PSCs due to its ability to attain high performance parameters. However, studies suggest that it actively participates within the process of degradation also it is expensive. The Cell-2 has CuI as HTM and device exhibited the best performance by attaining highest performance parameters among all. This work shows that the widely used Spiro-OMeTAD can be replaced using inorganic HTM such as CuI. This will not only help attaining high performance parameters but also will reduce the overall cost of the device.

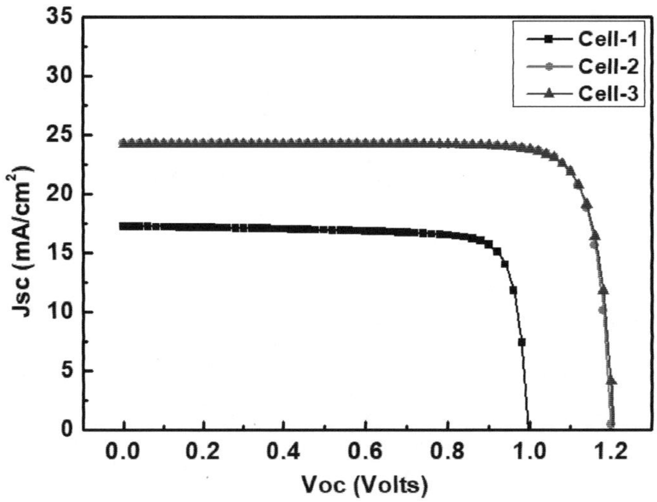


**Figure 3.** JV curve of the simulated cell.

**Table 2.** The attained performance parameters by the simulated cells.

| Cell No. | Voc(V) | Jsc(mA/cm$^2$) | FF(%) | PCE(%) |
|---|---|---|---|---|
| **Cell-1** | 0.9968 | 17.2450 | 82.14 | 14.12 |
| **Cell-2** | 1.2011 | 24.2167 | 84.2 | 24.5 |
| **Cell-3** | 1.2107 | 24.2176 | 83.42 | 24.46 |

*3.2 Performance analysis by varying HTL thickness*

The cell performance is influenced by HTL's thickness and material type. In this section, we study the influence of the HTL thickness variation from 50-150nm using organic (Spiro OMeTAD) and inorganic (CuI) HTMs. The thickness of HTLs not only affects the cell performance but also acts as a capping layer to avoid contact between cathode and absorber layer. Figure 4 shows the impact of HTL layer thickness (40-150nm) on the Fill factor (FF) of the simulated PSCs. It can be observed that, the Fill factor (FF) is not much affected for cell-2 (where CuI is being used as an HTM). However, Spiro OMeTAD based cell has exhibited a decrement in FF with increase in thickness of HTL. The organic material poses lower charge carrier mobility and conductivity as compared to the inorganic materials. As a result, when thickness of the organic HTM (Spiro OMeTAD) layer increases, the layer resistance increases which further lowers the performance of the cell. In case of inorganic HTL (CuI) the FF is not much affected by HTL thickness variation due to comparatively higher charge carrier mobility.

*3.3 Performance analysis using various contacts*

In this section, we simulate Sn-PSCs with and without HTM using different back contacts. The simulated device has configuration of FTO/ ZnO NP / $CsSn_{0.5}Ge_{0.5}I_3$/(Spiro-OMeTAD or CuI) /Back contacts. The different back contacts used are Pt (5.7eV), Ni(5.5eV), Au(5.1eV), C(5.0eV), Fe(4.81eV), Ag(4.74eV).

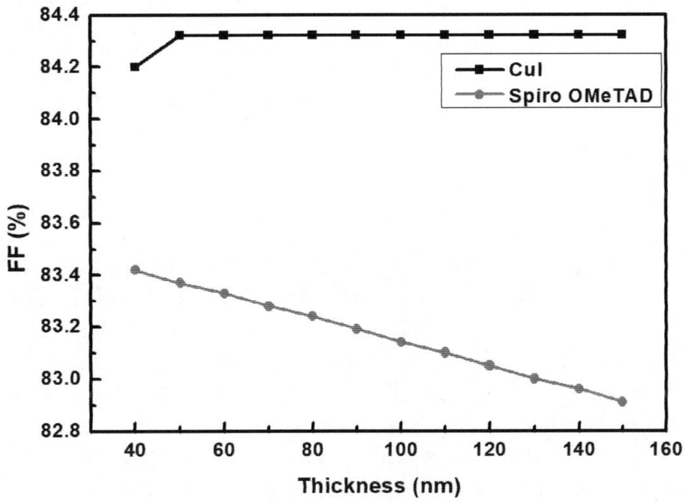


**Figure 4.** Impact of HTL layer thickness (40-150nm) on the Fillfactor (FF) of the simulated PSCs.

This analysis is carried out to understand the influence of different back contacts of Sn-PSCs performance using various HTLs. Figure 5 shows the variation of performance parameters with different back contacts on the Sn-PSC's performance with ZnO-NPs as ETM and various HTMs.

As the work function of the back contact tends to be higher than that of HTM's, it ensures proper hole collection and electron blocking. It can be said that, using a back contact with high work function will transfer holes efficiently leading to higher performance. It is observed that with increase in the contact work function the PCE initially increases and later gets saturated (Higher the value of bandgap of HTM earlier is the saturation point reached). The performance of Cell-1 (HTL free) seemed to be most dependent on the metal cathode. Cell-1 exhibited increment in performance with increase in contact work function till 5.4eV and later gets saturated. Cell-3 (using Spiro-OMeTAD) exhibited increment with increase in contact work function till 5.2 eV and later gets saturated. Cell-2 (using CuI) exhibited least metal cathode dependent behaviour by attaining the saturation point quiet early, around 4.84eV. It is because of the high charge carrier mobility of the used HTM (CuI). Therefore, it can be said that, in case of high charge carrier mobility metal contact with even lower function like Ag is capable of rendering good performance.

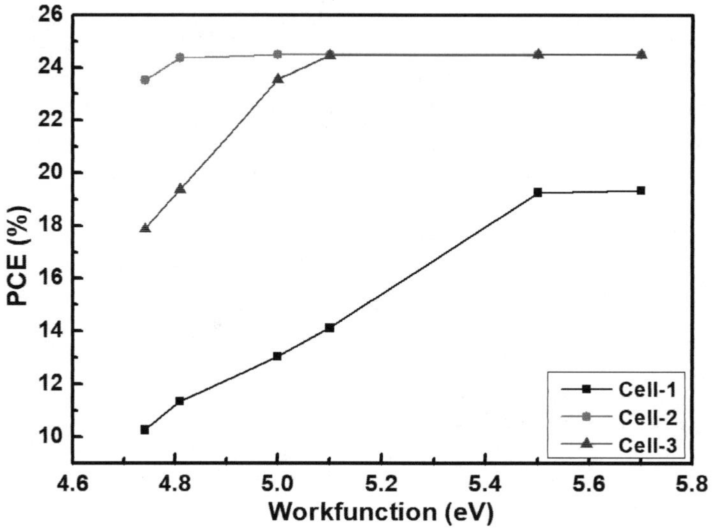


**Figure 5.** Energy-band diagram of the simulated cell.

## 4. Conclusions

In this work, we studied a hetrojunction $CsSn_{0.5}Ge_{0.5}I_3$ based PSCs with planar architecture using ZnO nanoparticles (NP) as an ETM. We use various perspectives of HTL and simulate HTL free, organic and inorganic HTMs to study its usage with $CsSn_{0.5}Ge_{0.5}I_3$ based PSCs. It was seen that, PSC with configuration: FTO/ZnO/ $CsSn_{0.5}Ge_{0.5}I_3$ /CuI/Au attained maximum performance with 24.5% PCE, 1.2011V $V_{oc}$, 82.4% fill factor and 24.2 mA/cm$^2$ $J_{sc}$. If CuI is used instead of the conventional Spiro-OMeTAD as HTM one can obtain good performance parameters along with reduction in overall device cost. Impact of HTL thickness variation on the performance of PSCs is also studied. The result signifies that HTMs carrier mobility and conductivity have influence on the HTL thickness selection. An analysis was conducted to study the performance using various metals as back contacts. It was observed that, presence and type of HTL has an impact on the metal cathode selection. This work also provides an alternative to use other cathode metal than widely used Au, thereby reducing the cost of the PSCs.

**Acknowledgement**

We thank Professor Marc Burgelman, University of Gent, Department of Electronics and Information Systems, for providing us SCAPS software package and allowing its use.

IOP Conf. Series: Materials Science and Engineering 1120 (2021) 012016 doi:10.1088/1757-899X/1120/1/012016

**References**


[1] NREL 2019 PV Research Cell Record Efficiency Chart

[2] Lyu M, Yun J-H, Chen P, Hao M and Wang L 2017 *Adv. Energy Mater.* **7** 1602512

[3] Roy P, Kumar Sinha N, Tiwari S and Khare A 2020 *Sol. Energy* **198** 665

[4] Shao S, Liu J, Portale G, Fang H-H, Blake G R, ten Brink G H, Koster L J A and Loi M A 2018 *Adv. Energy Mater.* **8** 1702019

[5] Chen M, Ju M-G, Garces H F, Carl A D, Ono L K, Hawash Z, Zhang Y, Shen T, Qi Y, Grimm R L, Pacifici D, Zeng X C, Zhou Y and Padture N P 2019 *Nat. Commun.* **10** 16

[6] Hao F, Stoumpos C C, Guo P, Zhou N, Marks T J, Chang R P H and Kanatzidis M G 2015 *J. Am. Chem. Soc.* **137** 11445

[7] Marc Burgelman, Koen Decock, Alex Niemegeers, Johan Verschraefen S D 2018 SCAPS manual

[8] Du H-J, Wang W-C and Zhu J-Z 2016 *Chinese Phys. B* **25** 108802

[9] Mandadapu U, Vedanayakam S V and Thyagarajan K 2017 *Indian J. Sci. Technol.* **10** 1

[10] Anwar F, Mahbub R, Satter S S and Ullah S M 2017 *Int. J. Photoenergy* **2017** 1

[11] Roy P, Kumar Sinha N and Khare A 2020 *Mater. Today Proc.*

[12] Roy P, Sinha N K, Tiwari S and Khare A 2020 *IOP Conf. Ser. Mater. Sci. Eng.* **798** 012020

[13] Raghvendra, Kumar R R and Pandey S K 2019 *Superlattices Microstruct.* **135** 106273

[14] Noh J H, Im S H, Heo J H, Mandal T N and Seok S Il 2013 *Nano Lett.* **13** 1764

[15] Mei A, Li X, Liu L, Ku Z, Liu T, Rong Y, Xu M, Hu M, Chen J, Yang Y, Gratzel M and Han H 2014 *Science* **345** 295

[16] Burschka J, Pellet N, Moon S-J, Humphry-Baker R, Gao P, Nazeeruddin M K and Grätzel M 2013 *Nature* **499** 316

[17] Hao F, Stoumpos C C, Cao D H, Chang R P H and Kanatzidis M G 2014 *Nat. Photonics* **8** 489

[18] Docampo P, Ball J M, Darwich M, Eperon G E and Snaith H J 2013 *Nat. Commun.* **4** 2761

[19] Mohamad Noh M F, Teh C H, Daik R, Lim E L, Yap C C, Ibrahim M A, Ahmad Ludin N, Mohd Yusoff A R bin, Jang J and Mat Teridi M A 2018 *J. Mater. Chem. C* **6** 682

[20] Mhamad S A, Mohammed A M, Aziz M and Aziz F 2019 Impact of Electron Transport Layers (ETLs) and Hole Transport Layer (HTLs) on Perovskite Solar Cells Performance *Nanostructured Materials for Next-Generation Energy Storage and Conversion* (Berlin, Heidelberg: Springer Berlin Heidelberg) p 227–46

IOP Conf. Series: Materials Science and Engineering 1120 (2021) 012017 doi:10.1088/1757-899X/1120/1/012017

# Simulation and analysis of schottky junction perovskite solar cells (SJPSCs)

**Numeshwar Kumar Sinha[1*], Priyanka Roy[1], Dhriti Sundar Ghosh[2] and Ayush Khare[1]**

[1]Thin Film Research Laboratory, Department of Physics, National Institute of Technology, Raipur – 492 010, India
[2]Department of Physics, Indian Institute of Technology, Bhilai, GEC Campus, Sejbahar, Raipur – 492 015, India

*Corresponding author's e-mail address:nksinha.phd2018.phy@nitrr.ac.in

**Abstract.** The Schottky junction solar cells are the simplest and single-layer structure of solar cells, offering low-cost and easy fabrication. In this study, we have simulated and analyzed the Schottky junction perovskite solar cells, having the device structure of FTO/$MAPbI_3$/Au. The simulation has done by Solar Cell Capacitance Simulator (SCAPS-1D). In this simulation, we have studied the effect of an absorber layer thickness, doping density, defect density, and the cell's temperature on the performance of the cells. We have further studied the effect of different back electrodes, like Au, Cu, Ag, Al, and Graphene in the performance of cells. The effect of series and shunt resistance has also been studied at optimized parameters. This simulation suggests that the SJPSCs show the highest optimized power conversion efficiency of greater than 14%, at actual series and shunt resistance of the cell. Simulating this simplest structure will help to further improvement in the performance of SJPSC. However, the use of lead makes the solar cells non-eco-friendly, but the proper encapsulation and disposal management can solve this problem.

## 1. Introduction

The Perovskite material, which is used for making solar cell devices, has properties that, it has high carrier mobility [1], long diffusion length [2], high charge carrier lifetime[3], high absorption coefficient [4] and tunable bandgap [5]. The Perovskite Solar Cells can be processed by solvent engineering techniques, which provide wide verities of cheap fabrication methods, like spin coating, dip coating, screen printing, etc. [6]. Due to such properties and easy fabrication techniques, the perovskite solar cells have been taken great interest and in just one decade the efficiency of PSCs has been accomplished to 25.5%, which very close to commercially available c-Si solar cells [7]. A Schottky junction solar cells are the metal-semiconductor interface, which offers, band, bending energy band diagram and that separates light generated carrier towards the electrodes. The Schottky junction solar cells (SJSCs) show their maximum performance when one contact is Ohmic and the other is barrier contact. The SJSCs are the simplest solar cells and beneficial for further reduction in the cost of solar cells. To form a Schottky barrier, the work function of the metal should be higher as compared to the electron affinity of the semiconductor for the case of n-type semiconductor, but for the p-type semiconductor, the work function of metal should be small. In our case, we have used the n-type $CH_3NH_3PbI_3$ ($MAPbI_3$) absorber layer, having an electron affinity of 3.9eV, with high work function metals [8]. In the present study, we have simulated SJPSCs with help of the Solar Cell Capacitance Simulator (SCAPS-1D)[9] and studied the effect of absorber thickness, doping density, defect density, and cell temperature on the performance of the cell. We have optimized all these parameters. The effect of different back electrodes and effect of series and shunt have also been studied. Our study reveals that, Au back electrode shows the maximum efficiency and Al is not suitable for the making of SJPSCs. Graphene, an interesting two-dimensional structure, has drawn intense interest due to its high stability, great charge carrier mobility and excellent conductivity[10][11]. In our simulation, we have used graphene layer as an electrode and achieved the efficiency of 14.14%. The use of graphene electrode can be suitable for further cost reduction of the device. We have achieved the optimized efficiency of 14.48% at series and shunt resistance of 3.18 $\Omega.cm^2$ and 758 $\Omega.cm^2$ respectively for FTO/n-type MAPbI3/Au structure.

## 2. Methodology and Simulation Parameters

For a simulation of the device, we have used SCAPS-1D simulation package, which is a one-dimension solar cell simulation program developed by the Department of Electronics and Information System, University of Gent, Belgium[12]. It is a completely computer-based software tool and adapted for analysis

of homojunction, heterojunction, multi-junction, and Schottky junction solar cells. SCAPS is based on solving the one-dimension Poisson's equation,

$$\frac{\partial^2}{\partial x^2}\varphi(x) = \frac{q}{\varepsilon}[n(x) - p(x) - N_D^+(x) + N_A^-(x) - p_t(x) + n_t(x)] \quad (1)$$

and continuity equation of electrons and holes.

$$\begin{aligned} \frac{\partial n}{\partial t} &= \frac{1}{q}\frac{\partial J_n}{\partial x} + G - R \\ \frac{\partial p}{\partial t} &= -\frac{1}{q}\frac{\partial J_p}{\partial x} + G - R \end{aligned} \quad (2)$$

Here, $\varphi$ is the potential, $q$ is the elementary charge, ε is the permittivity, $n$ is the density of free electron, $p$ is the density of free holes, $N_D^+$ is the ionized donor like doping density, $N_A^-$ is the ionized acceptor like doping density, $p_t$ is trapped hole density, $n_t$ is trapped electron density, $R$ is the recombination rate, $G$ is the generation rate, $J_n$ is current density due to electrons, $J_p$ is current density due to holes. The simulation parameters of different layers, chosen very carefully form different research, are given in table 1[13–15]. The simulated device structure is shown in figure 1(a) and the band diagram is shown in figure 1(b). In this SCAPS-1D simulation package, we can add up to seven layers. In such simulation packages, initial conditions are very important to find the exact solution. In the equilibrium state, where there is no light and no voltage applied, initial conditions are taken into account by assuming that the quasi-Fermi level in the system is at zero and no conceivable drops throughout the structure. The short circuit current density is determined as calculation begins after illumination and this is set to be the next initial guess. Layer by layer, this process continues and obtained the final optimized solution.

**Table 1.** Simulation Parameters of different layers used in the simulation.

| Parameters | $MAPbI_3$ | Graphene | FTO |
|---|---|---|---|
| **Thickness ($\mu m$)** | 0.1 – 1.2 | 0.05 | 0.05 |
| **Band-gap (eV)** | 1.5 | 0.8 | 3.5 |
| **Electron Affinity (eV)** | 3.9 | 4.8 | 4.4 |
| **Dielectric Permittivity (Relative)** | 30 | 7.1 | 9 |
| **CB effective density of states ($cm^{-3}$)** | $2.5x10^{19}$ | $3x10^{19}$ | $2x10^{18}$ |
| **VB effective density of states ($cm^{-3}$)** | $2.5x10^{19}$ | $3x10^{19}$ | $2x10^{18}$ |
| **Electron Thermal velocity (cm/s)** | $1x10^{7}$ | $1x10^{7}$ | $1x10^{7}$ |
| **Hole Thermal Velocity (cm/s)** | $1x10^{7}$ | $1x10^{7}$ | $1x10^{7}$ |
| **Electron Mobility ($cm^2$/Vs)** | 15 | $1x10^{1}$ | $2x10^{3}$ |
| **Hole Mobility ($cm^2$/Vs)** | 15 | $1x10^{5}$ | $2x10^{3}$ |
| **Donor density ($cm^{-3}$)** | $1x10^{13}$ – $1x10^{16}$ | 00 | $2x10^{19}$ |
| **Acceptor Density ($cm^{-3}$)** | 00 | $1x10^{16}$ | 00 |

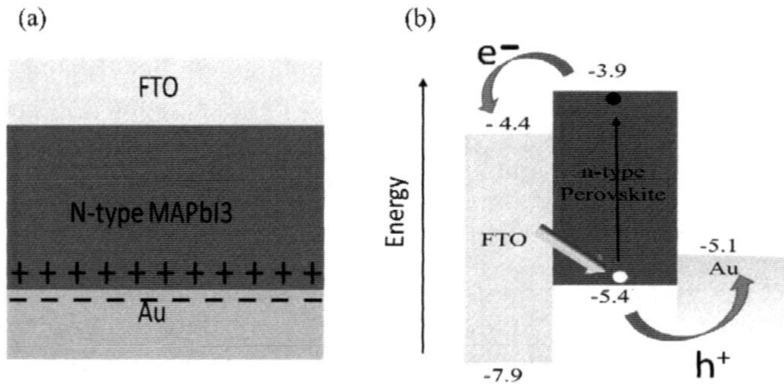


**Figure 1.** (a) Structure of simulated device and (b) band diagram of the device.

## 3. Results and Discussions

### *3.1. Effect of absorber thickness on the performance of the cells*

The performance of the cell varies with the thickness of an absorber layer as shown in figure 1. We have varied the thickness of an n-doped $CH_3NH_3PbI_3$ ($MAPbI_3$) from 0.1 μm to 1.2 μm. As increasing the thickness, the short circuit current density ($J_{sc}$) and open-circuit voltage ($V_{oc}$) increases as shown in the figure 2, it may be due to increasing the total number of electron-hole pairs as increasing the thickness[16]. The efficiency of the cell increases with increasing the thickness of an absorber layer and reached to optimum value at 0.75 μm. Further increasing the thickness, $J_{sc}$ of the cell get saturated and it may be due to the limiting the collection probability of light-generated charge carriers by the junction[17][18]. The $V_{oc}$ of the cell decreases, caused by the increase in the recombination as increasing the thickness, and thus the overall performance decreases[15]. For larger the lifetime of the charge carrier will increase the efficiency of the cell at a particular thickness of an absorber layer[19].

IOP Conf. Series: Materials Science and Engineering 1120 (2021) 012017 doi:10.1088/1757-899X/1120/1/012017

### *3.2 Effect of doping density of an absorber layer on the performance of the cells*

$MAPbI_3$ has properties that it can behave like n-type or p-type depending on the ratio of $MAI/PbI_2$[20]. Experimentally, it has been found that, if the film is MAI rich then it behaves as p-type whereas if the film is $PbI_2$ rich then the film work as n-type[21][22]. In our simulation, we have used an n-type $MAPbI_3$ perovskite material. We have studied the effect of an n-type doping concentration of an absorber layer on the performance of the cells by varying the doping density from $10^{13}cm^{-3}$ to $10^{18}cm^{-3}$. Increasing the n-type doping concentration increases the Fill factor (FF) and $J_{sc}$ of the cell because increasing the doping concentration increases the conductivity of the materials and increases collection of the light-generated charge carriers[23]. From Quantum Efficiency (QE) figure 4 (a), it clear that the charge collection slightly increases with increasing the doping density and but for larger the doping density QE get reduces. The increasing the doping density also increases the defect in the material which increases the recombination of the material[22]. As from figure 4 (b), it clear that the recombination rate in the material increases very when we increase the doping concentration from $10^{16}$ to $10^{18}$ $cm^{-3}$. That leads to decreasing in the $V_{oc}$. The overall efficiency of the device remains approximately the same for doping density $10^{13}$ to $10^{15}$ $cm^{-3}$ and decreases rapidly after doping density increases from $10^{15}$ to $10^{18}$ $cm^{-3}$.

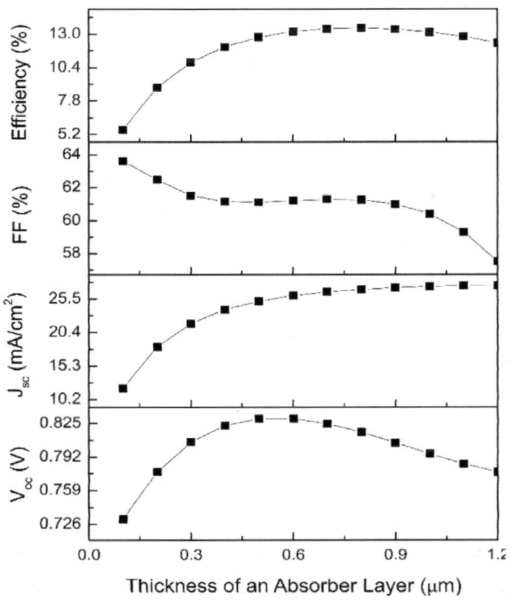


**Figure 2.** Effect of an absorber thickness on the performance parameters of the device.

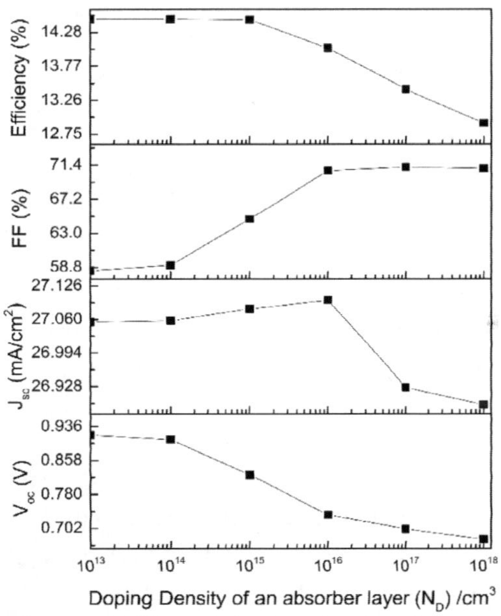


**Figure 3.** Effect of doping density of an absorber layer on performance parameters of the device.

### *3.3. Effect of defect density of an absorber layer on the performance of the cells*

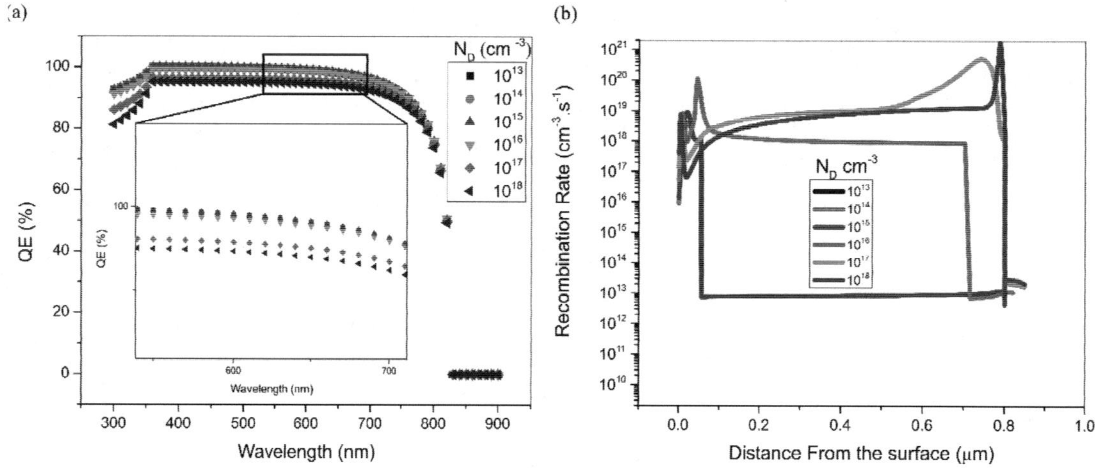


**Figure 4.** Effect of doping density of an absorber layer on (a) quantum efficiency and (b) recombination rate.

The defects in the solar cell's material plays a very important role. The defects create an intermediate energy level between the conduction band and valance band and work as a recombination center [24][25]. Thus, increasing the defects in the material will increase the recombination[26]. The defect reduces the diffusion length and lifetime of the charge carrier and thus reducing the collection probability of light generated carrier by the junction[27]. As we increase the defect density in the light-harvesting material from $10^{13}$ to $10^{15}cm^{-3}$, all the performance parameters get reduces tardily but when we further increase the defect density from $10^{16}$ to $10^{18}$ $cm^{-3}$ the performance parameters get reduces linearly approximately.

### *3.4. Effect of cell's temperature on the performance of the cells*

In India, at different seasons, the various regions have different temperatures[28] and the analysis of the influence of the temperature of the cells becomes more significant. The temperature of the cell has been varied from 300K to 350K. The increase in the temperature increases the recombination current or dark current and thus decreases the $V_{oc}$ of the device[29]. Increasing the temperature decreases the energy gap of the material by increasing the thermal energy of the charge carriers with temperature[30] and that leads to increasing the $J_{sc}$ as temperature[31] as shown in the figure 6. From the figure 6, it is clear that the rate of increasing of $J_{sc}$ is very small as compared to the rate decreasing of $V_{oc}$, thus the overall efficiency of the cell decreases with increasing the cell's temperature.

### *3.5. Effect of series and shunt resistance*

The series resistance comprises resistance of all the layers in the device whereas, shunt resistance is due to defects and poor quality of the cell. The series resistance predominately affects the $J_{sc}$ and Shunt resistance mainly impacts $V_{oc}$. The effect of series and shunt resistance on the cell's performance at optimized parameters (Thickness = 0.75μm, $N_D = 10^{15} cm^{-3}$, $N_t = 10^{13} cm^{-3}$, Temperature 300K) has been studied. The cell shows 16.98% efficiency by considering the cell as an ideal solar cell in which there is no series and shunt resistance. But in the actual device, the series and shunt resistance must present, and when including the series and shunt resistance the device shows 14.48% efficiency for series and shunt resistance of $3.15\Omega cm^2$ and $758\Omega cm^2$ respectively. These values of series and shunt resistance are corresponding to the actual device which we have taken from the experiment of another researcher[32].

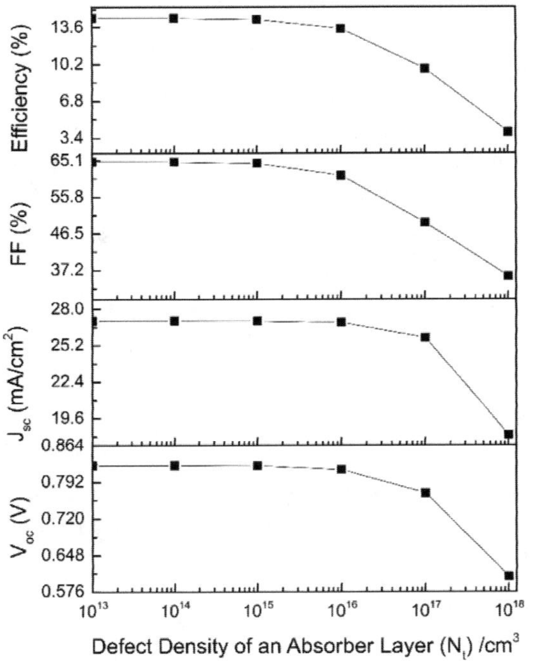


Figure 5. Effect of defect density of an absorber layer on the performance parameters of the device.

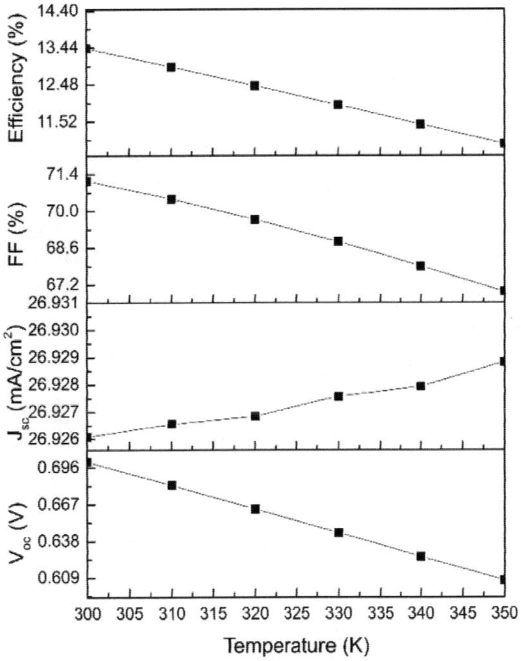


**Figure 6.** Effect of cell temperature on the performance of the cell.

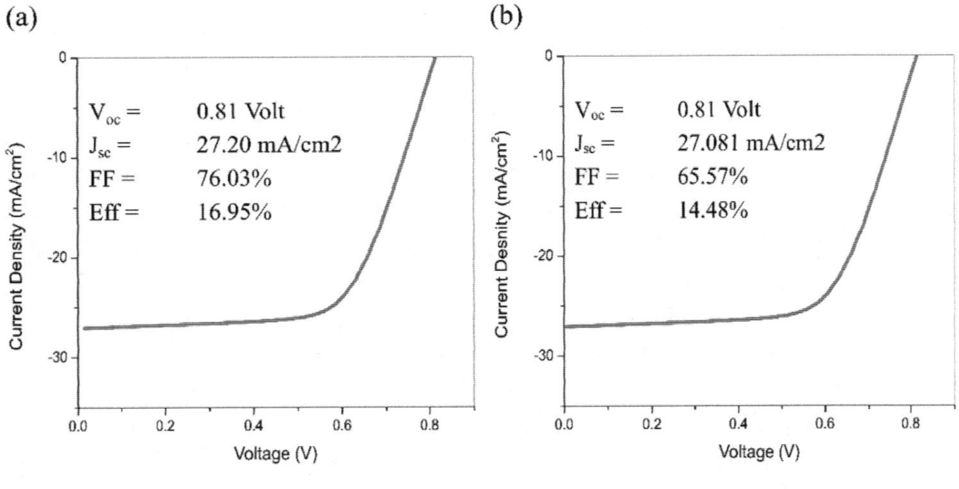


**Figure 7.** J-V curve of the cell for (a) an ideal case and (b) due to effect of series and shunt resistance.

### *3.6. Effect of different back electrodes*

The built-in-potential ($V_{bi}$) plays a vital role in the performance of solar cells. The built-in-potential provides the necessary action of separation of light-generated charge carriers. In the Schottky junction solar cells, the built-in-potential is defined as [22] in equation (3).

$$V_{bi} = \frac{\Phi_B-(E_C-E_F)}{q} \quad (3)$$

Where, $E_C$ is conduction band of n-type perovskite, $E_F$ is the fermi level of n-type semiconductor, $q$ is an electronic charge, and $\Phi_B$ is Schottky barrier height at metal/perovskite interface defined as

$$\Phi_B = \Phi_m - \chi \quad (4)$$

Here, $\Phi_m$ is the work function of metal and $\chi$ is the electron affinity of semiconductors. From equation (3), it is clear that the built-in-potential strongly depends on the work function of the metal. When we move from high work function back electrode to low work function back electrode, the strength of $V_{bi}$ decreases and thus decreasing the overall performance of the cells as shown in table 2. The device shows the maximum efficiency of 14.48% for the case of Au back electrode. For the case of Al back electrode, our device does not work. It may be due to a lower strength of $V_{bi}$, that might not sufficient to separate the light-generated charge carriers towards the electrode to generate the photovoltaic effect. The graphene has high conductivity[33] and high charge carrier mobility[34] and forms a Schottky junction between graphene/perovskite interface because of its high work function[35]. In our simulation, by adding a graphene layer above the perovskite layer, as parameters given in table 1, the device shows 14.14% efficiency. The use of graphene may be a good alternative of Au for making highly efficient SJPSCs and for further diminution of the cost of the cells.

**Table 2.** Effect of different back electrodes on the performance parameters of the device.

| Device Structure | $V_{oc}$ (V) | $J_{sc}$ mA/cm$^2$ | FF % | Efficiency % |
|---|---|---|---|---|
| **FTO/MAPbI$_3$/Au** | 0.815726 | 27.08188 | 65.5365 | 14.4779 |
| **FTO/MAPbI$_3$/Graphene** | 0.760943 | 25.62309 | 72.5078 | 14.1374 |
| **FTO/MAPbI$_3$/Cu** | 0.430565 | 26.97764 | 45.6396 | 5.3013 |
| **FTO/MAPbI$_3$/Ag** | 0.232208 | 24.85145 | 28.3431 | 1.6356 |
| **FTO/MAPbI$_3$/Al** | | **Cell do not work** | | |

## 4. Conclusions

In this study, we have simulated SJPSCs by using SCAPS-1D simulation package. our simulated device structure is FTO/n-type MAPbI3/back electrode. We have simulated and analyzed the effect of absorber's thickness, doping density, defect density, and cell temperature on the performance of the device and optimized all these parameters. The device shows the efficiency of 16.98% for the case of ideal solar cells and when we include the effect of series and shunt resistance, the device shows the efficiency of 14.48% at series and shunt resistance of 3.18 $\Omega.cm^2$ and 758 $\Omega.cm^2$ respectively. The effect of back electrodes has also been studied. The device shows a maximum efficiency of 14.48% for FTO/n-type_MAPbI$_3$/Au structure. Our simulation suggests that Al is not suitable for the making of SJPSCs because of its low work function. By using graphene as an electrode, we obtained an efficiency of 14.14% and maybe the promising alternative of gold.

## 5. Acknowledgment

We would like to thank the Department of Electronics and Information Systems (ELIS), University of Gent, Belgium, for providing the SCAPS-1D Simulation Package.

## References

[1] Herz L M 2017 *ACS Energy Lett.* **2** 1539
[2] Dong Q, Fang Y, Shao Y, Mulligan P, Qiu J, Cao L and Huang J 2015 *Science* **347** 967
[3] Kiermasch D, Rieder P, Tvingstedt K, Baumann A and Dyakonov V 2016 *Sci. Rep.* **6** 39333
[4] Davis M and Yu Z 2020 *J. Semicond.* **41** 041603
[5] Ju D, Dang Y, Zhu Z, Liu H, Chueh C-C, Li X, Wang L, Hu X, Jen A K Y and Tao X 2018 *Chem. Mater.* **30** 1556
[6] Aparna Thankappan 2018 *Perovskite Photovoltaics Basic to Advanced Concepts and Implementation* 1st edition pp 36
[7] Best Research-Cell Efficiency Chart | Photovoltaic Research | NREL https://www.nrel.gov/pv/cell-efficiency.html (accessed Feb 22, 2021).
[8] Gagandeep, Singh M and Kumar R 2019 *Dae Solid State Phys. Symp. 2018* **2115** 030548
[9] Burgelman M, Nollet P and Degrave S 2000 *Thin Solid Films* **361** 527
[10] Bouclé J and Herlin-Boime N 2016 *Synth. Met.* **222** 3
[11] Manzano-Ramírez A, López-Naranjo E J, Soboyejo W, Meas-Vong Y and Vilquin B 2015 *J. Nanomater.* **2015** 1
[12] Burgelman M, Decock K, Niemegeers A, Verschraegen J and Degrave S 2019 *SCAPS manual*
[13] Mandadapu U, Vedanayakam S V and Thyagarajan K 2017 *Indian J. Sci. Technol.* **10** 1
[14] Du H-J, Wang W-C and Gu Y-F 2017 *Chinese Phys. B* **26** 028803
[15] Bag A, Radhakrishnan R, Nekovei R and Jeyakumar R 2020 *Sol. Energy* **196** 177
[16] Liu D, Gangishetty M K and Kelly T L 2014 *J. Mater. Chem. A* **2** 19873
[17] Donolato C 2001 *J. Appl. Phys.* **89** 5687
[18] Honsberg, C.; Bowden, S. Collection probability https://www.pveducation.org/pvcdrom/solar-cell-operation/collection-probability (accessed Feb 8, 2021).
[19] Li S, Liu P, Pan L, Li W, Yang S E, Shi Z, Guo H, Xia T, Zhang S and Chen Y 2019 *Sol. Energy Mater. Sol. Cells* **199** 75
[20] Rudd P N and Huang J 2019 *Trends Chem.* **1** 394
[21] Wang Q, Shao Y, Xie H, Lyu L, Liu X, Gao Y and Huang J 2014 *Appl. Phys. Lett.* **105** 63508
[22] Lin L, Jiang L, Li P, Li X and Qiu Y 2020 *Energy Technol.* **8** 1901197
[23] Li Y, Li C, Yu H, Yuan B, Xu F, Wei H and Cao B 2020 *Front. Chem.* **8** 754
[24] Sherkar T S, Momblona C, Gil-Escrig L, Ávila J, Sessolo M, Bolink H J and Koster L J A 2017 *ACS Energy Lett.* **2** 1214
[25] Liu Y, Stradins P, Deng H, Luo J and Wei S-H 2016 *Appl. Phys. Lett.* **108** 022101
[26] Ball J M and Petrozza A 2016 *Nat. Energy* **1** 16149
[27] Astakhov O, Carius R, Finger F, Petrusenko Y, Borysenko V and Barankov D 2009 *Phys. Rev. B* **79** 104205
[28] India climate: average weather, temperature, precipitation, best time https://www.climatestotravel.com/climate/india (accessed Feb 8, 2021).
[29] SOLANKI C S 2016 Solar Photovoltaics Fundamentals, Technologies And Applications 3rd edition p 105
[30] Van Zeghbroeck, B. 2.2.5 Temperature dependence of the energy bandgap https://ecee.colorado.edu/~bart/book/eband5.htm (accessed Feb 8, 2021).
[31] Patel K and Tyagi P K 2017 *Carbon N. Y.* **116** 744
[32] Huang P H, Wang Y H, Ke J C and Huang C J 2017 **10** 1
[33] Mogi H, Bamba T, Murakami M, Kawashima Y, Yoshimura M, Taninaka A, Yoshida S, Takeuchi O, Oigawa H and Shigekawa H 2019 *ACS Appl. Electron. Mater.* **1** 1762
[34] Geim A K and Novoselov K S 2007 *Nat. Mater.* **6** 183
[35] Song Y, Osherov A, Bulović V and Kong J 2018 *Adv. Sustain. Syst.* **2** 1700106

IOP Conf. Series: Materials Science and Engineering 1120 (2021) 012018 doi:10.1088/1757-899X/1120/1/012018

# Effect of cadmium sources on the structural, morphological, and optical properties of CdS films for solar cell applications

**Tarkeshwar Sinha, Lekha Verma and Ayush Khare**

Department of Physics, National Institute of Technology, Raipur - 492010 India

*Corresponding author's e-mail address: ts22701@gmail.com

**Abstract**. The variations in the structural, morphological, and optical properties of three CdS films deposited with different cadmium ion sources namely cadmium acetate, cadmium chloride, and cadmium sulfate are reported in this article. All three CdS films possess cubic crystalline structure with the most prominent cubic phase (111). The CdS film deposited with cadmium acetate showed the best crystallinity among them. The presence of both Cd and S with a difference in the percentage of its existence depending upon the type of cadmium source chosen is obtained in the elemental analysis of the CdS films. CdS film deposited from acetate based chemical solution exhibited broad transmittance behaviour in the visible wavelength region due to its superior surface morphology. The energy band-gap values for CBD-CdS films deposited with cadmium acetate, cadmium chloride, and cadmium sulfate are found as 2.30eV, 2.23eV, and 2.32eV, respectively. The film deposited with cadmium acetate showed better IR response and more intense PL emission peaks as compared to the other two films. The efficiency of the CdTe/CdS cell fabricated with CdS window layer deposited with cadmium acetate is found to be 10.02%.

## 1. Introduction

Solar energy is very attractive, clean, reliable, and effective way to meet the expected rise in energy consumption globally. Cadmium telluride (CdTe)/Cadmium sulphide (CdS) thin film solar cell is one of the most popular photovoltaic (PV) devices. In the film deposition process of different layers of a PV cell by chemical deposition techniques, the selection of appropriate precursors is a very important decision as it eliminates the complexities of the chemical reactions and reduces the time and labor of a researcher. Every material has some specific characteristics that can support or disturb the reaction rate during their addition to the final solution. This affects the film quality and its subsequent applicability for solar cell applications. Not only in chemical but also in physical deposition techniques, the usage of good quality cadmium and sulfur source in an adequate molar ratio ensures the deposition of CdS film up to some extent and the other deposition parameters can be controlled more efficiently [1-2]. Thiourea is a very popular choice of a sulfur ion source for the preparation of chemical solution to deposit CdS film as it offers flexibility in usage with different types of cadmium ion sources including cadmium acetate, cadmium sulfate, cadmium chloride, etc. [3-4].

## 2. Experimental details

In the present work, three CdS films were deposited by chemical bath deposition (CBD) technique using three different cadmium ion sources cadmium acetate ($Cd(CH_3COO)_2{\cdot}2H_2O$), cadmium chloride ($CdCl_2$), and cadmium sulfate ($CdSO_4$) on fluorine doped tin oxide (FTO) glass substrates. Thiourea

($H_2NCSNH_2$) and ammonium hydroxide ($NH_4OH$) were used as the sulfur ion source and complexing agent, respectively. Three different solutions were prepared using these three different cadmium ion sources separately, in which each of the chemical solutions contained 0.5M cadmium source material and 1M thiourea ($H_2NCSNH_2$). In each beaker, one FTO coated glass substrate was dipped vertically. Then beakers were placed for 1h in water bath which was already maintained at 70°C. After deposition, the substrates were rinsed with distilled water and dried in air. The deposited CdS films were annealed at 200°C for 1h in a hot air oven.

## 3. Results and discussions

*3.1 Structural analysis*
CdS thin films grown with different cadmium sources are in cubic crystalline structure as shown in the X-Ray diffraction (XRD) pattern (Figure 1). The XRD pattern of the first CdS film deposited with cadmium acetate exhibit peaks corresponding to diffraction angles 26.6°, 30.7°, and 43.8° which are characteristics of cubic phases (111), (200), and (220). In the XRD pattern of CdS film deposited with cadmium chloride, two characteristic peaks of CdS have been obtained which are situated at diffraction angle 26.6° and 43.8° respectively. These peaks are characteristics of cubic phases (111) and (220) [5]. In the XRD pattern of CdS film deposited with cadmium sulfate only one peak corresponding to diffraction angles 26.5° has been obtained which is a characteristic of cubic phase (111) [6]. The diffraction peaks corresponding to FTO are marked by *. In XRD spectra, the intensities of the preferential orientation peak (111) along with the other two peaks are higher for the CdS films deposited with cadmium acetate as compared to the CdS films deposited with the other cadmium source materials. Thus, the film deposited with cadmium acetate shows far better crystalline behaviour than the films deposited with cadmium chloride and cadmium sulfate.

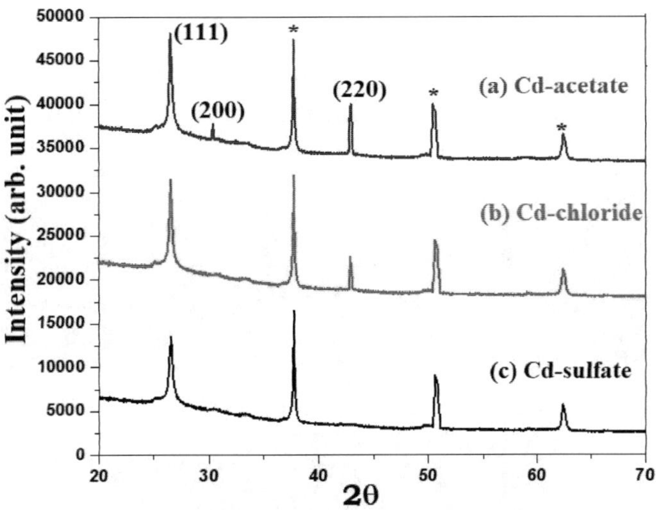


**Figure 1.** The XRD patterns of CBD-CdS films deposited with different cadmium sources.

*3.2 Morphological analysis*
The morphologies of the CdS films deposited with different cadmium sources were investigated by scanning electron microscopy (SEM) and the SEM micrographs are shown in figure 2. The SEM analysis reveals that the morphologies of CdS films were sensitive to the change in the cadmium ion source and an interesting change has been observed in the morphology. As shown in the SEM images, there is a regular growth pattern of fine CdS particles on the surface of the film deposited with cadmium acetate and cadmium chloride. The surface of the film deposited with cadmium sulfate has some cracks that disturb the morphology and quality of CdS film. Here, surface of film deposited with cadmium

acetate is best among all three films as it is more compact and smooth. Also, its surface is covered very well by the particles having defined grain boundaries that make it a more suitable choice for solar cell applications than the other films deposited with cadmium chloride and cadmium sulfate.

**Figure 2.** SEM images of CdS films deposited with (a) cadmium acetate, (b) cadmium chloride, and (c) cadmium sulfate.

The energy dispersive X-ray (EDX) spectra of CdS films deposited with different cadmium sources is shown in figure 3. The images in figure 3 confirm the presence of cadmium and sulfur in all three CdS films deposited with cadmium acetate, cadmium chloride, and cadmium sulfate.

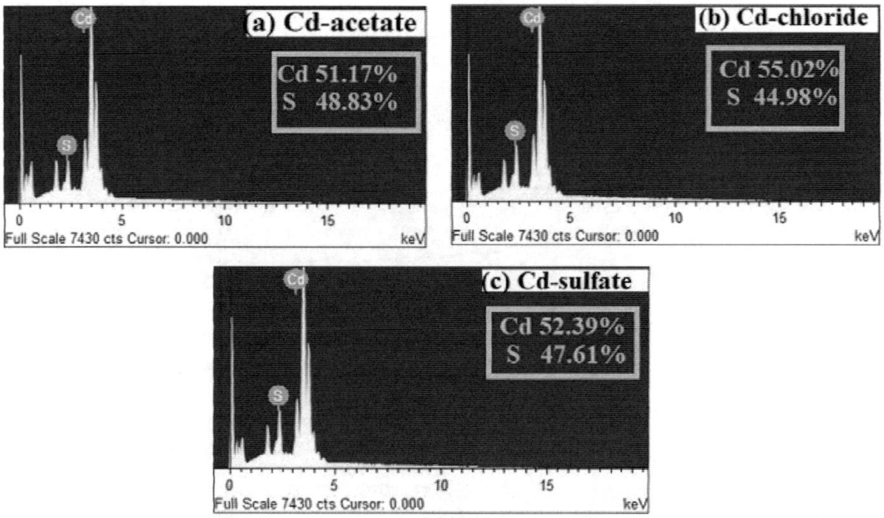


**Figure 3.** EDX spectra of CdS films deposited with different cadmium sources.

The presence of both Cd and S with a difference in the percentage of its existence is found in the EDX spectra, which depends on the different cadmium sources and the Cd/S concentrations taken during the preparation of CdS films [7]. The EDX analysis confirms that the elemental composition is not disturbed drastically for any of the CdS film suggesting that all three cadmium sources (cadmium acetate, cadmium chloride, and cadmium sulfate) can make valid pair with thiourea (sulfur source) for the deposition of a CdS film.

*3.3 Optical analysis*

The Optical characteristics of the window layer play an important role in determining the performance of solar cells [8]. Optical transmittance spectra of the CBD-CdS films were measured by UV–VIS spectrophotometer and it was recorded in wavelength range 400–700 nm. Figure 4 presents the transmittance spectra of the CdS films deposited with different types of cadmium ion sources. Here, the CdS films deposited with cadmium acetate and cadmium chloride show 60-90% optical transmission, but the film deposited with cadmium sulfate exhibit significantly lesser transmittance (less than 60%). In this case, the transmittance spectra complement the SEM investigation also as the CdS film deposited with cadmium acetate, which had the best surface morphology, shows the highest transmittance among all three films.

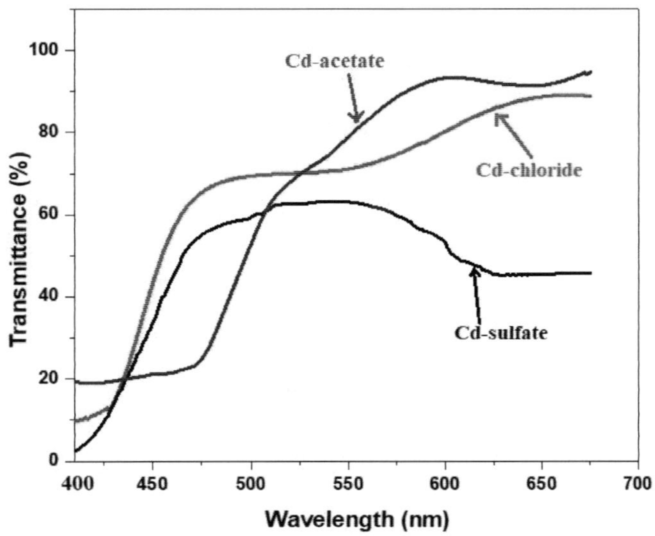


**Figure 4.** Transmittance spectra of CdS films deposited with different cadmium sources.

The energy band-gap ($E_g$) of the CdS films deposited with different cadmium sources is determined by Tauc's plot (Figure 5).

As per the extrapolation of the linear part of the curves plotted between $(\alpha h\nu)^2$ versus $(h\nu)$ to the intercept on the horizontal axis, the band-gap values of CdS films deposited with cadmium acetate, cadmium chloride, and cadmium sulfate are found as 2.30eV, 2.23eV, and 2.32eV, respectively. These values are in good agreement with the literature [9].

*3.4 PL study*

Photoluminescence (PL) spectroscopy has been used to study the luminescence spectra of CBD-CdS films with an excitation wavelength of 435nm. The PL spectra of the deposited CdS films have been recorded in the wavelength range from 400nm to 650nm (Figure 6). As the PL spectra indicate, the CdS films exhibited two distinct PL bands but the peak positions and PL intensities of the films were different for films deposited by different cadmium sources. The film deposited with cadmium acetate had PL emission peaks centered around 470nm and 546nm. The film deposited with cadmium chloride had PL emission peaks centered around 446nm and 549nm. The film deposited with cadmium sulfate and cadmium chloride had PL emission peaks centered around 442nm and 546nm. The PL intensities of all bands are higher for CdS film deposited with cadmium acetate than the rest of the two films.

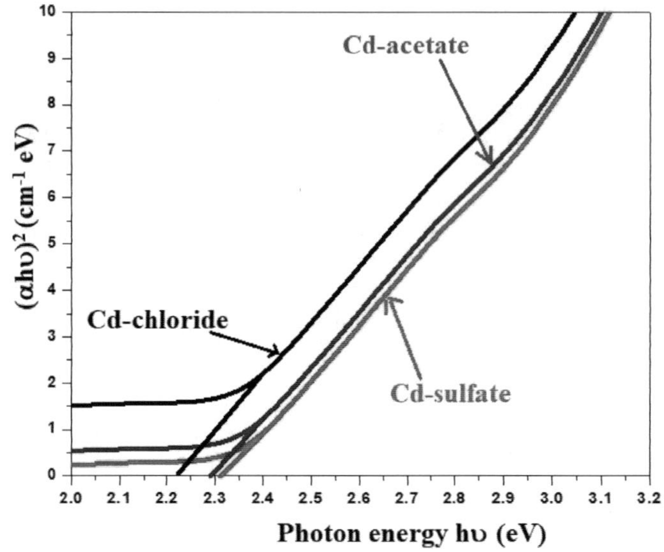


**Figure 5.** Tauc's plots of CdS films deposited with different cadmium sources.

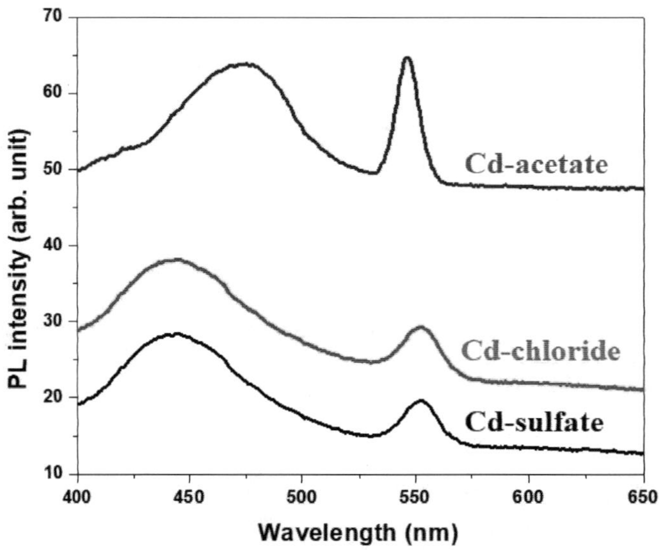


**Figure 6.** PL spectra of CdS films deposited with different cadmium sources.

*3.5 FT-IR study*

Figure 7 depicts the Fourier transform infrared (FT-IR) spectra of deposited films are recorded on a Bruker FT-IR spectrometer. The broad absorption band around 1300 $cm^{-1}$ is due to the presence of C-S stretching band, while bands around 1450 $cm^{-1}$ and 1960 $cm^{-1}$ are assigned to the presence of triple-bonded S–C–N vibrations. The narrow and weak bands around 1645 $cm^{-1}$ are due to the bending vibrations of CdS. The other weak bands observed in CBD-CdS films between 700 $cm^{-1}$ and 900 $cm^{-1}$ are due to the presence of C-S and Cd-S stretching band of CdS. The characteristics peaks obtained for

CdS films deposited with cadmium acetate are in good accordance with the literature [10]. However, in the IR spectra of CdS film deposited with cadmium sulfate, some of the characteristic peaks are missing and the positions of characteristic peaks were found to be scattered in the IR spectra of the film deposited with cadmium chloride. Comparatively, the intensities of these bands vary significantly for films deposited with different cadmium sources, which shows that the IR spectra of the films depend on the choice of a suitable cadmium source.

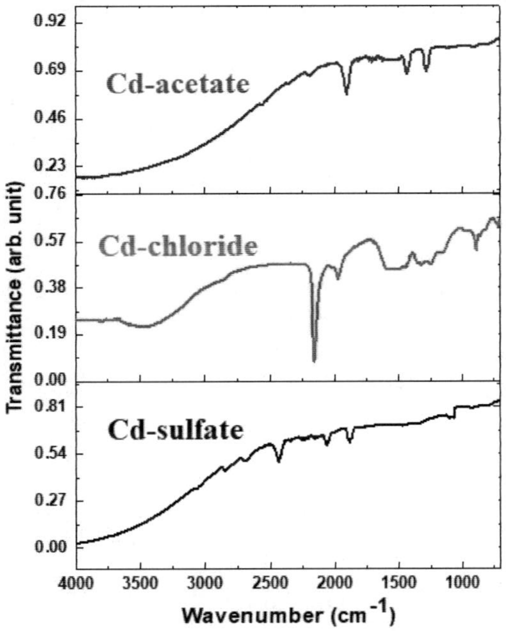


**Figure 7.** FT-IR spectra of CdS films deposited with different cadmium sources.

*3.6 Determination of PV parameters of the CdTe/CdS solar cell*

CdS film deposited with cadmium acetate has good crystallinity, better morphology, and higher optical transmittance. So it has been used for the fabrication of the CdTe/CdS solar cell. For this, a CdTe absorber layer is deposited on the pre-deposited CBD-CdS window layer by electrodeposition technique. To complete the cell structure a silver back contact was used. Figure 8 shows I–V curve obtained at room temperature for CdTe/CdS solar cell having a CdS window layer deposited with cadmium acetate.

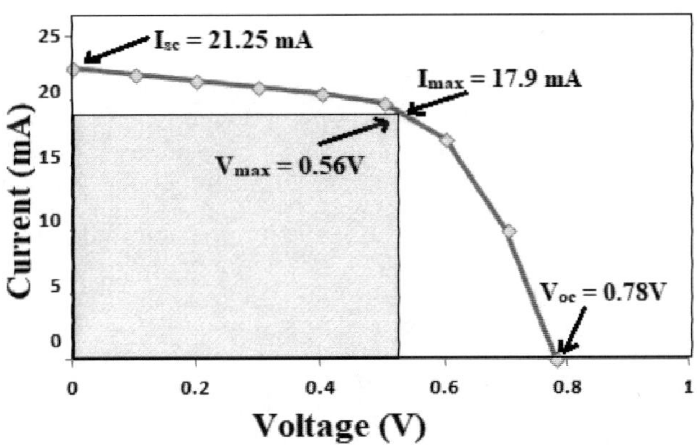


**Figure 8.** I–V characteristic of CdTe/CdS solar cell.

The PV parameters of the cell were measured using current-voltage (I-V) curve under the light illumination of 100mW/cm$^2$ by class AAA Solar Simulator. The fabricated CdTe/CdS solar cell demonstrates a short circuit current ($I_{sc}$) = 21.25 mA, open circuit voltage (Voc) = 0.78 V, a fill factor = 0.60 or 60%, and an efficiency (η) = 10.02%.

## 4. Conclusions

The variations in the structural, morphological, and optical properties of three CdS films deposited with different cadmium ion sources are studied extensively. All three CdS films possess cubic crystalline structure and the CdS film deposited with cadmium acetate exhibited best crystallinity. CdS film deposited using cadmium acetate showed very good transmittance behaviour due to its superior surface morphology. Energy band-gap values for CBD-CdS films deposited with different cadmium sources are found in the range of 2.23eV to 2.32eV. Film deposited with cadmium acetate showed better IR responsce and PL behavior. The efficiency of the CdTe/CdS cell is found to be 10.02%.

## References

[1] Pandya S and Raval K 2017 *J. Mater. Sci.: Mater. Electron.* **28** 18031
[2] Anbarasi M, Nagarethinam V S and Balu A R 2015 *J. Appl. Sci. Eng. Res.* **4** 135
[3] Kamble A, Sinha B, Agawane G, Vanalakar, Kim I Y, Kim J Y, Kale S S, Patil P and Kim J H 2016 *Phys. Chem. Chem. Phys.* **18** 28024
[4] Senthamilselvi V, Saravanakumar K, Begum N J, Anandhi R, Ravichandran A T, Sakthivel B and Ravichandran K 2012 *J. Mater. Sci.: Mater. Electron.* **23** 302
[5] Zhao X H, Wei A X, Zhao Y and Liu J 2013 *J. Mater. Sci.: Mater. Electron.* **24** 457
[6] Islam S, Hossain T, Sarwar H and Rashid 2020 J. Theor. Appl. Phys. **14** 265
[7] Mohammed I M, Gubari G M, Huse N P, Dive A S, Han S H and Sharma R 2020 *J. Mater. Sci.: Mater. Electron.* **31** 9989
[8] Basak A, Singh U P, Mishra S S and Mondal A 2017 *J. Mater. Sci.: Mater. Electron.* **28** 6560
[9] Werta S Z, Echendu O K, Dejene F B, Urgessa Z N and Botha J R 2018 *Appl. Phys. A* **124** 576
[10] Sinha T, Verma L and Khare A 2020 *Appl. Phys. A* **126** 867

# Effect of pH on structural, optical, and morphological properties of ED-CdTe film

**Lekha Verma and Ayush Khare**

Department of Physics, Thin Film Research Laboratory, National Institute of Technology, Raipur, 492010, India

*Corresponding author's e-mail address: lekhaverma.85@gmail.com

**Abstract**. Cadmium telluride (CdTe) thin films have been deposited by electrodeposition (ED) technique on fluorine doped tin oxide (FTO) coated glass substrates using an aqueous electrolyte bath solution. The pH-value of the electrolyte solution is one of the most important deposition parameter that should be optimized to achieve high quality photoconductive material. The films have been deposited at different pH values ranging from 1.5 to 3. The structural, optical, and morphological properties of the samples were characterized using X-ray diffraction (XRD), UV–Visible (UV-Vis) spectroscopy, spectroflurophotometer, scanning electron microscopy (SEM), and energy dispersive X-ray analysis (EDX) techniques respectively. The XRD analysis shows that the ED-CdTe films possess the cubic crystalline structure and the diffraction peak intensity increased significantly after changing the pH values. The direct energy band-gap ($E_g$=1.44eV) is obtained in optical absorption spectra. The SEM analysis shows very good morphological growth by changing the pH values and the film deposited with lower pH showed more intense PL emission peaks as compared to the other two films. CdTe film deposited at pH 1.5 will be better choice as an absorber layer of a CdTe/CdS solar cell.

## 1. Introduction

For the fabrication of solar cell devices, semiconductor thin films are the very promising and popular solar cell materials. Cadmium telluride (CdTe) is one of the most capable p-type semiconducting materials due to its suitable band-gap of 1.5 eV, high chemical stability, high absorption coefficient, and very good optical properties. The theoretical efficiency of CdTe thin film based solar photovoltaic (PV) devices with n-type cadmium sulfide (CdS) window layer 28%-30% and these cells have shown constant long-term performance. The first solar has introduced the new world record for CdTe photovoltaic (PV) solar cell efficiency of 22.1% and commercially available CdTe/CdS solar modules are available having an efficiency of 18.2% [1]. Here, the role of CdS window layer is to permit the energetic shorter wavelength photons to be transferred for the incidence at the hetero junction with minimum possible absorption loss. On the other hand, the CdTe absorber layer should absorb most of the solar radiation incident near the hetero-junction with minimum recombination losses. The fundamental requirement for the CdTe layer for good device efficiency is the minimization of bulk recombination and interface density [2]. CdTe have been deposited by variety of methods including chemical bath deposition (CBD) [3], electrodeposition (ED) [4], closed space sublimation (CSS) [5], radio frequency sputtering (RF) [6], laser ablation [7] etc. Electrodeposited CdTe thin film layer is the best absorber layer for polycrystalline CdTe/CdS thin film solar cell [8]. Additionally, ED is quite easy, simple and inexpensive technique. This method provides good control over preparatory conditions such as stirring rate of the solution, deposition time, growth temperature, concentration of the solution, and pH of the electrolyte. In the present work, CdTe thin films have been deposited on FTO substrate using 2-electrode ED technique at different pH of electrolyte bath ranging from 1.5 to 3.5. The main purpose of this work is to determine the optimum pH where the PV activity of CdTe layers can be achieved for development in thin film solar cells.

## 2. Materials and methods

CdTe electrolyte solution is prepared with cadmium sulphate ($CdSO_4$), tellurium dioxide ($TeO_2$), and sulphuric acid ($H_2SO_4$). Here, very low concentration of Te (~1.0–5.0mM) and high concentration of Cd (0.5–2.0M) are maintained. For this, 1000 ppm of cadmium chloride ($CdCl_2$) is added to the aqueous solution. Three different CdTe solutions are prepared at different pH values 1.5, 2.5, and 3.5. The solution is heated at 65°C for 10 min and meanwhile the solution was stirred constantly at 300 rpm using a hotplate magnetic stirrer. A saturated graphite electrode and a transparent glass/Fluorine doped tin oxide (FTO) having sheet resistance (7Ω/square) are employed here as anode and cathode, respectively. CdTe films were washed by distilled water after the completion of deposition and dried in air. With the help of a hot air oven, all three films were then annealed for 10 min at 150°C. Resulting samples were analyzed to reveal the difference in their structural, optical, and morphological properties in accordance with the different pH values.

Optical absorption measurements have been performed using the Shimadzu make UV-1800 spectrophotometer and the band-gap energy values of ED-CdTe films are determined by extrapolating the linear portion of the $(\alpha hv)^2$ vs. $hv$ plot. The surface morphological characteristics of the FTO/CdTe layer are studied using the scanning electron microscope (Zeiss EVO 18). The structural properties of the samples have been studied by PANalytical X'pert multifunctional powder X-ray diffractometer using Cu-Kα radiation (λ=1.54 Å) in the scanning range from 20° to 90°. Luminescence behaviour of the films is studied using spectroflurophotometer (Shimadzu made) instrument. The film deposition process is governed by the following chemical reactions [9]:

$$HTeO_2^{+} + 3H^{+} + 4e^{-} \longrightarrow Te + 2H_2O \quad (1)$$

$$Cd^{2+} + 2e^{-} \longrightarrow Cd \quad (2)$$

$$HTeO_2^{+} + Cd^{2+} + 3H^{+} + 6e^{-} \longrightarrow CdTe + 2H_2O \quad (3)$$

## 3. Results and Discussions

### *3.1 UV-Visible (UV-Vis) studies*

Information related to optical properties such as band-gap energy ($E_g$), transmission and absorption behaviour of deposited films is very important for the fabrication of a good quality PV device. Fig. 1 shows the graph plotted between $(\alpha hv)^2$ and photon energy ($hv$). The band-gap values for CdTe films deposited at pH values of 1.5, 2.5, and 3.5 were estimated from intercepts on $hv$ axis, which are obtained as 1.44 eV, 1.48 eV, and 1.53 eV, respectively. From this graph it can be noticed that when pH value of electrolyte solution increases the band-gap of the CdTe film increases. The above results are close to the band-gap value of CdTe documented in the literature [10-11].

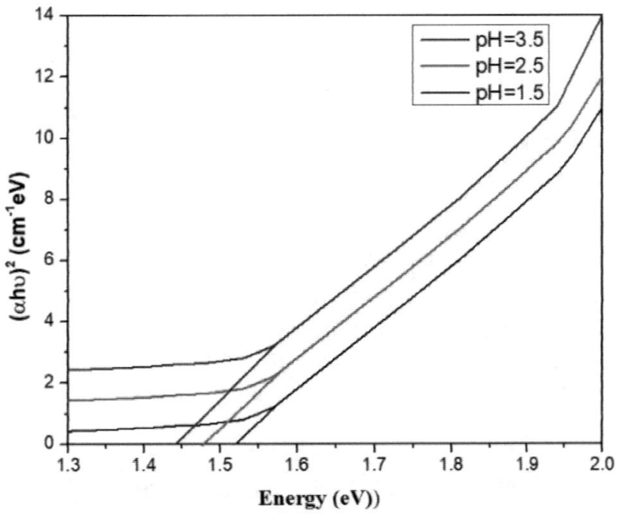


**Figure 1.** Tauc's plot of ED-CdTe films deposited at different pH values.

*3.2 Photoluminescence (PL) studies*

The emission of light caused by the irradiation of a material with other light is known as PL. Fig. 2 shows the PL spectra of ED-CdTe layers deposited at different pH values. In this PL study, the source of excitation was laser beam (He-Ne) of wavelength 632 nm (1.96 eV). PL band observed at 800nm ($E_g$ = 1.55eV) is related to band to band transition. The weak PL band obtained at 840nm ($E_g$ = 1.47eV) is related to the band to band transition or recombination of holes and electrons through surface/defect states. These findings are in good agreements with the the previous studies [12]. Both of these PL bands are obtained for the films deposited at pH 1.5 and 2.5, but the intensity of second PL band situated at 840 is very low for the film deposited at pH 2.5. Also, only first PL band (800nm) is seen in the PL spectra of CdTe film deposited at pH 3.5. This result indicates that pH of 1.5 is suitable for growing stoichiometric CdTe layer to be used as absorber layer in a solar cell.

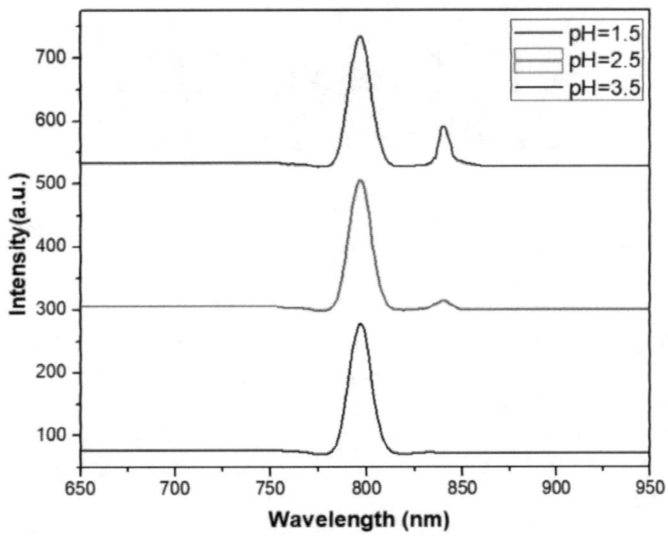


**Figure 2.** PL spectra of ED-CdTe films deposited at different pH values.

### *3.3 Scanning electron microscopy (SEM) studies*

In the development of thin film solar cells, fundamental knowledge about the surface morphology of constituting layers is very important. SEM is a microscope that produces images of a sample by scanning the surface with focused beam of electrons. SEM images are recorded using 20KV electron beam voltage and 20000 magnifications (Fig. 3). Fig. 3(a) is the SEM image of FTO/CdTe layer having pH value of 1.5. The grains exhibit large in size and cauliflower like structure. Fig. 3(b) and (c) are the SEM images of CdTe layer having pH value of 2.5 and 3.5, respectively. It shows the agglomeration of crystalline consisting of many smaller size grains and these films looked somewhat blurred as their morphology got disturbed by increasing pH concentration [13]. Another study found that the better morphological behavior has been obtained for the film deposited at Low pH as it helps to prevent the formation of phases of cadmium hydroxide and improves the solubility of Tellurium in electrolyte solution [14]. Also, the rate of deposition of CdTe layer became slow at higher pH value due to the less availability of $H^+$ ions [15] thus the film deposited at low pH values are more preferable choice for PV application than the film deposited on higher pH value.

**Figure 3.** SEM micrograph of ED-CdTe films deposited at different pH values.

### *3.4 X-Ray Diffraction (XRD) studies*

XRD is a process in which atoms of a crystal cause an interference pattern generated by incident X-ray beams depending upon the nature of their spacing. XRD is a popular characterization technique used for the identification of various issues related to the crystal structure of solids [16]. Fig.4 represents the XRD diffractograms of ED-CdTe films deposited at pH 1.5, 2.5 and 3.5. In the XRD pattern of CdTe film deposited at pH 1.5, four diffraction peaks (111), (002), (022) and (113) related to cubic phase situated at diffraction angles 24.4, 26.3, 37.9, and 46.0, respectively are seen. In the XRD pattern of CdTe film deposited at pH 2.5, all of the four peaks are observed but the intensities of the peaks decreased significantly. But, for the CdTe film deposited at pH 3.5 only two cubic phases (111) and

(113) were found. Some other peaks were found in the XRD images due to FTO substrates. It is clear from the above discussion that the CdTe film deposited at pH 1.5 has best crystallinity.

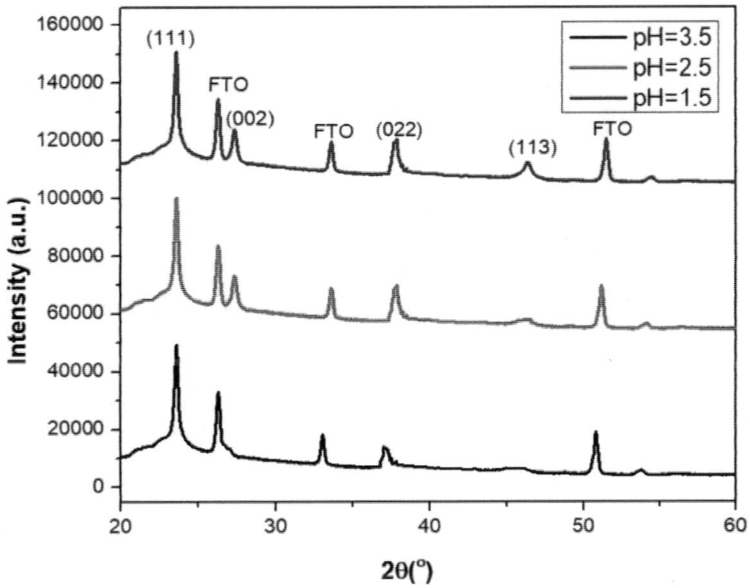


**Figure 4.** XRD patterns of ED-CdTe films deposited at different pH values.

## 4. Conclusions

CdTe films have been successfully on FTO substrates at different pH value of the solutions using a 2-electrode ED setup. The influences of pH of aqueous electrolytes on structural, optical and morphological properties are studied. The results obtained from the optical absorption measurement show that FTO/CdTe layer deposited at pH=1.5, 2.5 and 3.5 have band-gap energy ($E_g$) 1.44 eV, 1.45 eV, and 1.52eV, respectively. All three films are cubic crystalline in nature but the film deposited at pH 1.5 showed best crystalinity. The surface morphology of the films is found to improve for the lowest pH value 1.5. The film deposited with lower pH showed more intense PL emission peaks as compared to the other two films The better material properties can be achieved at pH 1.5 and CdTe film deposited at pH 1.5 will be a good selection as an absorber layer of a CdTe/CdS solar cell.

## Acknowledgements

The authors are thankful to the Science and Engineering Research Board (SERB) project no. EMR/2017/000431, New Delhi, India for their financial assistance and to the Department of Metallurgical Engg., NIT Raipur, India, for their help in performing XRD and SEM studies at their center.

## References

[1] Fang Z, Wang X C, Wu H C and Zhao C Z 2011 *Int. J. Photoenergy* https://doi.org/10.1155/2011/297350
[2] Sharma R K, Jain K and Rastogi A C 2003 *Curr. Appl Physics* **3** 199
[3] Olvera-Felix C, Ramirez-Bon R, Ochoa-Landín R, Ruvalcaba-Manzo S G and Castillo S J 2020 *J. Electron Mater* **49** 1257
[4] Sali D P and Chaure N B 2020 *Mater. Today* https://doi.org/10.1016/j.matpr.2020.07.475
[5] Doroody C, Rahman K S, Abdullah S F, Harif M N, Rosly H N, Tiong S K and Amin N 2020 *Results Phys* **18** 103

[6] Kulkarni R, Rondiya S, Pawbake A, Waykar R, Jadhavar A, Jadkar V, Bhorde A, Date A, Pathan H and Jadkar S 2017 *Energy Procedia* **110** 188
[7] Ghosh B, Hussain S, Ghosh D, Bhar R and Pal A K 2012 *Physica B* **407** 4214
[8] Wang J, Zhou X, Lv P, Yang L, Ding D, Niu J, Liu L, Fu W and Yang H 2016 *Electrochim Acta* **202** 32
[9] Fathy M, Elyamny S, Mahmoud S and Kashyout AEHB 2015 *Int. J. Electrochem. Sci.* **10** 6030
[10] Ojo A A and Dharmadasa I M 2017 *J. Mater. Sci: Mater Electron* **28** 7231
[11] Echendu O K, Okeoma K B, Oriaku C I and Dharmadasa I M 2016 *Adv. Mater. Sci.* **146** 4477
[12] Sinha T, Lilhare D and Khare A 2019 *Rare Metals* **38** 1
[13] Murase K, Watanabe H, Mori S, Hirato T and Awakura Y 1999 *J. Electrochem. Soc.* **146** 4477
[14] Salim H I, Patel V, Abbas A, Walls J M and Dharmadasa I M 2015 *J. Mater. Sci: Mater. Electron.* **26** 3119
[15] Madhu U, Mukherjee N, Bandyopadhyay N R and Mondal A 2007 *Indian J. Pure Appl. Phy.* **45** 226
[16] Miyake M, Murase K, Hirato T and Awakura Y 2003 *J. Electrochem. Soc.* **150** C 413

IOP Conf. Series: Materials Science and Engineering 1120 (2021) 012020 doi:10.1088/1757-899X/1120/1/012020

# Observation of coronal mass ejections in association with sun spot number and solar flares

**Preetam Singh Gour[1], Nitin P Singh[1], Shiva Soni[1] and Sapan Mohan Saini[2*]**

[1] Department of Physics, Jaipur National University, Jaipur, India
[2] Department of Physics, National Institute of Technology, Raipur, India

*Corresponding author's e-mail address: smsaini.phy@nitrr.ac.in

**Abstract.** The sun's atmosphere is frequently disrupted by coronal mass ejections (CMEs) coupled with different solar happening like sun spot number (SSN), geomagnetic storms (GMS), solar energetic particle and solar flare. CMEs play the important role in the root cause of weather in earth's space environment among all solar events. CMEs are considered as the major natural hazardous happening at the surface of sun because this event can cause several other phenomena like solar flare and many more. In this work, we report a statistical observation for the relationship of CMEs having linear speed >500 km/s with SSN and solar flares that were registered during the period 1997-2015. Test results of the annual correlation between SSN and CMEs is 0.77 and can be represented by a linear regression equation. We concluded that SSN and solar flares are well correlated with CMEs in most of the yearly compiled data for this time span.

## 1. Introduction

Coronal mass ejections (CMEs) consist of large structures containing plasma and magnetic fields that are expelled from the Sun into the atmosphere. They are of importance for both scientific and technological reasons. Scientifically they are of significance because they remove built-up magnetic energy and plasma from the solar corona [1] and technologically they are of significance because they are liable for the most extreme space weather effects at Earth [2] as well as at other planets and spacecraft throughout the atmosphere. Maximum ejected material comes from the low corona, although cooler and denser material probably from chromosphere or photosphere is also involved sometimes. The CMEs plasma is entrained on an expanding magnetic field, which has commonly the form of helical field lines with changing pitch angles. Up to the early years of this century, the images of CMEs had been made near the Sun firstly by coronagraphs on board spacecraft. Coronagraphs sight the outward flow of density structures flowing from the Sun by observing Thomson-scattered sunlight from the free electrons in coronal and heliospheric plasma. This outflow has an angular dependence which must be noted in the measured brightness [3-5]. CME related phenomena such as solar flares and prominences have been observed since the late 19th century, and energetic particles [6] type II and IV radio bursts [7] and interplanetary shocks [8] have been known since the 1940s, 50s and 60s.

On the positive side, there is a concord on the origination of CMEs by a loss of equilibrium of a magnetic configuration in the solar corona [9] and the conversion of magnetic energy to radiative energy and kinetic energy that accelerates the plasma cloud into the heliosphere. The magnetic reconnection process play a fundamental role ends up in a reconfiguration of the disrupted magnetic fields. Several models have been proposed to explain the initial phase of eruption, particularly the "mass-loading model" [1] the "catastrophe model" [10] the "breakout model" [11] and the "shearing arcade model" [12], see [13] for a review. The identification and measurement of CMEs remain difficult due to their complexity and the variety of shapes [14] further complicated by their temporal evolution and the effects of viewing geometry.

Solar cycle is a key parameter in the helioshperic phenomena which takes approximately 11 years on the average. With this time, the sunspot number (SSN) starts from a minimum value and increases which marks the start of the solar cycle. Then, the SSN tends to reach a maximum value, then decreases to a minimum value again which marks the end of a solar cycle. The reason for this phenomenon is due to the continuous fluctuation of the solar magnetic field, and this credence is strongly supported by the observed fact that a sunspot maintains a strong magnetic field compared to that of the sun's surface [15]. Sunspots are the most apparent feature on the disturbed surface of the photosphere above the solar atmosphere and appear to play a key role in major solar and terrestrial disturbances [16]. Many researchers study the occurrence of CMEs in relation to sunspot number and found that sometimes the correlation is quite weak between them during the maximum phase of solar cycle [17-18].

## 2. Data Sources

During the rising phase of solar cycles 23 and 24 more than thousands CMEs were observed by SOHO/LASCO (Solar and Heliospheric observatory/large angle spectrometric coronagraph) spacecraft LASCO has three telescope C1, C2, C3. Linear speed of CMEs should be equal or greater than 500km/sec. The CME data was collected from the CDAW data centre of SOHO/LASCO online catalogue of CMEs [19]. X-ray solar flare data was collected from online catalogue of GOES (geostationary environmental satellite) [20]. The SSN was extracted from NOAA National Geophysical Data Center [21] shown in table-1.

**Table 1.** Occurrence of average SSN and CMEs for 1997-2015

| Year | SSN | No. of CMEs |
|---|---|---|
| 1997 | 249.6 | 2375 |
| 1998 | 739 | 9486 |
| 1999 | 1151.8 | 7488 |
| 2000 | 1479.3 | 11679 |
| 2001 | 1478.9 | 11921 |
| 2002 | 1313.6 | 14840 |
| 2003 | 789.1 | 9226 |
| 2004 | 519.9 | 9968 |
| 2005 | 372.4 | 11512 |
| 2006 | 184 | 5582 |
| 2007 | 101.4 | 1527 |
| 2008 | 29 | 0 |
| 2009 | 33.8 | 0 |
| 2010 | 186 | 3153 |
| 2011 | 601.6 | 9157 |
| 2012 | 633 | 8714 |
| 2013 | 729 | 11783 |
| 2014 | 896.4 | 8963 |
| 2015 | 553.4 | 10637 |
| Total | 12041.2 | 148011 |

## 3. Analysis Result and Discussion

In this study we find out the linear relationship of coronal mass ejections with X-ray solar flares and sunspot number. For this study we extracted CMEs data from CDAW data center [19] at angular width $360^{o}$ and having acceleration 2000 $m/s^{2}$ which are associated with solar flare in the period 1997 to 2015. The relation between coronal mass ejection and intensity of solar flares which is calculated here is during 23 & 24 solar magnetic cycle (1997 to 2015).

The correlation coefficient is a degree of association, denoted by r. It is sometimes called Pearson's correlation coefficient after its originator and is a measure of linear association. The correlation coefficient is measured on a scale that varies from + 1 through 0 to – 1. Complete correlation between two variables is expressed by either + 1 or -1. When one variable increases as the other increases the correlation is positive; when one decreases as the other increases it is negative. Complete absence of correlation is represented by 0.

The calculation of the correlation coefficient with x representing the values of the independent variable (CMEs) and y representing the values of the dependent variable (Solar flares). The formula to be used is:

$$r = \frac{\Sigma(x-\bar{x})(y-\bar{y})}{\sqrt{[\Sigma(x-\bar{x})^2(y-\bar{y})^2]}} \qquad (1)$$

*3.1 Coronal mass ejection and solar flares*

We have selected those CMEs events whose velocity will be equal or greater than 500km/sec. we have no observed CMEs events for 2007, 2008 & 2009. During 1997-2015 we have found 412 CMEs events which were associated with different categories of X-ray solar flares (A, B, C, M, & X). No CMEs events observed for A-class solar flares, out of 412 CMEs events 55, 160, 126 and 71 events are associated with B, C, M and X-class X-ray solar flares respectively. The relation between flares and CMEs are positive as well as negative. There are total five types of solar flares A, B, C, M, & X. when speed of CMEs are correlated by intensity of solar flares via correlation coefficient whose value vary from -1 to +1.

To understand the correlation between CMEs and B-class solar flares draw a scatter plot from the trend line of the plot there is weak negative correlation between these two events with correlation coefficient -0.3(figure 1).

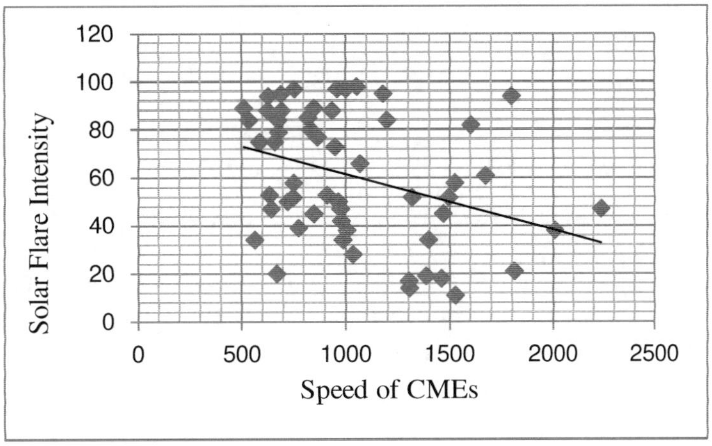


**Figure 1.** Scatter plot between CMEs speed and B class flares with correlation coefficient -0.35.

From the trend line of the scatter plot shown in figure 2 we have no correlation between CMEs and C class solar flares with correlation coefficient – 0.0 between these two events.

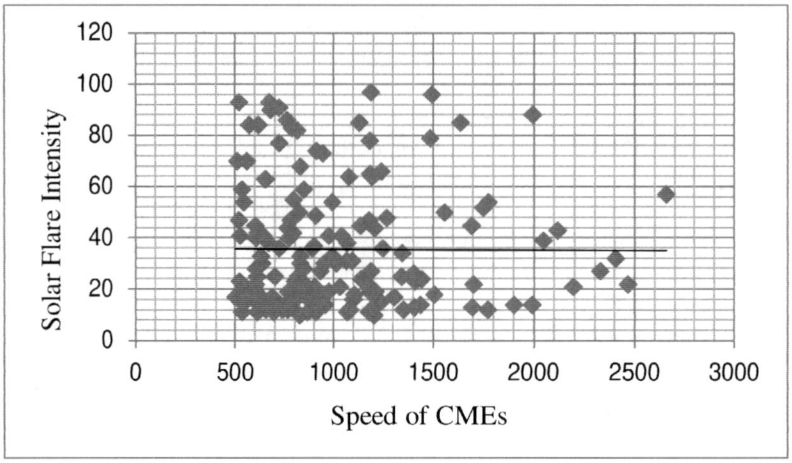


**Figure 2.** Scatter plot between CMEs speed and C class flares with correlation coefficient -0.00.

Further to understand the statistical behavior between CMEs and M class X-ray solar flares draw a scatter plot and from the trend line of the scatter plot we have found that there is a weak positive correlation between these two events with correlation coefficient 0.3 (figure 3).

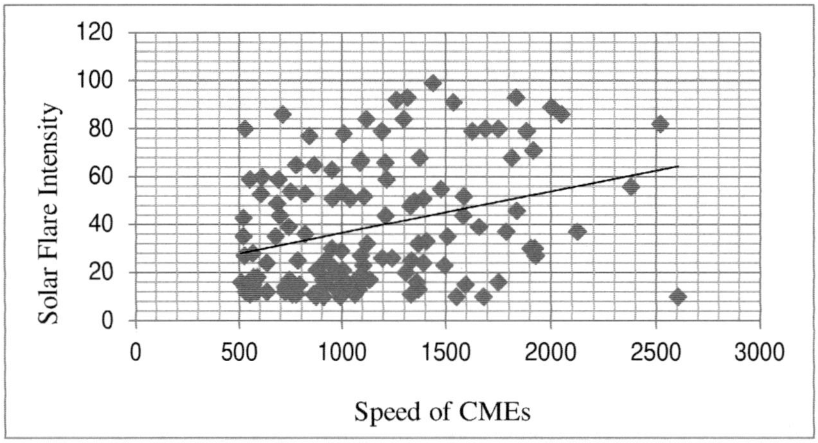


**Figure 3.** Scatter plot between CMEs speed and M class flares with correlation coefficient 0.31.

To understand the statistical behavior between CMEs and X class X-ray solar flares draw a scatter plot and from the trend line of the scatter plot we have found that there is no correlation between these two events with correlation coefficient 0.09 (figure 4).

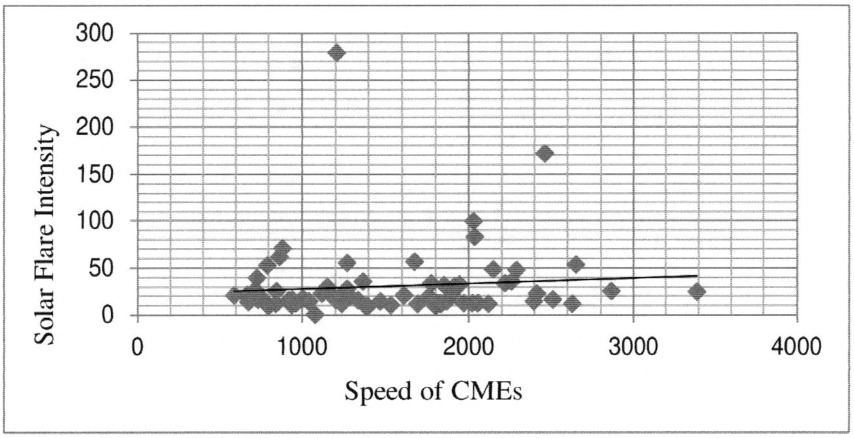


**Figure 4.** Scatter plot between CMEs speed and M class flares with correlation coefficient 0.09.

*3.2 Coronal mass ejection and sunspot number*

The statistical analysis was conducted to derive the correlation between SSN and CMEs for the annual time during the period from 1997-2015 through solar cycles 23 and 24 to investigate the behaviors of these parameters. A total of 12041 sun spot number and 148011 CMEs of velocity over 500 km/s were observed during the period from 1997-2015 through solar cycles 23 & 24. Table-2 shows the values of the observed SSNs and CMEs for the selected period. The number of occurrence of CMEs slightly increases towards solar maxima and slightly decreases towards solar minima. We have found that in the years 2008 and 2009 didn't record even single CMEs over the speed of 500 km/sec. On the other side sunspots have shown regularity even in the declining phase of the solar cycle. It can be seen that in 2008 and 2009 doesn't show any relation between CMEs and sunspots. Since 2008 and 2009 went through the declining phase of solar cycle; the record of CMEs has kept low.

In this article, we find that the linear regression equation is $y = 0.08252x - 9.1293$ and the correlation coefficient is 0.77, which measures the degree of a linear relation. A very good relationship rate with the positive slope is observed between the occurrence rate of CMEs and SSN, as shown in the correlation chart of figure 5.

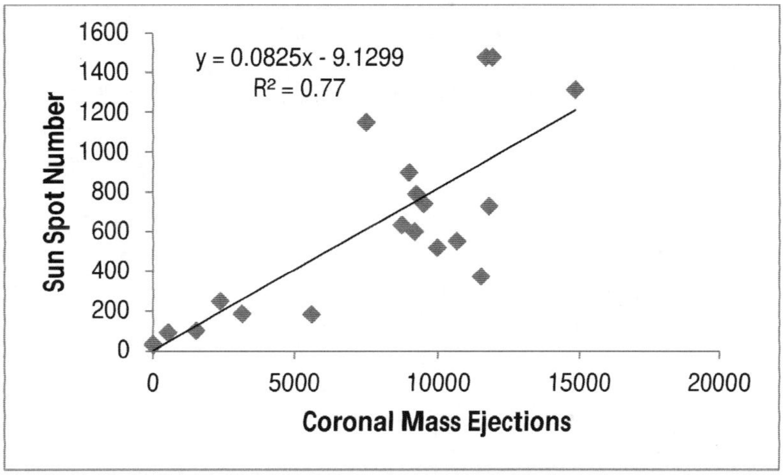


**Figure 5.** Correlation between CMEs and SSN for 1997-2015 with correlation coefficient 0.77.

## 4. Conclusion

In this study, the statistical analysis was conducted to investigate the behaviors of CMEs with solar flares and SSN and derive the mutual correlation between these parameters for the annual time during the period from 1997 to 2015 through solar cycle 23 & 24. Although the solar cycle number 24 started in December 2008, data were taken before that to increase the sample number which gives better statistics in the curve fitting. The solar cycle 24 initially have activity less than cycle 23. Emissions of CMEs can occur at any time during a solar activity but increase with solar activity and solar energy. The present results foreshowed a high affiliation relation between CMEs and SSNs as expected since both generate from active regions of the sun. A good relationship with a positive slope exists between SSN and the occurrence rate of CMEs. The correlation coefficient between CMEs and SSN is 0.77. The present results for CMEs and solar flares is not as per expectation, only M class X-ray solar flare is weakly related to CMEs other type of solar flares are having no correlation with CMEs.

**Acknowledgement**

We are very thanking full to this CME catalog is generated and maintained at the CDAW Data Center by NASA and The Catholic University of America in cooperation with the Naval Research Laboratory. SOHO is a project of international cooperation between ESA and NASA. Also thank full to NOAA and solar geophysical data report for making the data freely available.

**References**

[1] Low B C 1996 *Solar Phys.* **167** 217
[2] Baker D N, Balstad R, Bodeau J M, Cameron E, Fennell J F, Fisher G M, Forbes K F, Kintner P L et al 2008 Severe Space Weather Events – Understanding Societal and Economic Impacts: A Workshop Report, The National Academies Press, Washington DC
[3] Billings D E 1966 A Guide to the Solar Corona Academic Press, New York; London
[4] Vourlidas A and Howard R A 2006 *Astrophys. J.* **642** 1216
[5] Howard T A and Tappin S J 2009 *Space Sci. Rev.* **147** 31
[6] Forbush S E 1946 *Phys. Rev.* **70** 771
[7] Wild J P, Murray J D and Rowe W C 1954 *Aust. J. Phys.* **7** 439
[8] Sonnett C P, Colburn D S, Davis Jr L, Smith E J and Coleman Jr P J 1964 *Phys. Rev. Lett.* **13** 153
[9] Priest E R 1988 *Astrophys. J.* **328** 848
[10] Forbes T G 2000 *J. Geophys. Res.* **105** 23153
[11] Antiochos S K, DeVore C R and Klimchuk J A 1999 *Astrophys. J.* **510** 485
[12] Mikic Z and Linker J A 1994 *Astrophys. J.* **430** 898
[13] Zhang M and Low B C 2005 *Annu. Rev. Astron. Astrophys.* **43** 103
[14] Lamy P L, Floyd O, Boclet B, Wojak J, Gilardy H and Barlyaeva T 2019 C *Space Sci Rev* **215 (39)** 1
[15] U L V K, Varghese B S and Kurian P J 2017 *International Journal of Engineering and Applied Sciences* **4** 68
[16] Pandey S K, Shrivastava R, Borkar L K, Dubey S C et al 2010 *Current science* **98** 1496
[17] Gopalswamy N 2010 Coronal Mass Ejections from Sunspot and Non-Sunspot Regions. In Siraj H, **Rutten** R J (Eds.) *Magnetic Coupling between the Interior and Atmosphere of the Sun* (1st ed., pp 289-307) Astrophysics and Space Science Proceedings, Springer-Verlag Berlin Heidelberg
[18] Gour P S, Joshi H and Soni S 2020 *Int. J. Innovat. Res. Growth* **9** 92
[19] http://cdaw.gsfc.nasa.gov./CME_list/index.html
[20] http://www.ngdc.noaa.gov.stp./space-weather/solar-data/solar-features/solar-flares/x-rays/goes/xrs/
[21] http://.ngdc.noaa.gov/stp

IOP Conf. Series: Materials Science and Engineering 1120 (2021) 012021 doi:10.1088/1757-899X/1120/1/012021

# Heat conduction mechanism in $Al_xGa_{1-x}N$ ultraviolet light emitting diode

**Priyanka Sahare and Bijay Kumar Sahoo**

Department of Physics, National Institute of Technology, Raipur-492010

*Corresponding author's e-mail address: psahare.phd2018.phy@nitrr.ac.in

**Abstract**. $Al_xGa_{1-x}N$ ultra violet light emitting diode (UV LED) has wavelength ranges from 200 to 400nm.UV light has wide range of applications from food processing, health sector, environment and security to space exploration. The energy of one photon of UV light ranges from 3.1 to 6.5 eV which is enough to raise temperature of the active region of the diode above room temperature. The issue is termed as self-heating which shrinks band gap of $Al_xGa_{1-x}N$ semiconductor and decreases the emission efficiency of LED. Efficiency droop directly depends on thermal transport in active region of the $Al_xGa_{1-x}N$ UV LED. Thus, thermal properties of $Al_xGa_{1-x}N$ UV LED should be studied to explore strategies to minimize the issue. In this work we discussed heat conduction mechanism in $Al_xGa_{1-x}N$ UV LED. The heat flow in active region is studied by solving heat flux equation in the active region for different material parameter and thermal conductivity. Our study finds that high thermal conductivity of $Al_xGa_{1-x}N$ can minimize the self-heating issue and can improve the efficiency of $Al_xGa_{1-x}N$ UV LED.

## 1. Introduction

$Al_xGa_{1-x}N$ ultra violet light emitting diode (UV LED) has wavelength ranges from 200 to 400nm. UV light has wide range of applications from food processing, health sector, environment and security to space exploration. $Al_xGa_{1-x}N$ based devices are mostly used due to their superior characteristic like high light emission efficiency (11%), high mobility (> 2430 $cm^2/V$), high frequency, high electric field ($10^6$meV/cm) and high sheet carrier density ($10^{12}cm^2$) [1-3]. Despite these feature $Al_xGa_{1-x}N$ UV LED is not able to achieve the ideal result due to self-heating problem. Self-heating is arising due to rise in temperature (from 300 – 500K) in the active region of $Al_xGa_{1-x}N$ UV LED [4]. Rise in temperature reduces band-gap and enhances non-radiative processes. It affects device reliability and performance [5]. This reduces efficiency. Thus, one of the reasons of efficiency droop is self-heating which is a function of temperature. Self-heating is caused due to low thermal conductivity (*k*) of active region as superlattice (SL) possesses low *k*. High current increases thermal energy of electrons which helps to leak to p-side without recombination. Several researches have been done to minimize self-heating. Self-heating in AlGaN/GaN high electron mobility transistor (HEMT) has been studied by Gaska for different substrates [6]. Self-heating has been studied for doped and un-doped channel of AlGaN/GaN heterstructure field effect transistor (HFET) [11]. To minimize self-heating and modify performance of the LED, it is needed to control the heat flow in the device. To do thermal management in the active region of LED, heat conduction mechanism in the LED must be understood. Heat conduction mechanism is directly depends on thermal conductivity of $Al_xGa_{1-x}N$/GaN SL UV LED. If the thermal conductivity of SL is high, heat energy will faster exhaust from the active region to outer from the

$Al_xGaN$ UV LED. In this work we discussed the heat conduction mechanism in $Al_xGa_{1-x}N$ UV LED. To obtain the effect of self-heating we used 2D or 3D heat flux equation. In this paper we simulate self-heating effect on $Al_xGa_{1-x}N$ SL for different structure and geometry.

## 2. Theoretical Model

The self-heating effect in SLs can be calculated by heat flow equation. To determine heat flow Wachutka's thermodynamically model of lattice heating is used [7]. The lattice heat flow equation is solved by Poisson and continuity equations. The heat flow equation in 3D can be written as

$$H = \rho C \frac{\partial u}{\partial t} - k\nabla^2 u \tag{1}$$

In above equation $k$ is thermal conductivity (W $m^{-1}$ $K^{-1}$), H is the heat energy (W $m^{-3}$), C is the heat capacity (J K $m^{-3}$), and T is the temperature (K). The thermal conduction equation is basically a partial differential equation that can be solved by the 2D or 3D finite element method.The time derivative is solved by the central difference method [10].In eq. (1)

$$\frac{\partial u}{\partial t} = \frac{u_{(i+1)} - u_{(i-1)}}{2h} \tag{2}$$

$$\nabla^2 u = \frac{\partial^2 u}{\partial x^2} + \frac{\partial^2 u}{\partial y^2} + \frac{\partial^2 u}{\partial z^2}$$
$$= \frac{u_{i+1} - 2u_i + u_{i-1}}{\partial x^2} + \frac{u_{i+1} - 2u_i + u_{i-1}}{\partial y^2} + \frac{u_{i+1} - 2u_i + u_{i-1}}{\partial z^2} \tag{3}$$

For boundary $u|_{left,right,top,bottom} = 0$. This equation represents conservation of energy in form of continuity equation for energy variations of thermal mass at point (x, y, z) in space at the specific time and the divergence of heat flux per unit area. The term $k\nabla u$ ($k\nabla u = Q = k(T - T_0)$) is heat current that represents heat transfer rate of device. In heat flow equation mass density, heat capacity, and heat current all are constant term except thermal conductivity $k$. Thus, thekey factor in heat flow equation isthermal conductivity of material. The total thermal conductivity of material can be denoted as $k = k_e + k_p$ here $k_e$ and $k_p$ is thermal conductivity of semiconductor due to electron and phonon contribution. In the calculation $k_p = 10^{-3} k_e$; so, $k_e$ is neglected due to low doping ($\sim 10^{16} cm^{-3}$) [9]. The thermal conductivity becomes $k = k_p$. This $k_p$ is calculated by Callaway's formula i.e.

$$k = k_p = [\frac{k_\beta^{\,4}\, T^3}{2\pi^2 \hbar^3 v} \int_0^{\theta_D/T} \frac{\tau_c\, y^4 e^y}{(e^y - 1)^2} dy\,] \tag{4}$$

In above equation, $y = \hbar\omega / k_\beta T$ , $k_\beta$ is Boltzmann constant, $\hbar$ is plank's constant, v is group velocity, T is temperature, $\tau_c$ is combined relaxation time. The combined relaxation time can be calculated by using normal scattering $\tau_{nr}^{-1}$, umklapp phonon scattering ($\tau_u^{-1}$), dislocation scattering ($\tau_d^{-1}$), boundary scattering ($\tau_b^{-1}$), point defect scattering ($\tau_p^{-1}$), phonon-electron scattering.

$$\tau_C^{-1} = \tau_{nr}^{-1} + \tau_u^{-1} + \tau_d^{-1} + \tau_b^{-1} + \tau_p^{-1} + \tau_{ph-e}^{-1} \tag{5}$$

## 3. Result and Discussion

2D heat conduction equation has been solved for x-axis, y-axis and temperature is found. We have computed the rise in tempearture in the active region (AlGaN/GaN SL). In this work, H = 0, and $T_0$ =300K are taken for bottom of the SL substrate. We have used constant value of thermal conductivity (*k*), mass density (ρ ) Specific heat C: k = 2.7 $Wcm^{-1}K^{-1}$, ρ = 3266*x + 6150(1-x) Kg $cm^{-3}$, C = 748*x+431*(1-x), for x = 0.1, Lx = 0.1, Ly = 0.1[10]. We used this value to calculate temperature distribution in active region of AlGaN/GaN UV LED. In Figure1 right side colour bar is used for temperature scale. It was observed that temperature risesafter 350K. In figure 1 boundary of different colour specify different temperature region. We can see that temperature is high at length 0.1 nm in active region of device. It was observed active region of the SL has high temperature as shown in figure 1. Some of researchers have reported the self-heating effect for AlGaN/GaN HFET for doped and un-doped region for active region thickness 50 nm [11] which is 5 time thicker than we consider here.

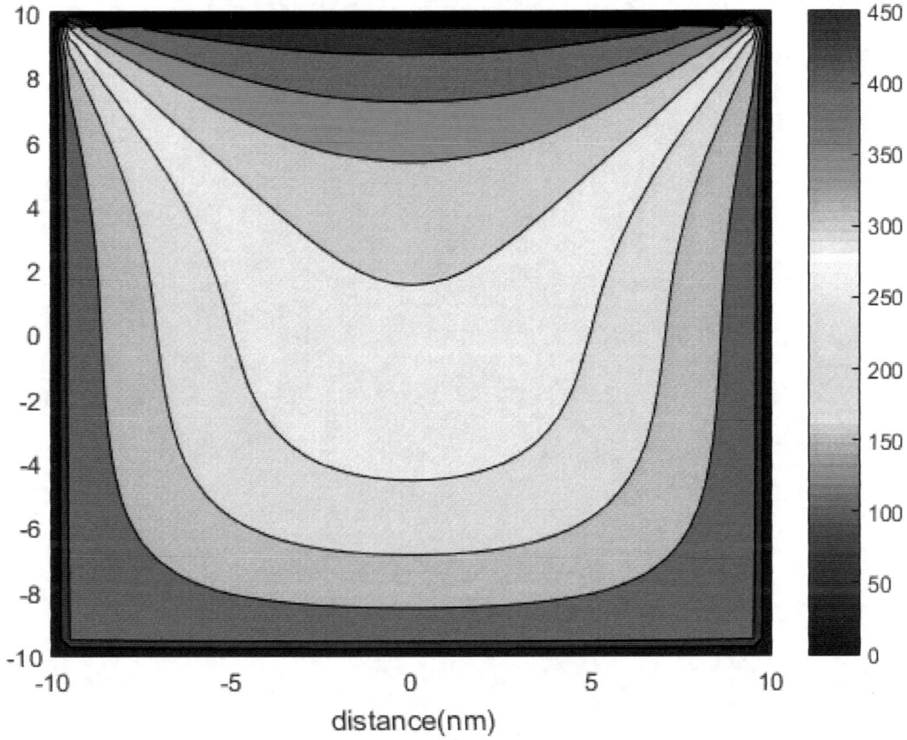


**Figure 1.** Temperature contour in AlGaN/GAN SL.

## 4. Conclusion

In this work, we have investigated the heat conduction mechanism in$Al_xGa_{(1-x)}N$/GaN UV LEDs. It is found thatmost of the heat unable to move from active region of device to outside due to low k. The heat conduction depends on k; and heat will transfer faster when k is high. Phonon engineering will improve k of AlGaN/GaN UV LED. Removal of heat from the active region minimizes rise in temperature. Minimization in temperature will reduce strength of self-heating process.

**Acknowledgments**
Author (PS) acknowledges with thanks to Department of Science and Technology (DST) Govt. of India for financial support through awarding WOS-A Fellowship; project File No. SR/WOS-A/PM-97/2018.

## References

[1] Pandey A and Shin W J 2020 *Photonic Research* **8** 331
[2] Dalpati P and Urata S 2020 *Superlattice and Microstructure* **147** 106648
[3] Khan A and Balakrishan K 2008 *Nat Photonics* **2** 77
[4] Wu Y R and Singh J 2007 *J. Appl. Phys.* **101** 113712
[5] Chattopadhyay M K and Tokekar S 2008 *Microelectronics Journal* **39** 1181
[6] Gaska 1998 *IEEE Electron Device Letters* **19** 89
[7] Wachutka G K 1990 *IEEE Trans.Computer-Aided Design* **9** 1141
[8] Zou J 2002 *J. Appl. Phys.* **92** 2534
[9] Florescu D I and Asnin V M 2000 *J. Appl. Phys.* **88** 3295
[10] Mitterhuber L and Hammer R 2020 *Energies* **13** 2363
[11] Fillippov and Balandin A 2003 *Nanotech* **1 3** 333

IOP Conf. Series: Materials Science and Engineering 1120 (2021) 012022 doi:10.1088/1757-899X/1120/1/012022

# A study of enhanced structural, microstructural and dielectric behaviour of aliovalent ions doped $BaZr_{0.05}Ti_{0.95}O_3$ ceramic

**G Nag Bhargavi[1], Tanmaya Badapanda[2], Ayush Khare[3] and M Shahid Anwar[4]**

[1] Department of Physics, Govt. Pt. Shyamacharan Shukla College Dharsiwa Raipur-493221, India
[2] Nanophotonics Laboratory, Department of Physics, C.V. Raman College of Engineering, Bhubaneswar-752054, India
[3] Department of Physics, National Institute of Technology, Raipur-492010, India,
[4] Colloids & Materials Chemistry, Institute of Minerals and Materials Technology, Bhubaneswar-751013, India

*Corresponding author's e-mail address: bhargavi.nag24@gmail.com

**Abstract.** The lead-free ceramics are considered to be the best substitutes for lead based ceramics which have tremendous harmful effects regarding environmental pollution. In this concern, the lead-free ceramics have been widely attracted by research communities. In the present study we have prepared lead free $BaZr_{0.05}Ti_{0.95}O_3$perovskite ceramics with various doping concentrations of $Gd^{3+}$ ions by the conventional solid-state reaction method. The perovskite phase formation was studied by x-ray diffraction which indicate the transformation of crystal symmetry from orthorhombic to tetragonal structure for the $Gd^{3+}$ ions doped samples. The scanning electron microscopy studies revealed the modification in grain size on doping Gd ions also the energy dispersive X-ray spectra have been obtained to study the compositional variations. The dielectric and loss studies have been performed in the large range of temperature and frequency. The low dielectric loss and high dielectric constant suggests possible applications in memory devices and ceramic capacitors. Temperature coefficient of capacitance plots have also been plotted.

## 1. Introduction

$BaTiO_3$(BT) a commonly known perovskite was commercially used for acoustic and ultrasonic generation and detection though its Curie temperature was quite low (i.e. 120 °C) [1-6]. Soon it was replaced by another perovskites like $PbTiO_3$ and Pb(Zr, Ti)$O_3$ [7-13]. Though the lead based ceramic compounds have proved to be extremely good as far as the industrial needs are concerned but, the toxicity of Pb restricts its practical applications. Thus, in recent years the need for lead free perovskite ceramics caused a reconsidered interest in $BaTiO_3$ based ceramics [14-18]. The structural and dielectric properties of $BaTiO_3$ can be modified to a large extent by the addition of some aliovalent ions (heterovalent dopants) such as $La^{3+}$, $Ce^{3+}$, $Nb^{5+}$, $Nd^{3+}$, $Pr^{3+}$, $Eu^{3+}$, $Er^{3+}$, $Gd^{3+}$, $Mg^{2+}$, $Sn^{4+}$ etc. on $Ba^{2+}$ or $Zr^{4+}$ on $Ti^{4+}$ sites that forms the solid solution [19-26]. Among various compositions $Zr^{4+}$ with ~15% substitution in Ba(Zr, Ti)$O_3$(BZT) undergoes three structural phase transitions (rhombohedra to orthorhombic, orthorhombic to tetragonal and tetragonal to cubic) like $BaTiO_3$, also the Curie temperature decreases [27-30]. After thorough investigation of literature, out of the several rare earth ions our focus was on Gd whose ionic radius is 0.91Å. The incorporation of rare-earth ($Gd^{3+}$) ions into the BT host system was studied by Borah et al. in 2014 [31]. The results of their study revealed that $Gd^{3+}$ ions greatly manifest dielectric relaxation and carrier conduction mechanisms. The dielectric constant showed a decreasing trend with doping content and with increasing frequency. However, in the low-frequency region due to the combined result of orientational polarization and electrical conduction the loss tangent (tan δ), was found to be quite high in the doped samples as compared to their un-doped counterpart. In another study the amphoteric nature of $Gd^{3+}$ ions were discussed by Li et al.in 2012 [32]. According to the study because of the mediocre ionic radius of $Gd^{3+}$ ion, it may replace the A-site ions or B-site ions. After the analysis of XRD patterns they revealed that different substitution sites of $Gd^{3+}$ ions could be affected by amount of $Gd_2O_3$ doping. $Gd^{3+}$ tended to occupy Ba-site when $Gd_2O_3$ concentration was less than 0.25 mol%. On further increasing concentration of $Gd_2O_3$ doping amount to 0.3 mol%, $Gd^{3+}$ions get substituted into both the Ba- and Ti-sites as a result of which an improvement on dielectric constant of $BaTiO_3$-based ceramics is seen. They used self-compensation model to discuss the dielectric behaviour. Reddy et al. in 2011[33] gave a detailed study of dielectric studies properties when $Gd^{3+}$ ions are doped at A site of $BaZr_{0.1}Ti_{0.9}O_3$ ceramic. The microstructural investigation on the sintered ceramics showed that Gd doping significantly reduced the grain size of pure BZT ceramics. Change in the Gd concentration had minor influence on the grain size and on morphology. The Curie temperature ($T_C$) significantly decrease with an increase in the Gd content. However, the maximum value of dielectric constant at $T_C$ was seen for 2 mol% Gd and on further increase in Gd content the dielectric constant at $T_C$ decreased. They reported that the dielectric constant was significantly improved as compared to that of undoped BZT ceramic. Hence it is concluded that the tunable dielectric materials with good dielectric properties can be prepared by doping BZT with Gd.

In the present study we have chosen 5 mole % $Zr^{4+}$ doped BT i.e. $BaZr_{0.05}Ti_{0.95}O_3$ as base material. This composition of BZT shows normal ferroelectric behavior and also the presence of morphotropic phase boundaries give special features by improving the piezoelectric properties. Further we have chosen $Gd^{3+}$ as an heterovalent ion which can replace $Ba^{2+}$ ions at A-site as according to the formula $Ba_{1-x}Gd_{2x/3}Zr_{0.05}Ti_{0.95}O_3$. Thus, the objective of the present work is to study the influence of $Gd^{3+}$ ions doping on the dielectric properties of $BaZr_{0.05}Ti_{0.95}O_3$.

## 2. Materials and methods

The perovskite samples of pure and Gd doped Barium Gadolinium Zirconium Titanate with formula $Ba_{1-x}Gd_{2x/3}Zr_{0.05}Ti_{0.95}O_3$ (x=0.00, 0.01, 0.02) were prepared by conventional high temperature solid state reaction method. For the sake of convenience, we will mention the samples as BZT, BGZT1 and BGZT2 for x= 0.00, 0.01 and 0.02 respectively. The starting raw materials were $BaCO_3$, $TiO_2$, $ZrO_2$ and $Gd_2O_3$ (all from Merck India) all were more than 99% pure. The stoichiometric ratio of all the chemicals were thoroughly mixed in an agate mortar in dry and wet mixing with appropriate amount of Acetone for 5 hours. After proper mixing the samples were calcined at 1200°C for 4 hours at a heating rate of 5 °C per minute. A small amount of polyvinyl alcohol was added to the calcinated powders for the fabrication of pellets which were sintered at 1250 °C. The circular disc shaped pellets were prepared by applying uniaxially a pressure of 200 MPa. A preliminary study on compound formation and structural parameters was carried out using an X-ray diffraction (XRD) technique with an X-ray powder diffractometer. The XRD pattern of the calcinated powder was recorded at room temperature PANAlytical X'pert pro with Cu-$K_\alpha$ radiation (1.5405 Å) in a wide range of Bragg's angles 2θ ($20 \leq 2\theta \leq 80$). The dielectric constant and dielectric loss were measured on N4L-NumetriQ LCR meter (model PSM1735).

## 3. Results and Discussions

### *3.1 Structural characterization*

The X-ray diffraction patterns of BZT, BGZT1 and BGZT2 ceramic specimen are shown in figure1. After a keen analysis by X'pert Highscore software, the Perovskite structure was identified for all the specimens at room temperature. The peaks of BGZT1 and BGZT2 specimens seems to move slightly towards right (higher 2θ). The lattice parameters from the XRD data were calculated using Checkcell software shown in inset. The structural evolution of the specimens with these compositional changes can be readily understood in terms of the ionic radii of the cations at A-site ($r_{Ba2+}$ = 1.61 Å and $r_{Gd3+}$ = 0.91Å) and B-site ($r_{Ti4+}$ = 0.605 Å and $r_{Zr4+}$ = 0.72 Å). The splitting of peaks at 2θ=45° for BGZT1 and BGZT2 show the tetragonal symmetry indicating the ferroelectric phase at room temperature. However, for the BZT sample the crystal symmetry was observed to be orthorhombic at room temperature. The inset shows the variation of d-spacing and cell volume both these values decrease with increasing concentration of Gd.

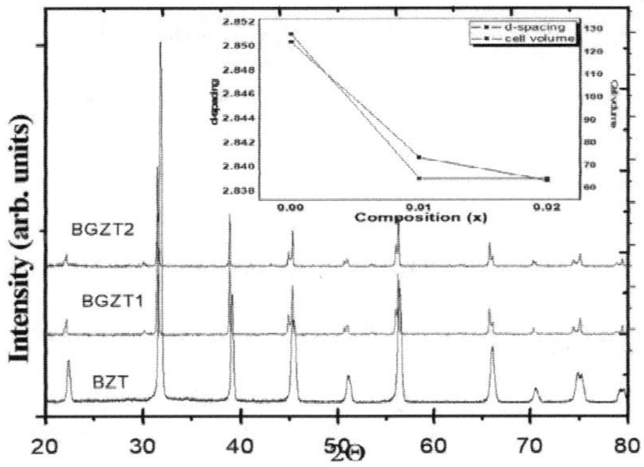


**Figure 1.** XRD plots of BZT, BGZT1 and BGZT2 at room temperature. The inset is variation of d-spacing and cell volume with composition.

*3.2 Microstructural studies*

Figure 2(a)-(c) reveal the scanning electron microscopic images of BZT, BZGT1, BZGT2 samples. The non-uniform distributions of the irregular shaped grains are observed in all the ceramics. The microstructural modifications in terms of grain size can be seen in the pictures. The average grain sizes were decreased sharply with Gd addition accompanied with uniform grains. As significant from the figures the average grain size for BZT is ~20μm whereas for BGZT1 and BGZT2 it reduces to ~2μm. However, no remarkable change in the density of the ceramics with the variation of Gd concentration in the BZT system. The compositional characterization was done by Energy dispersive X-ray spectroscopy (EDX) which is shown in inset of figure 2. Table 1 shows the calculated and EDX derived composition of BZT, BGZT1 and BGZT2 ceramics. There exists small variation in both the data which is possibly due the deficiency of oxygen during sintering at high temperature.

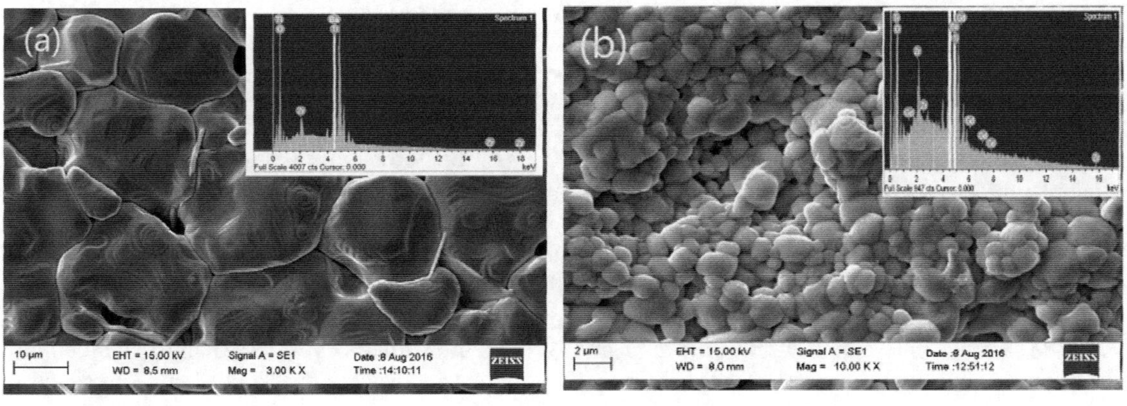

**Figure 2.** SEM micrographs of (a) BZT (b) BGZT1 (c) BGZT2 ceramics. The insets show EDX spectrum showing the elemental composition of the ceramics.

**Table 1.** Calculated and EDX derived composition of $BaGdZr_{0.05}Ti_{0.95}O_3$ ceramic.

| Element | BZT | | BGZT1 | | BGZT2 | |
|---|---|---|---|---|---|---|
| | Wt.% calculated | Wt.% from EDX | Wt.% calculated | Wt.% from EDX | Wt.% calculated | Wt.% from EDX |
| **Ba** | 57.59 | 57.65 | 56.39 | 56.45 | 56.17 | 56.11 |
| **Zr** | 1.48 | 1.59 | 1.50 | 1.55 | 1.52 | 1.46 |
| **Ti** | 18.52 | 18.12 | 18.57 | 18.55 | 18.60 | 18.46 |
| **O** | 22.41 | 22.64 | 22.96 | 22.85 | 22.90 | 23.12 |
| **Gd** | - | - | 0.58 | 0.60 | 0.81 | 0.85 |

*3.3 Dielectric studies*

We have done the dielectric studies and dielectric loss studies of Gd doped BZT as a function of temperature at various frequencies. But these results were reported in our earlier publication [34] so here, we only report the dielectric study as function of frequency at various temperatures. Figure 3 (a-c) illustrated the frequency dependent dielectric studies of BZT, BGZT1 and BGZT2 ceramics at various temperatures from room temperature to temperature above phase transition. We have noticed an inverse relation between dielectric constant and frequency i.e. ε' decreases with increasing frequency. All the ferroelectric materials follow this behaviour. At lower frequencies the values of dielectric constant are higher while in the higher frequency region the dielectric constant drops and attains a saturation. To understand this behaviour Debye gave a modified law for the dipoles oscillating in AC fields [35]:

$$\varepsilon^* = \varepsilon_\infty + \frac{\varepsilon_s - \varepsilon_\infty}{[1+(i\omega\tau)^{1-\alpha}]} \quad (1)$$

where $\varepsilon_s$ and $\varepsilon_\infty$ are the low and high-frequency values of ε, ω (=2π*f*) is the cyclic frequency, *f* is the frequency of measurement, τ is the relaxation time and α measures the distribution of relaxation time. According to this relation at lower frequencies the relaxation time is quite high so the dipoles get sufficient time to follow the electric field resulting in high value of ε'. With increasing frequency, the dipoles begin to lag behind the field as a result of which decrease in ε' values are seen. At much higher frequencies due very short relaxation times the dipoles could not orient themselves in the direction of the applied electric filed and hence reduction in the dielectric values are seen.

The inset in figure 3 shows the dielectric loss as function of frequency for the samples. The dielectric loss of the samples is less than 0.2 and for BZT and BGZT2 the value is even less than 0.1 which show good thermal stability for practical applications. The dielectric loss is attributed to the presence of oxygen vacancies which are commonly seen in the oxide perovskites. The dielectric loss curves follow the similar trend as seen in the dielectric constant vs. frequency plots. The curves are more dominant at lower frequencies than higher frequencies. At higher frequencies the dipoles could not follow the applied AC fields because of infinitesimally small relaxation time. In BZT at some frequencies between $10^2$-$10^4$ Hz a small hump is seen indicating dielectric relaxation in the sample however, in BGZT1 and BGZT2 samples this kind of relaxation is not seen.

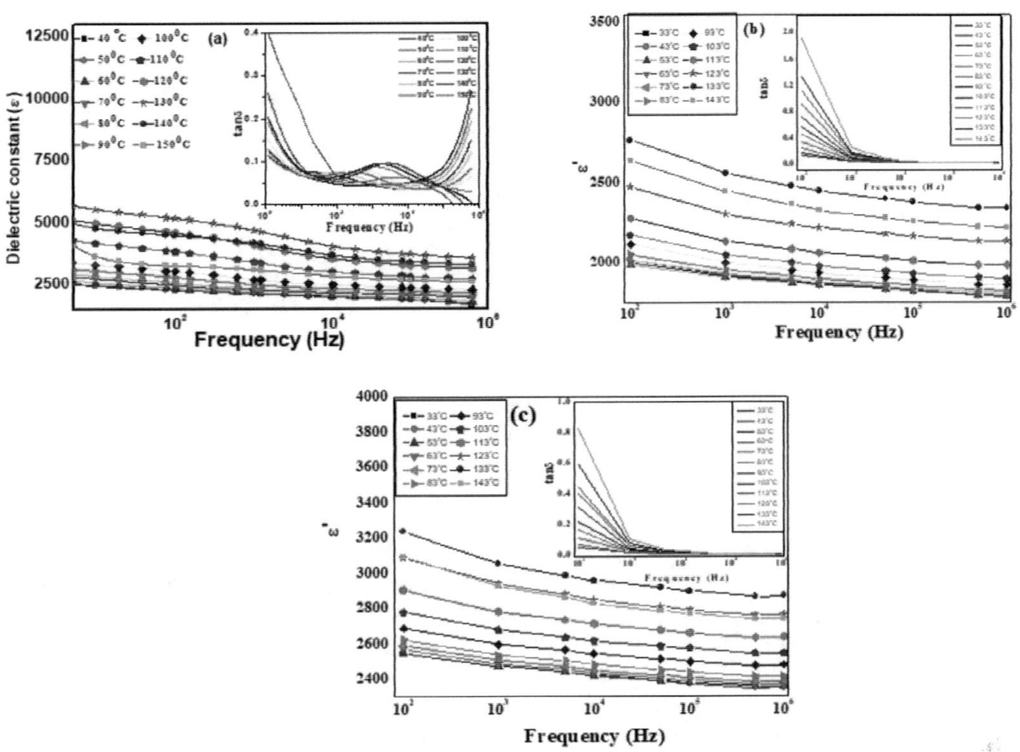


**Figure 3.** Variation of dielectric constant with frequency at various temperatures from room temperature to Tc. The inset shows the corresponding dielectric loss values. (a) BZT (b) BGZT1 (c) BGZT2.

The capacitance and dissipation factor of ceramic capacitors are significantly affected by doping concentrations. The permittivity/capacitance of ferroelectrics is generally temperature dependent and varies especially near to the Curie temperature. In order to measure the degree of capacitance variation with temperature, the temperature coefficients of capacitance (TCC) near phase transition temperature were calculated using the formula [36]:

$$TCC(\%)=\frac{C(T)-C(T_c)}{C(T_c)}x100\% \qquad (2)$$

Where C(T) is capacitance at the temperature T and C(Tc) the reference value of the capacitance at phase transition temperature (Tc). The TCC(%) measured at 100kHz frequency is represented as a function of the temperature for BZT, BGZT1 and BGZT2 in figure 4. In case of doped samples, the variation of capacitance is less than 25% from 40°C-160°C. The corresponding temperature interval of 120°C is considered as a ceramic with good stability which is a fundamental property of all devices for energy storage application. Analysis of the plots suggests that the TCC of BGZT2 sample is almost close to the temperature stability specifications for X8R and X9R capacitors, which state that the capacitance change should fit within ±15% of the tolerance limit for long temperature range.

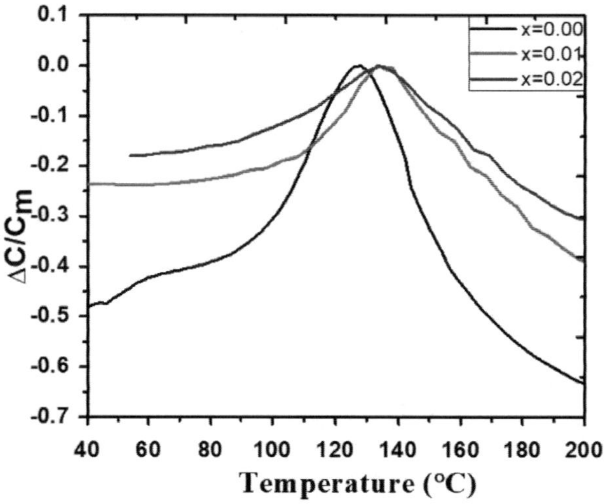


**Figure 4.** Temperature coefficient of capacitance with temperature.

*3.4 Conductivity studies*

Figure 5 shows the variation of AC conductivity of BZT, BGZT1 and BGZT2 samples with temperature (40°C-160°C that corresponds to the temperature nearby phase transition) measured at 100kHz. Here, ac conductivity was evaluated from dielectric data by employing the following relation [37]:

$$\sigma_{ac} = \omega\varepsilon_0\varepsilon'\tan(\delta) \quad (3)$$

where, $\sigma_{ac}$ is the ac conductivity, ω =2πf, is the angular frequency and $\varepsilon_0$ is the absolute permittivity. It isapparent from the figure that in low temperature regime the ac conductivity
depends significantly on the frequency. However, the increase in temperature, where marked dielectric relaxation takes place, reduced gradually the frequency dependence of $\sigma_{ac}$so that the conductivity was mainly determined by temperature. At still higher temperatures (near phase transition) there was a deviation probably due to the onset of dielectric phase transition. The ac conductivity found to increase with temperature at lower frequencies however, at higher frequency we could not find much variation ac conductivity. Near Curie temperature, the domain structure breaks up, carriers become free and take part in conduction mechanism. Interpretation of different theoretical models concludes that ac conductivity originates from migration of ions by hopping between neighbouring potential wells at lower temperatures which eventually give rise to dc conductivity at high temperatures. Oxygen vacancies are the most mobile charge carriers in oxide ferroelectrics and play an important role in the conduction process in most dielectric ceramics. Around the oxygen vacancy, long range potential wells may be formed, there can be large number of titanium centres within each potential well surrounding the oxygen vacancy. Conduction electrons created by the ionization of oxygen vacancies can cause hopping of electrons between $Ti^{4+}$ and $Ti^{3+}$ [38-40]. The figure also indicates that the introduction of $Gd^{3+}$ ions has slightly increased the conductivity probably due to the presence of oxygen vacancies.

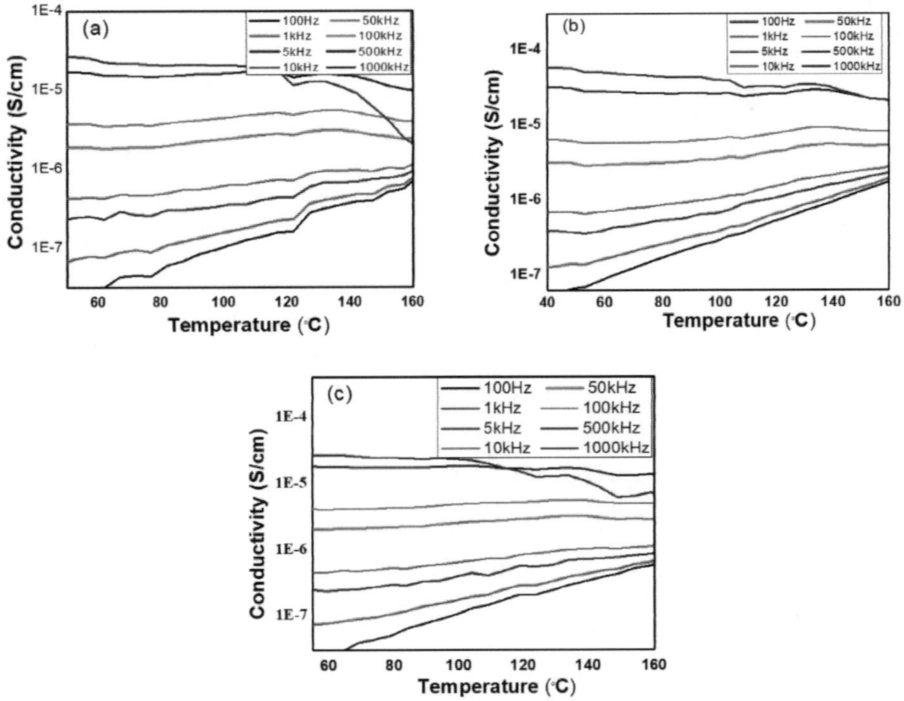


**Figure 5.** AC conductivity as function of temperature at various frequencies from 100Hz-1000kHz (a) BZT (b) BGZT1 (c) BGZT2.

## 4. Conclusion

The performance characteristics of ceramic capacitors are greatly determined by the properties of the dielectric material. Thus, from the present study we conclude that lead free ceramic samples of Gadolinium doped Barium Zirconium Titanate samples have been synthesized successfully by the conventional solid-state reaction method. The XRD technique has confirmed the perovskite phase with orthorhombic structure for BZT and tetragonal for BGZT1 and BGZT2. The microstructural studies reveal that the shape, size and distribution of BZT has got modified by addition of $Gd_2O_3$ in it and reduced from 20μm to 2μm. The dielectric values found to decrease for the doped samples but the dielectric loss for these samples is little larger. Also, the variations in temperature effected capacitance and dissipation factor of ceramic capacitors are found. The conductivity values are increased for doped samples.

## References

[1] Jiang B, Iocozzia J, Zhao L, Zhang H, Harn Y W, Chen Y and Lin Z 2019 *Chemical Society Reviews* **48** 1194
[2] Bisen S, Mishra A and Jarabana K M 2016 In *AIP Conference Proceedings* (Vol. 1731, No. 1, p. 050070). AIP Publishing LLC
[3] Sidorkin A, Nesterenko L, Gagou Y, Saint-Gregoire P, Vorotnikov E and Popravko N 2018 *Materials* **11** 1436
[4] Acosta M, Novak N, Rojas V, Patel S, Vaish R, Koruza, J, Rossetti Jr.G A and Rödel J 2017 *Applied Physics Reviews* **4** 041305
[5] Li Y J, Xu M, Feng J Q and Dang Z M 2006 *Applied Physics Letters* **89** 072902
[6] Kuo D H, Chang C C, Su T Y, Wang W K and Lin B Y2001 *J. Eur. Ceram. Soc.* **21** 1171
[7] Kour P and Sinha S 2012 *Digest Journal of Nanomaterials and Biostructures* **7** 1327

[8] Bochenek D, Niemiec P and Dercz G 2020 *Materials* **13** 1996
[9] Pérez de la Cruz J, Joanni E, Vilarinho P M and KholkinA L 2010 *J. Appl. Phys.* **108**114106
[10] Gupta R, Verma S, Singh V and Bamzai K K 2015 Processing and Application of Ceramics **9** 1
[11] Zhan X L, Chen Z X, Cross L E and Schulze WA 1983 *J. Mat. Sci.* **18** 968
[12] Zhang Q M, Zhao J and Cross L E 1996 *J. appl.Phys.* **79** 3181
[13] Lü C, Zhang Y, Zhang H, Zhang Z, Shen M and Chen Y 2019 *Energy Conversion and Management* **182** 34
[14] Ahmad M M, Alismail L, Alshoaibi A, Aljaafari A, Kotb H M and Hassanie R 2019 *Results in Phys.* **15** 102799
[15] Kuang S J, Tang X G, Li L Y, Jiang Y P and Liu Q X 2009 *Scripta Materialia* **61** 68
[16] Puli V S, Pradhan D K, Riggs B C, Chrisey D B andKatiyar R S 2014 *J. alloys and compds.* **584** 369
[17] Thanachayanont C, Yordsri V, Kijamnajsuk S, Binhayeeniyi N and Muensit N 2012 *Materials Lett.* **82** 205
[18] Xue D, Gao J, Zhou Y, Ding X and Ren X 2015 *J. Appl. Phys.* **117** 124107
[19] Park I J, and Han Y H 2014 *Metals and Mater. Inter.* **20** 1157
[20] Narang S B and Kaur D. 2009 *Int.Ferroelectr.* **105** 87
[21] Hernández LaraJ P, Pérez Labra M, Barrientos Hernández F R, Romero Serrano JA, Ávila Dávila O, Thangarasu P, and Hernández Ramirez A 2017 *Mater.Resrch.* **20** 538
[22] Huang X, Zhang W, Xie J, Xu Q, Zhang L, Hao H, and Cao M 2017 *J. Mater. Sci.: Mater. in Electr.* **28** 4204
[23] Billah M, Ahamed A A, Sen A and Rahman M M 2015 International Conference on Mechanical, Industrial and Materials Engineering, ICMIME 2015
[24] Karuppanan A, Athikesavan V, Bhargav P B and Ramasamy P 2020 *J. Mater. Sci.: Mater. in Electr.* **1**10
[25] Canu G, Bottaro G, Buscaglia M T, Costa C, Condurache O, Curecheriu L and Armelao L 2019 *Sci. reports.* **9** 1
[26] Zhang Q, Chen J and Che M 2020 *Ferroelectrics.* **566** 30
[27] Moura F, Simões A Z, Stojanovic B D, Zaghete M A, Longo E and Varela J A 2008 *J. Alloys and Compds.* **462** 129
[28] Mahesh M L V, Prasad V B and James A R 2013 J. Mater. Sci.: Mater. Electron. **24** 4684
[29] Muensita S and Binhayeeniyi N 2010 In 2010 3rd International Nanoelectronics Conference (INEC) (p 877-877). IEEE.
[30] Garbarz-Glos B, Bormanis, K, Kalvane A, Jankowska-Sumara I, Budziak A, Suchanicz W and Śmiga W 2011 *Int. Ferroelectrics*, **123** 130
[31] Borah M and Mohanta D 2014 Appl. Phys. A. **115** 1057
[32] Li L, Wang M, Guo D, Fu R and Meng Q 2013 *J. Electroceram.* **30** 129
[33] Reddy S B, Rao K P and Rao M R 2011 *J. alloys and compds.* **509** 1266
[34] Bhargavi G N, Khare A, Badapanda T and Anwar M S 2018 *Appl. Phys. A* **124** 746
[35] Wilson J N, Frost J M, Wallace S K and Walsh A 2019 *APL Materials* **7** 010901
[36] Borderon C, Nadaud K, Coulibaly M, Renoud R and Gundel H 2019 *Int. J. Adv. Resrch. Phys. Sci.* **6** 2349
[37] Reddy P S and Rao TS 2014 *Int. J. Eng. Resrch. & Tech.* **3** 857
[38] Funke K 1993 *Porgs. Sol. Stat. Chem.* **22** 111
[39] Jonscher A K 1996 Universal Relaxation Law, *Chelsea Dielectrics Press Ltd*, London, ISBN 0950871125
[40] Ngai K L 1993 *Phys. Rev. B* **48** 13481

IOP Conf. Series: Materials Science and Engineering 1120 (2021) 012023 doi:10.1088/1757-899X/1120/1/012023

# Study of electrical conduction behaviour of $Y_{6-x}Ba_4(SiO_4)_6O_2$ oxyapatite doped with ytterbium ions

**Ritu Gupta and Sadhana Agrawal***

Department of Physics NIT Raipur G.E. road 492010, NIT Raipur, India

*Corresponding author's e-mail address: sagrawal.phy@nitrr.ac.in

**Abstract.** This paper reports the electrical conduction behaviour of Yttrium Barium Silicate Oxyapatite $Y_{6-x}Ba_4Yb_x(SiO_4)_6O_2$, synthesized by solution combustion method where x varies from 0.01 mol% to 0.05 mol%.. The scanning electron microscopy was done for morphological analysis which shows the spheroidal shaped grains with average grain size of 35μm. The dielectric behaviour of the material was studied in the frequency range of 100Hz – 100KHz and temperature varying from 50ºC - 400ºC. The frequency dependent dielectric constant and dielectric loss has been plotted for different temperatures. Decrease in dielectric constant and dielectric loss with increase in frequency has been observed for YBaSiO:Yb compounds.

## 1. Introduction

Oxyapatite materials doped with rare earth materials have attracted researchers due to the good electrical conductivities and low activation energies of oxygen ions present in the compound [1]. These materials show optical, biological and electrical applications because of their good thermal and chemical stabilities. They are used as electrolytes for solid oxide fuel cells SOFCs [1-2], through electrochemical reactions with excellent environmental compatibilities. The silicate oxyapatites also have good luminescent properties [3-4].

The general formula of silicate oxyapatite is $A_4B_6(SiO_4)_6O_2$, where A and B are cationic sites and allow various substitutions of alkali, alkaline earth and transition elements providing structural flexibility to the compound. The movement of these cations and oxygen (anion), are responsible for the electrical conduction behaviour of oxyapatite materials [5-6]. Various electrical properties can be examined like grain and grain boundary effect using complex impedance spectroscopy [7]. Response of a. c. conductivity to frequency can also be analysed using resistor – capacitor model [8]. In lower frequency region, the value of a. c. conductivity decays due to the slow orientation of charge carriers and in high frequency region, hopping of charge carriers results in rise in a.c. conductivity.

The influence of rare earth ions on the dielectric properties of various apatite materials such as $Sr_{10}(PO_4)_6(OH)_2$, $Ca_{10}(PO_4)_6(OH)_2$ etc has been widely researched [9-10]. Among various rare earth ions, Ytterbium has good chemical stability and mechanical strength [11]. The $Yb^{3+}$ ions when coated on surface forms a protective layer against corrosive properties. It has been reported that when $Yb^{3+}$ ions are doped in $La_9YbSi_6O_{27}$ shows enhanced dielectric constant [12] and a. c. conductivity [13]. Keeping this in view, the present study has been carried out on $Yb^{3+}$ ions doped $Y_{6-x}Ba_4(SiO_4)_6O_2$ oxyapatite compound.

Solution combustion method is used for synthesizing the oxyapatite compound, $Y_{6-x}Ba_4(SiO_4)_6O_2$: xYb with varying concentration of x ranging from 0.01 mol% to 0.05 mol%.

## 2. Material synthesis

Solution combustion method is one of the promising method due to cost effectiveness, simple equipment, low temperature requirement, less time consumption and also the particle size is obtained in the nano meter region. The flow chart for the synthesis of $Y_{6-x}Ba_4(SiO_4)_6O_2$: xYb compound is shown in figure 1. All the reagents i.e. $Y(NO_3)_3.6H_2O$, $Ba(NO_3)_2$, $Yb(NO_3)_3 \cdot 5H_2O$, SiO2 and citric acid, were weighed according to the nominal composition. After weighing the chemicals were individually dissolved in 10ml of distilled water and then all the reagents were mixed together to form a solution (50 ml). The solution was placed in a magnetic stirrer with 500 rpm to facilitate thoroughly mixing of solution along with heat supply. Meanwhile, citric acid was added drop wise and at sufficient temperature combustion took place, producing a powdered sample which after subsequent mixing placed in a programmed muffle furnace for 5 hours at 1200°C for calcination and then ground into a fine powder.

Then, the fine powders were mixed with binding agent and given the form of pellet using KBr press machine and sintered at 900°C.

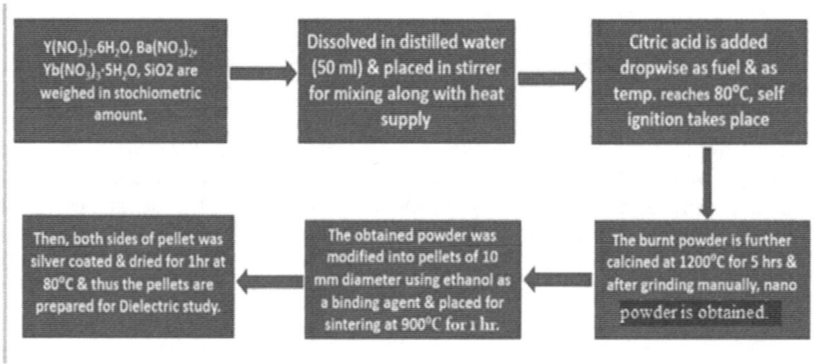


**Figure 1**. Flow chart for solution combustion synthesis.

## 3. Results and discussions

### *3.1 SEM*

SEM micrograph for $Y_{6-x}Ba_4Yb_x(SiO_4)_6O_2$ (x = 0.05 mol%) compound is shown in figure 2. SEM image show the formation of spheroidal grains with higher grain connectivity. The average grain size is calculated to be 35μm.

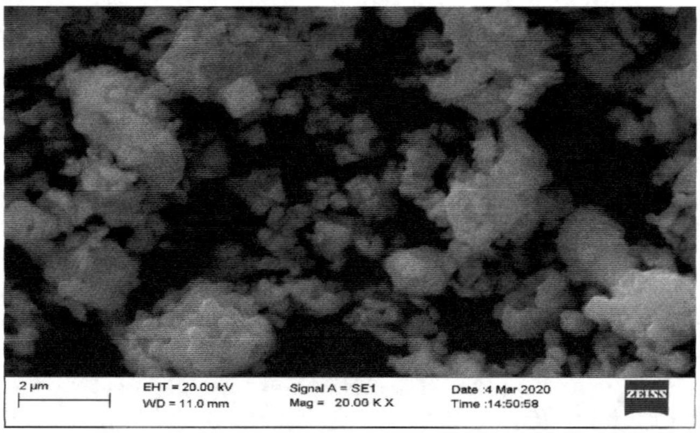


**Figure 2**. SEM micrograph of $Y_{6-x}Ba_4Yb_x(SiO_4)_6O_2$ (x = 0.05 mol%) compound.

*3.2 Dielectric studies*

Dielectric studies were done in the frequency range from 100 Hz to 100000 Hz at temperatures varying from 50 °C to 400 °C using HIOKI-IM3533 LCR Meter.

The dielectric constant (€) also known as permitivitty is a measure of degree to which it can oppose the flow of charge. Figure 3 shows the change in value of dielectric constant with applied frequency for different values of temperature. The graph shows decreasing pattern of € with increase in frequency. This is due to the reason that at lower frequency, the orientation of molecules take place due to different polarizations i.e. electronic, ionic and space charge polarization[14] where as at higher frequency the dipoles could not orient in the direction of applied a.c. field [15].

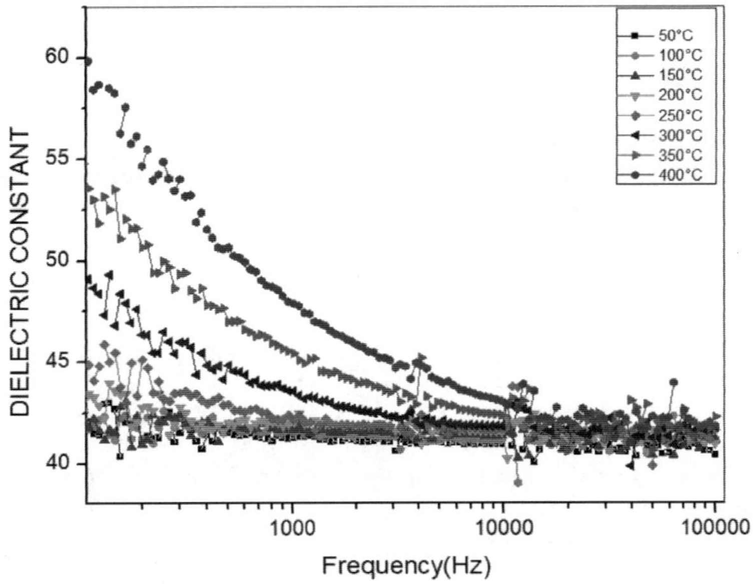


**Figure 3**. Plot of dielectric constant Vs frequency for $Y_{6-x}Ba_4Yb_x(SiO_4)_6O_2$ (x = 0.05 mol%) compound.

The dielectric loss (tanδ) also shows similar variation as dielectric constant with respect to frequency, i.e. at lower frequency dielectric loss is maximum and with rise in frequency it decreases as shown in figure 4. The tanδ can be defined as the loss of energy due to heating of the material in the presence of electric field. The change in value of dielectric loss with frequency is due to the reason that the ytterbium ions in the compound creating metastable states. In the metastable state due to relaxation phenomena and time lag, the emission energy is found to be less than the absorbed energy [15]. The loss curve recorded at highest temperature is maximum.

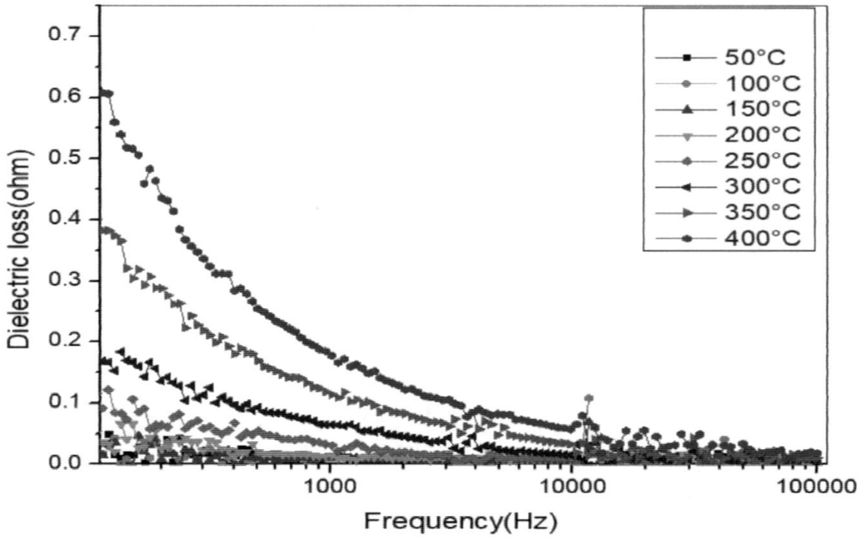


**Figure 4**. Plot of dielectric loss Vs frequency for $Y_{6-x}Ba_4Yb_x(SiO_4)_6O_2$ (x = 0.05 mol%) compound.

Figure 5 shows the variation of dielectric constant with temperature for $Y_{6-x}Ba_4Yb_x(SiO_4)_6O_2$ (x = 0.05 mol%) compound. At lower temperature the dipoles do not follow the direction of applied field hence dielectric constant shows saturated value. However, with increase in temperature the dipoles start orienting in the direction of applied field and the value of dielectric constant increases [16]. The change in dielectric loss with respect to temperature for $Y_{6-x}Ba_4Yb_x(SiO_4)_6O_2$ (x = 0.05 mol%) compound is shown in figure 6. As temperature increases dielectric loss also increases due to the rapid motion of charge carriers present in the compound [17].

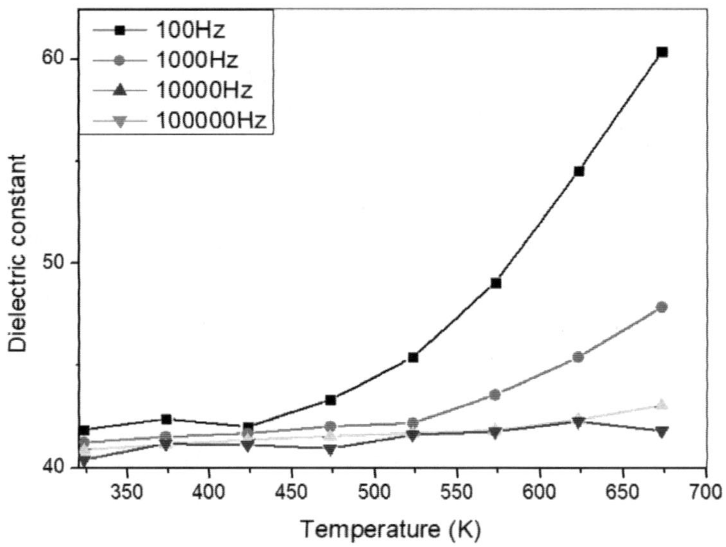


**Figure 5.** Plot of Dielectric constant against Temperature for $Y_{6-x}Ba_4Yb_x(SiO_4)_6O_2$ (x = 0.05 mol%) compound.

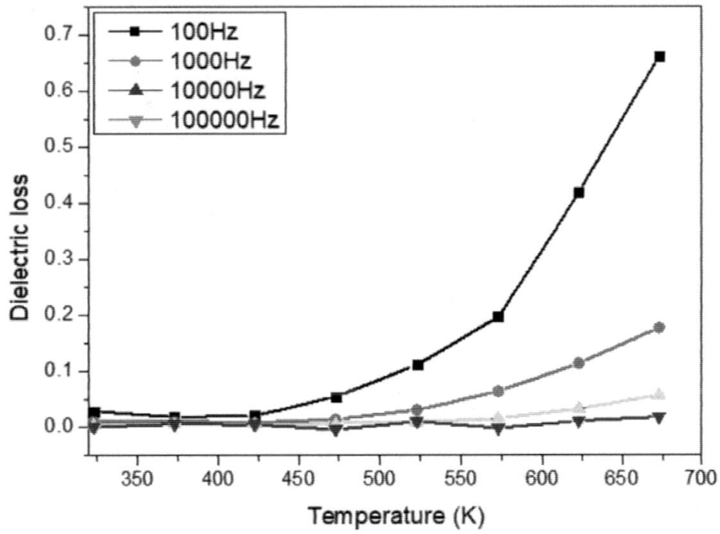


**Figure 6.** Plot of Dielectric loss against Temperature for $Y_{6-X}Ba_4Yb_x(SiO_4)_6O_2$ (x = 0.05 mol%) compound.

## 4.Cole-cole plot

The relation between the real (Z') and imaginary (Z'') part of impedance can be studied by Complex Impedance Spectroscopy (CIS) [18]. The CIS investigation of the prepared material is done with frequency from 100 to 100000 Hz and temperature varying from 50ºC - 400ºC.

Figure 7, shows the Cole-Cole plot for the compound $Y_{6-X}Ba_4Yb_x(SiO_4)_6O_2$ (x = 0.05 mol%). The value of Z' and Z'' increases with rise in frequency. Cole – Cole plots govern the nature of sample whether it is Debye or Non – Debye. The synthesized compound showed Debye nature. It also tells the effect of grains and grain boundaries.

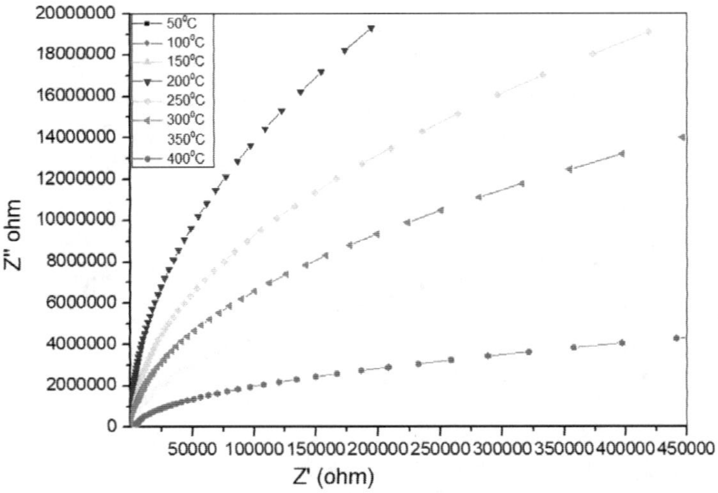


**Figure 7.** Imaginary part of impedance Vs real part of impedance for $Y_{6-X}Ba_4Yb_x(SiO_4)_6O_2$ (x = 0.05 mol%) compound.

## 5. A. C. conductivity analysis

The response of a. c. conductivity in accordance with frequency has been recorded and is depicted in the graph below (figure 8). At lower values of frequency, the variation of conductivity is almost saturated but with rise in frequency, it increases showing the conduction behaviour of $Y_{6-x}Ba_4Yb_x(SiO_4)_6O_2$ (x = 0.05 mol%) compound. This AC conductivity rise is almost linear with frequency for all temperatures, following Johnscher's power law which is given as, [19]

$$\sigma(\omega) = \sigma(0)+A\omega^n \quad (1)$$

where $\omega$ is angular frequency, $n$ is a constant taking values from 0 to 1 and $\sigma(0)$ is the frequency independent conductivity. The low frequency conductivity is due to the motion of charge carriers through the channel connecting the cationic and anionic sites present in the compound. The relation used for the calculation of a. c. conductivity is given as,

$$\sigma = \frac{z'}{z'^2+z''^2} \times \frac{t}{A} \quad (2)$$

where t is the thickness of pellet while A is the area of it.

The curves recorded at low temperature regions were overlapping and to get a clear view of the response curve, the scaling has been reset and embedded in figure. 6

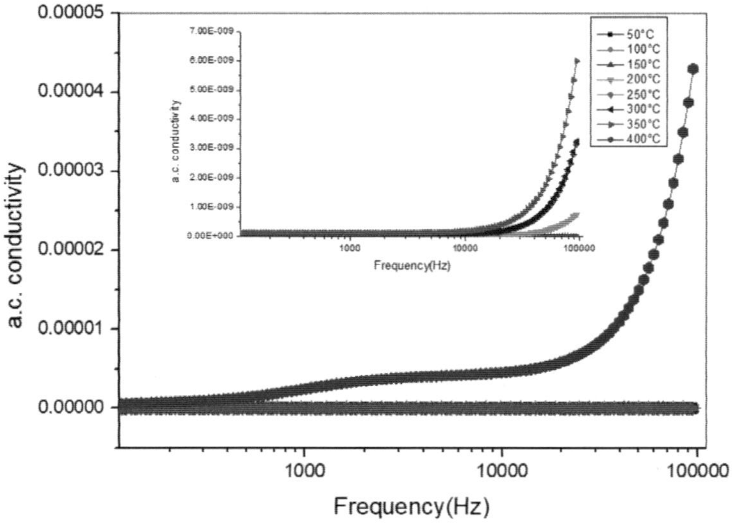


**Figure 8**. The graph for a. c. conductivity Vs frequency for $Y_{6-x}Ba_4Yb_x(SiO_4)_6O_2$ (x = 0.05 mol%) compound

## 6. Conclusions

Synthesis of Yttrium Barium Silicate Oxyapatite doped with Ytterbium ions has been done by solution combustion technique. SEM micrograph showed spheroidal grains with higher grain connectivity. The dielectric studies of YBaSiO: Yb compound showed inverse relation with frequency. The temperature dependent dielectric studies showed that as temperature increases dielectric parameters increases due to increase in polarization. The synthesized compound showed increase in a. c. conductivity with increase in applied frequency. The debye nature of sample has been observed from the Cole-Cole plot.

## References

[1] Reina L L, Losilla E R et al. 2005*Chem. Mater.* **17** 596
[2] Béchade E, Julien I et al. 2008 *Jour. Of the Euro. Cer. Soc.* **28** 2717
[3] Steigerwald D A, Bhat J C et al. 2002 *Jour. of Sel. Top. in Quant. Electro.* **8** 310
[4] Kharton V V, Shaula A L et al. 2004 *Jour. of the Electro. Soc.* **151** 1236
[5] Slater P R, Sansom J E H and Tolchard J R 2004 *Chemi. Rec.* **4** 373
[6] Yao H C, Wang J S et al. 2010 *Sol. Sta. Ioni.* **181** 41

[7] Tank K P, Sharma P et al. 2011 Crys. Res. and Tech. **46** 1309
[8] Naddari T, Savariault J M et al. 2002 *Jour. of Soli. Stat. Cir.* **166** 237
[9] Wafi R A, Jafer R et al. 2017 *Cera. Interna.* **43** 17153
[10] Valdes J J P, Rodriguez V et al. 1995 *Journ. of Mate. Res.* **10** 2174
[11] Ganesan M, Berube C D et al. 2002 *Organometallics* **21** 1707
[12] Samad R, Rather M U D et al. 2018 *J. Mater. Sci. Mater. Electron.* **29** 4226
[13] Xiang J, Liu Z G et al 2012 *Electro. Act.* **65** 251
[14] Arends J, Royce B S H et al. 1968*Phys. Lett.* **27** 720
[15] Rout A and Agrawal S 2020 *Cer. Inter.* **46** 7912
[16] Rout A and Agrawal S 2021 *Cer. Inter.* **47** 7032
[17] Schüller J, Mel'nichenko Y B et al. 1994 *Phys. Rev. Lett.* **73** 2224
[18] Joshi M2017 *Mecha. Mate. Sci. and Eng.* **9** 261
[19] Gittings J P, Bowen C R et al 2009 *Act. Biomater.* **5** 743

# Electrical conduction mechanism of barium doped strontium hydroxyapatite synthesized by co – precipitation method

**Ananya Rout and Sadhana Agrawal***

Department of Physics, National Institute of Technology Raipur, 492010, India

*Corresponding author's e-mail address: sagrawal.phy@nitrr.ac.in

**Abstract.** The crystalline samples of Barium (Ba) doped $Sr_{10-x}Ba_x(PO_4)_6(OH)_2$ with x = 0, 0.5 and 1 mol% were prepared by co – precipitation method. The microstructural (SEM) image shows the distribution of densely packed grains over the surface of prepared sample. The nature of cole – cole plots confirm that the conduction mechanism is governed only by the grains with no contribution of grain boundaries for all prepared samples. The resistance of grains decreases with increase in temperature and molar (mol%) concentration, which shows a negative temperature coefficient resistance (NTCR) behaviour for all prepared samples. The a.c. conductivity increases with increase in frequency which shows non – debye nature of the prepared samples.

## 1. Introduction

Hydroxyapatite compounds have attracted the interest of researchers due to their large band gap, semiconducting properties, electrical conduction mechanism, high optical strength and high luminescence efficiency [1]. The general formula of hydroxyapatite is $X_{10}(YO_4)_6(OH)_2$ where X can be substituted by any divalent and trivalent cation and Y can be substituted by halogen atom [2]. It has a flexible structure due to which different ions can be easily incorporated into the lattice.

Strontium bearing apatites are considered well known hosts with various applications in optical and electronic devices [3]. It has application in biological field as it can be incorporated in human bones [4]. It also has suitable application as nuclear waste storage materials [5]. Many investigations on bone grafting, dental implants is done through electrical conduction phenomena in hydroxyapatite.
The electrical conduction analysis is done through complex impedance spectroscopy [6], which is a potential tool to get information regarding the electrical and relaxation behaviour of compounds. In addition, it also gives information regarding the grain and grain boundary effect, real and imaginary part of impedance, resistive and capacitive elements in the compounds [7].

The present paper reports the microstructural, dielectric, impedance and a.c conductivity studies of Ba doped $Sr_{10-x}Ba_x(PO_4)_6(OH)_2$compound. The a.c impedance studies reveal that the prepared compound obeys Jonscher's power law and appears non – debye in nature.

IOP Conf. Series: Materials Science and Engineering 1120 (2021) 012024 doi:10.1088/1757-899X/1120/1/012024

## 2. Experimental method

All the staring materials $Sr(NO_3)_2.4H_2O$, $Ba(NO_3)_2.4H_2O$ and $(NH_4)_2H(PO_4)$ supplied by Sigma Aldrich were taken in stoichiometric amount. The materials were dissolved in 50 ml of distilled water and kept on a heater attached magnetic stirrer with 500 rpm for 1 hour. In between $NH_4OH$ was added dropwise to set the PH at 10. After 1 hour of vigorous stirring the solution was filtered. The filtered precipitant was dried in an oven at $80^0C$ for 3 hour. The dried precipitant was grond in agate mortar and pestle for 1 hour and placed in programming furnace at $900^0C$ for 4 hours. The calcined powders were pressed with 200 MPa pressure to form pellets. The pellets were sintered at $900^0C$ for 1 hour as shown in figure 1.

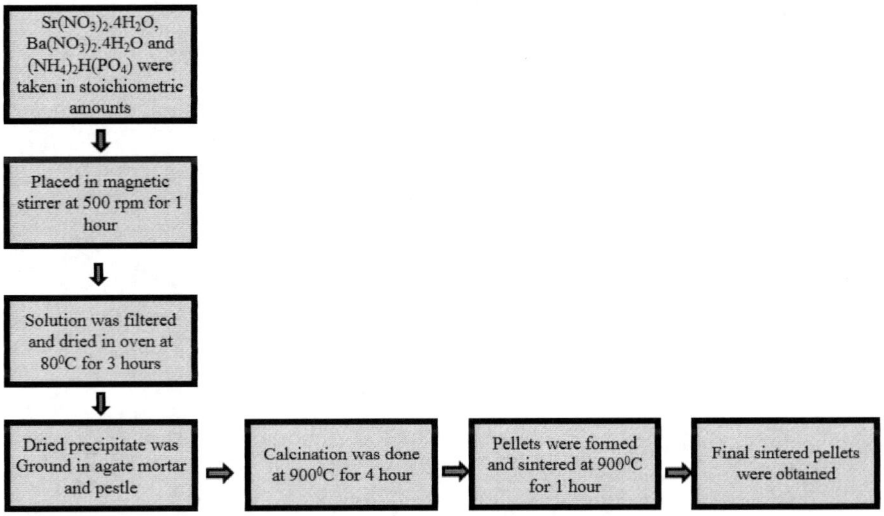


**Figure 1.** Flow chart for synthesis of $Sr_{10-x}Ba_x(PO_4)_6(OH)_2$ compound.

## 3. Results and discussions

### *3.1. Morphological and Structural Analysis*

The scanning electron microscopy images of $Sr_{10-x}Ba_x(PO_4)_6(OH)_2$ (x = 0.5 mol%) are shown in figure 2(a) and 2(b) respectively. The SEM images show the formation of clusters of highly densified and agglomerated spheroidal grains with varying grain size which may be due to sintering of samples at high temperature ($1200^0C$). The powder diffraction X-ray analysis of $Sr_{10-x}Ba_x(PO_4)_6(OH)_2$ compound confirms the formation of single phase hexagonal structure with space group P63/m [8].

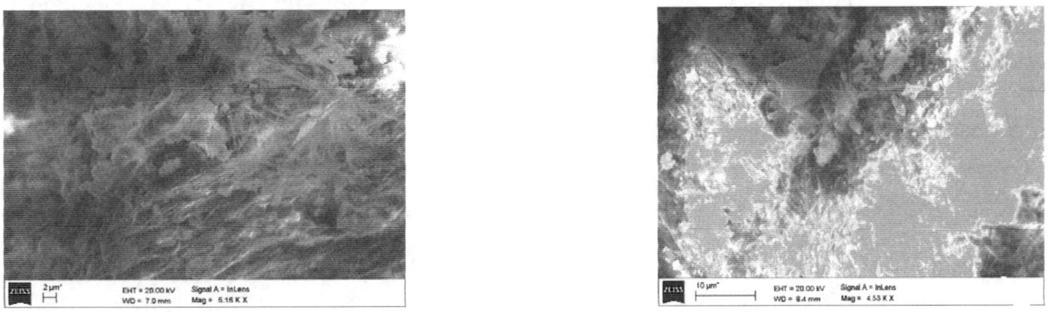

(a) (b)

**Figure 2.** SEM images of $Sr_{10-x}Ba_x(PO_4)_6(OH)_2$ (x = 0.5 mol%) compound (a) 6.16 KX (b) 4.33 KX.

### *3.2. Dielectric studies*

Dielectric studies were performed in the frequency range of 1000 Hz to 100000 Hz and temperature varying from 30$^0$C to 200$^0$C with 5$^0$C sweep mode. The change in value of dielectric constant (Ɛ) with applied frequency is given in figure 3. It has been noticed that Ɛ decreases with increase in frequency which shows the ceasing effect of polarization for higher frequency [9]. At low frequency region the dipoles are oriented in the applied electric field direction, as the frequency rises the dipoles lag to oscillate in the track of applied electric field and dielectric constant decreases.

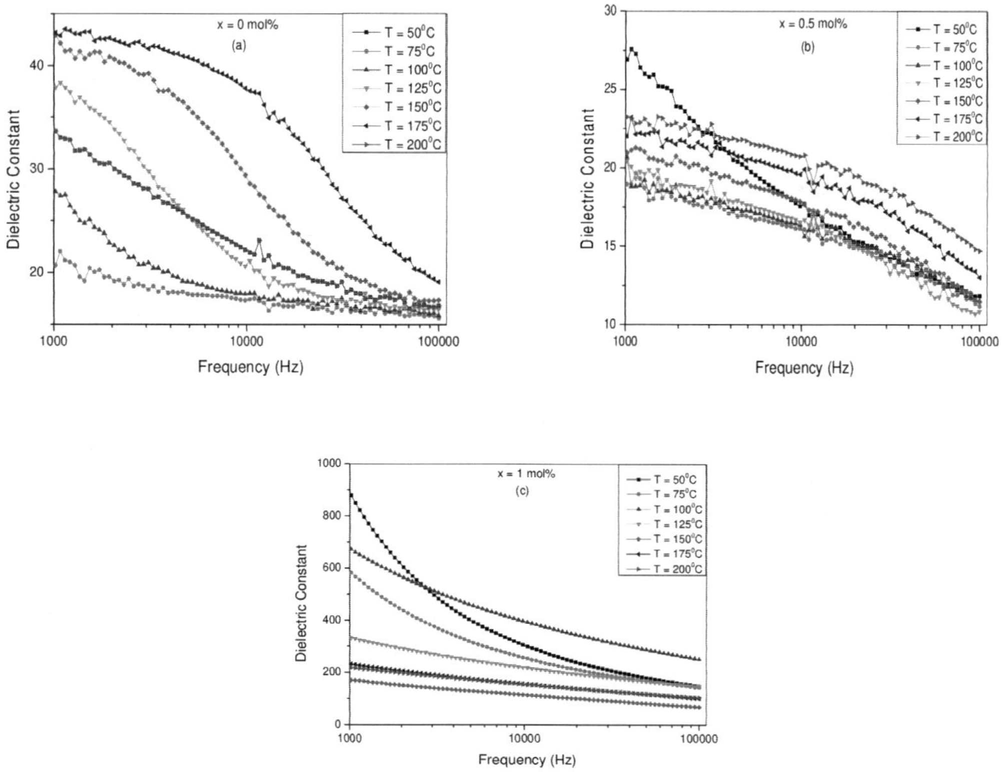


**Figure 3.** Dielectric constant plots of $Sr_{10-x}Ba_x(PO_4)_6(OH)_2$ compound for different x values (a) 0 mol% (b) 0.5 mol% (c) 1 mol%.

Figure 4 shows the variation of dielectric constant with varying temperature. At low temperature the dipoles do not orient in the direction of applied field hence, dielectric constant shows a constant value. As the temperature increases the dipoles get sufficient energy to orient themselves in the direction of applied field and hence the dielectric constant increases [10].

IOP Conf. Series: Materials Science and Engineering 1120 (2021) 012024 doi:10.1088/1757-899X/1120/1/012024

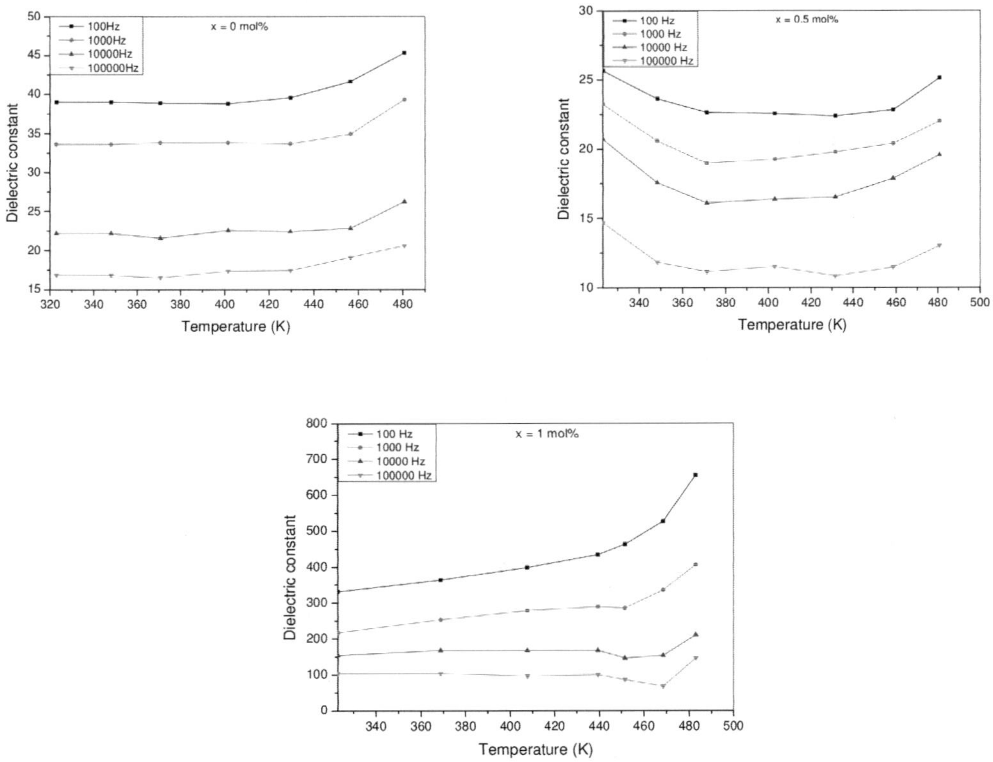


**Figure 4.** Temperature dependent dielectric constant plots of $Sr_{10-x}Ba_x(PO_4)_6(OH)_2$ compound for different x values (a) 0 mol% (b) 0.5 mol% (c) 1 mol%.

The change in the value of dielectric loss (tanδ) with applied frequency is shown in figure 5. The dielectric loss value up surges with rise in temperature and as the doping concentration increases tanδ value decreases in the high frequency side [11]. It shows the hopping mechanism of charge carriers in the $Sr_{10-x}Ba_x(PO_4)_6(OH)_2$ compound. For x = 0 mol% and x = 0.5mol% the dielectric loss increases with increase in frequency due to hopping of charge carriers in the direction of applied electric field. With further increase in dopant concentration to x = 1mol% the hopping of charge carriers lag behind the frequency of applied electric field as a result dielectric loss decreases with increase in frequency [12].

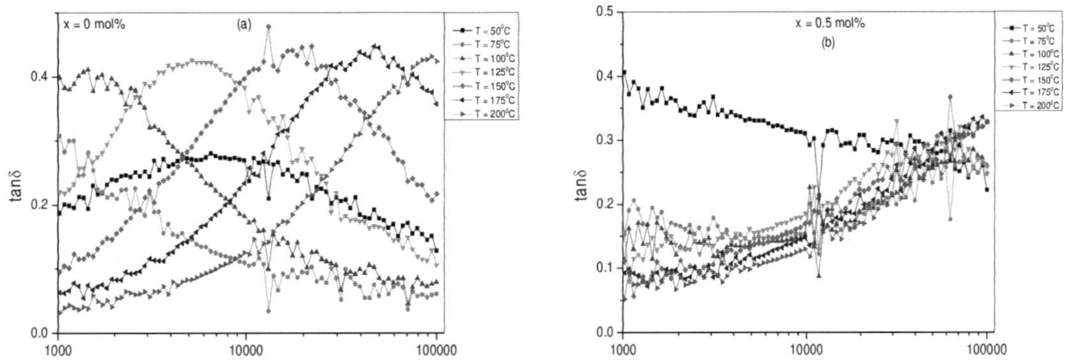

IOP Conf. Series: Materials Science and Engineering 1120 (2021) 012024 doi:10.1088/1757-899X/1120/1/012024

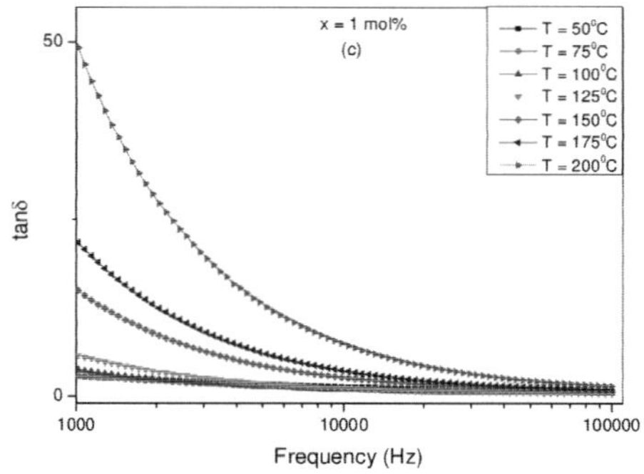


**Figure 5.** Dielectric loss plots of $Sr_{10-x}Ba_x(PO_4)_6(OH)_2$ compound for different x values (a) 0 mol% (b) 0.5 mol% (c) 1 mol%.

*3.3 Cole – Cole Plots*

The cole – cole plots for $Sr_{10-x}Ba_x(PO_4)_6(OH)_2$is shown in figure 6. It can be seen that with rise in temperature the semi-circular arc decreases and bends towards real part of impedance (Z') which shows increase in a.c conductivity of the samples [13].

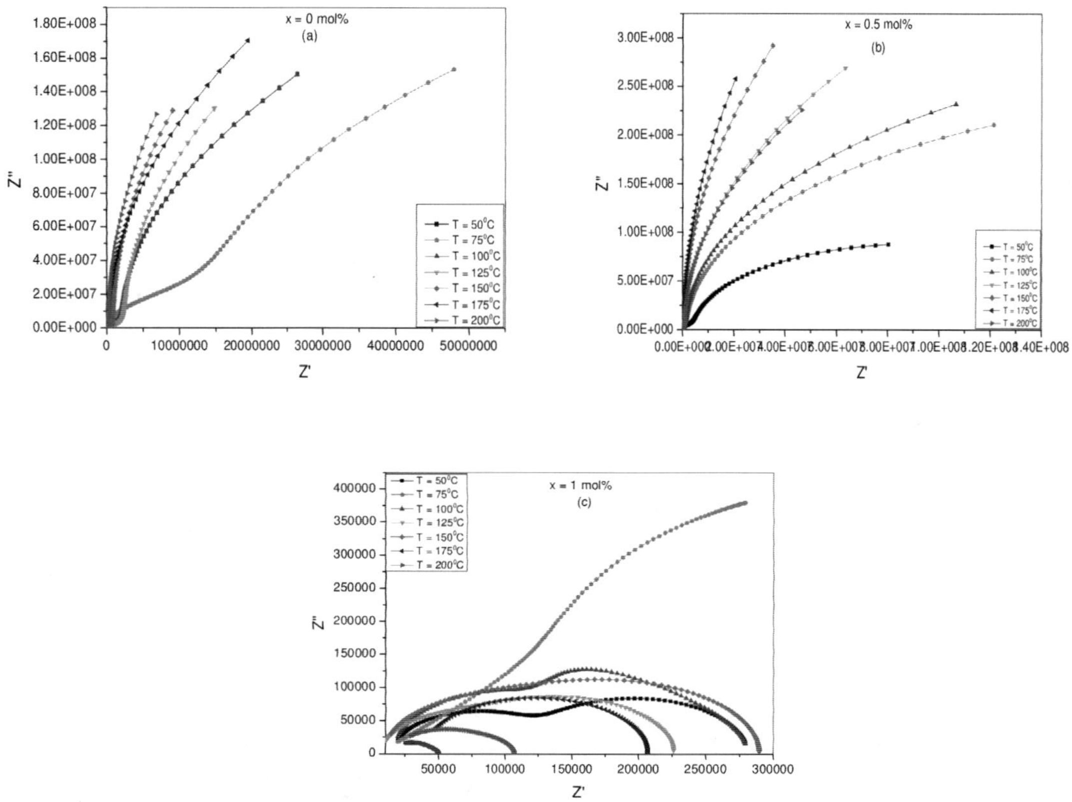


**Figure 6.** Cole – Cole plots of $Sr_{10-x}Ba_x(PO_4)_6(OH)_2$ compound for different x values (a) 0 mol% (b) 0.5 mol% (c) 1 mol%.

### *3.4 A.C Conductivity Analysis*

Figure 7 shows the frequency dependent conductivity study at different temperatures for $Sr_{10-x}Ba_x(PO_4)_6(OH)_2$ compound. It is observed that at low frequency a plateau region is observed corresponding to d.c conductivity [14]. As the frequency increases the conductivity increases which shows that the compound obeys Jonscher's power law.

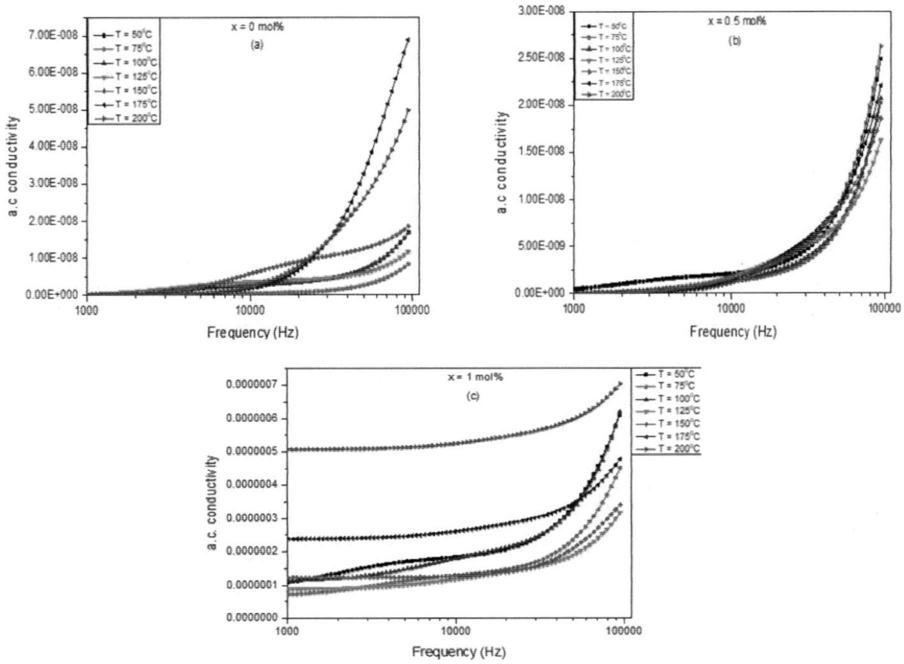


**Figure 7.** A.C conductivity plots of $Sr_{10-x}Ba_x(PO_4)_6(OH)_2$ compound for different x values (a) 0 mol% (b) 0.5 mol% (c) 1 mol%.

## 4. Conclusion

The Ba doped $Sr_{10-x}Ba_x(PO_4)_6(OH)_2$ with x = 0 -1 mol% synthesized by co – precipitation technique. SEM images show the distribution of high densified grains over the surface of samples. The dielectric constant decreases with increase in frequency which shows the ceasing effect of polarization for higher frequency. Dielectric loss follows the same trend and decreases at high frequency region which shows the hopping of charge carriers of the prepared samples. Cole – Cole plots show the presence of grains effect of the samples. A.C conductivity plots obey Jonscher's power law and appears non – debye in nature.

## References

[1] Ahmad Z, Thian E S et al 2008 *J. Mater. Sci.* **19** 3093
[2] Munir G, Huang J, Edirisinghe M J et al 2011 *J. R. Soc.* **8** 678
[3] Sudarsanan K and Young R 1980 *Acta Crystallogr., Sect. B: Struct. Crystallogr. Cryst. Chem.* **36** 1525
[4] Zhang W, Shen Y et al 2011 *Acta Biomater.* **7** 800

[5] Maji B K, Jena H, Krishnaiah M and Asuvathraman R 2016 *J. Therm. Anal. Calorim.* **124** 857
[6] Yakuphanoglu F, Yahia I S, Barim G, Filiz B S 2010 *Synth. Met.* **160** 1718
[7] Ahn J H, Lee J U, Kim T W 2007 *Current Appl. Phys.* **7** 509
[8] Ananya R, Sadhana A 2020 *IOP Conf. Ser.* **798** 2020
[9] Mah S B, Jang N W, Park J H, Paik D S, and Park C Y 2003 *Mat. Res. Bull.* **35** 1113
[10] Okazaki K and Nagata K 1973 *J. Am. Ceram. Soc.* **82** 56
[11] Aburba D et al 2018 *J. All. Comp.* **739** 729
[12] Ravish G, Paramjeet S et al 2013 *Int. Jour. Eng. Res.* Dev. **5** 9
[13] Ananya R and Sadhana A 2020 *Cer.Int.***46** 7912
[14] Jonscher A K 1977 *Nature* **267** 673

# A Review on characterization and application of fly ash cenosphere

**T Gupta [1] and P S Bokare[2]**

[1] OP Jindal University Raigarh, India
[2] OP Jindal University, Raigarh, India

*Corresponding author's e-mail address: tulika.gupta@opju.ac.in

**Abstract.** The Fly Ash Cenospheres (FAC) or alumino-silicate is one of the most valuable by-products of fly ash generated by coal thermal power plants. It is an industrial by-product of high commercial value. It is found that FAC is hollow spherical microspheres having sizes varying 5-500 μm and chemically inert particles and make them an ideal material to replace fine aggregate from cement concrete. This study presents a systematic review of observational studies on the properties and application of FAC. Different industries use FACs due to their unique properties such as hollow spherical shape and lightweight, less specific gravity, high compressive strength, high thermal resistance, and acoustic insulation, etc. The addition of FAC to concrete has found to reduce the density of concrete without compromising the strength. Other properties such as compressive strength, flexural strength, toughness, ductility, micro-structure, etc are also documented in this study. The gap in the literature is identified and reported at the end of the research report.

## 1. Introduction

The Fly Ash Cenosphere (FAC) or aluminosilicate is the by-product of coal fly ash emerging as a waste product of thermal power generation plants. The FAC is a value-added product, having many applications in industries as well as in construction. FAC is hollow, lightweight, spherical in shape, having a specific gravity less than water [1]. A general view of FAC is presented in figure 1.

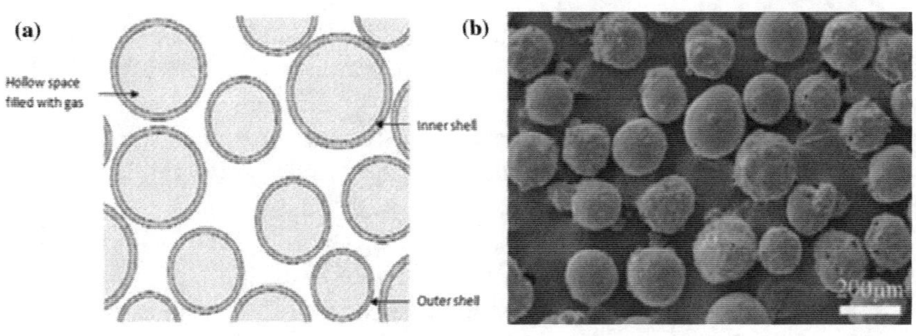


**Figure 1.** General view of FAC [2].

The FAC is recovered from the Fly Ash by the Dry Separation and Wet Separation method. A combination of these methods is generally used for the recovery of FAC from fly ash. A brief description of the recovery methods is presented in the next section.

### *1.1 Recovery Methods of FAC*

*(a) Dry Separation Process*
It comprises fluidized bed separation followed by screening and pneumatic separation in a free-fall air chamber. The mechanical activator assists the Fluidised bed separation to prevent agglomeration. This reduces the portion of material that requires further treatment. The FAC is then, individually subjected to pneumatic separation [3].

*(b) Wet Separation Process:*
This the most widely accepted and used method for the separation of the FAC from fly ash. Wet separation is carried out by using the liquid medium and based on the density difference between floaters and the liquid [4]. The FAC is light in weight as compared to liquid. Due to this, the FAC floats on the liquid. The wet separation method is popularly known as the "Sink and Float" method.

### *1.2 Properties of FAC*

Various researchers have reported the properties of FAC as lightweight, with high compressive strength, low thermal conductivity, hollow, etc. Therefore, the FAC is an ideal additive to the concrete to reduce the unit weight without compromising the strength of concrete. Typical properties of FAC are presented in Table 1.

**Table 1**. Typical properties of FAC [5].

| Sr. No | Property of FAC | Value |
|---|---|---|
| 1 | Size | 0.005 mm to 0.035 mm |
| 2 | Mean Diameter | 0.117 mm |
| 3 | Compressive Strength | 7.38 MPa |
| 4 | Composition | Silica, Alumina, Ferrous (Approx. 90%) |
| 5 | Electrical Resistance | R= 2.26×108 Ω |
| 6 | Mass Loss at 1000 C | Negligible |
| 7 | Wall Thickness | 10% of Diameter |

These properties (presented in Table 1) of FAC render it as an ideal application in replacement of fine aggregate in concrete, replacement of cement in concrete, acoustics, synthetic foam, paints, plaster, etc. In addition, the chemically inert nature of FAC is useful in many applications since it is chemically inert in nature.

The chemical characteristics (item no 4 of Table 1) indicate that the FAC is rich in silica, alumina, and ferrous. These constituents impart the physical properties to FAC which are highly useful as commercial application.

Though various researchers studied and reported the properties of FAC, they have used Fly Ash originated from different sources of coals, for studying the properties of FAC. Hence there is a large variation in reported properties of FAC. Another, these efforts are not documented in one place to provide the overall scenario of the characteristics and application of FAC. This report, therefore, summarizes the available research reports on the properties of FAC.

This paper is divided into three major sections dealing with methods of separation of FAC, characterization of FAC, and application of FAC followed by conclusions and the selected references.

## 2. Methods of Separation of FAC

The FAC is obtained by use of two methods, the dry separation method and the wet separation method. The wet separation method is popularly used by researchers. This section presents a brief account of the observations that are put forward by the researchers, world over, on both the separation methods.

### *2.1 Dry Separation Method.*

In the dry separation method, the researchers reported two types of separation methods. The first one uses the sieving by appropriate aperture sieve for physical separation of FAC from fly ash. In the second one, the centrifuge is used for the physical separation of FAC from fly ash. A typical flow diagram for dry separation is presented in figure 2.

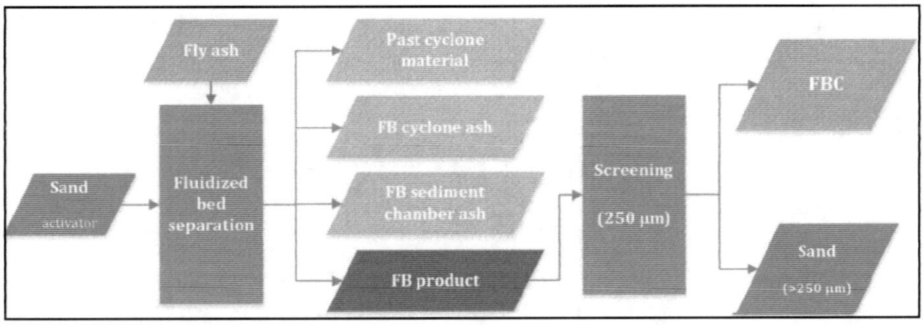


**Figure 2.** Typical flow diagram of dry separation process [3].

Author [3] carried out the dry separation method for FAC and further tried the wet separation using a fluidized bed. They recovered the FAC exceeding 81%. Author reported a 66% recovery of FAC using the dry separation method [6]. They tried the same sample separation by the wet method and resulted in a recovery of 80%.

### *2.2 Wet Separation Method*

Author reported that the most efficient separation method of FAC is wet separation [7]. They tried to recover the FAC in two plants of air classifier with 90% and 81% recovery, respectively. The difference in yields is reported due to different sources of coal. Table 2 presents a summary of the FAC yield reported by various authors.

**Table 2.** FAC yield observed by various researchers.

| Sr. No. | Author | % of FAC Obtained |
|---|---|---|
| 1 | Wrona et al. (2020) [3] | 81 |
| 2 | Acar et al. (2016) [7] | 81 |
| 3 | Petrus et al. (2011) [6] | 80 |
| 4 | Hajima et al. (2010) [8] | 66 |

Table 2 indicates that the yield of FAC is similar in both, dry and the wet separation method.

## 3. Characterization of FAC

This section presents the review of research reports published by researchers on the characterization of FAC collected from the fly ash by using different methods of separation. The various author investigated the properties of FAC by collecting samples from different coal-based power plants. On reviewing various research reports, the characteristics of FAC can be categorized as:

(i) Chemical and Mineralogical Characteristics
(ii) Morphological Characteristics
(iii) Physical Characteristics

The following discussion presents the summary of reports on these characteristics.

### *3.1 Chemical and Mineralogical Characteristics*

The FAC primarily contains silica and alumina [5, 9-22]. Various techniques such as X-Ray Diffraction Analysis (XRD), Scanning Electron Microscopy (SEM), X-Ray Fluorescence (XRF), etc. are used by researchers to determine various characteristics of FAC [2, 4, 5, 9-12, 14, 15, 19, 21, 23, 25-29].

Author used XRF and XRD to evaluate the characteristics of FAC [30]. Table 3 presents the gist of characteristics reported by them.

**Table 3.** Chemical composition of FAC [30].

| Country | $SiO_2$ | $Al_2O_3$ | $Fe_2O_3$ | CaO | $SO_3$ | $Na_2O$ | $K_2O$ | $P_2O_5$ | $TiO_2$ | MgO |
|---|---|---|---|---|---|---|---|---|---|---|
| India | 50-60 | 25-35 | 2-8 | 1-6 | 0-1 | 0-1 | 1-3 | 0-1 | 1-5 | 0-1 |
| China | 40-65 | 15-35 | 2-8 | 1-6 | 0-1 | 0-2 | 1-3 | 0-1 | 1-5 | 0-1 |
| Australia | 45- 60 | 25- 30 | 5-10 | 1-6 | 0-1 | 0-1 | 0-3 | 0-1 | 1-5 | 0-1 |
| USA | 30- 60 | 25- 35 | 2-15 | 1-15 | 0-1 | 0-4 | 0-4 | 0-1 | 1-5 | 0-1 |

Note: All figures are in %

Table 3 indicates that FAC mostly contains $SiO_2$ followed by $Al_2O_3$ irrespective of occurrence and location. The highest % of $SiO_2$ and $Al_2O_3$ in FAC is observed in India and China followed by Australia and the USA. The difference in the concentration of $SiO_2$ and $Al_2O_3$ can be contributed to the composition of coal in respective countries.

Adesina 2020 [31] reviewed various research efforts on mineralogical characteristics of the FAC as presented in Table 4.

**Table 4.** Chemical composition of FAC.

| Sr. No | Authors | $SiO_2$ | $Al_2O_3$ | $Fe_2O_3$ | CaO | $K_2O$ | $Na_2O$ |
|---|---|---|---|---|---|---|---|
| 1 | Wang et al. 2011 [32] | 62.2 | 29.7 | 3.5 | 0.9 | 1.7 | - |
| 2 | Sarkar et al. 2008 [33] | 54.6 | 36.9 | 1.8 | 0.39 | 2.8 | 0.3 |
| 3 | Ngu et al. 2007 [33] | 62.4 | 26.2 | 0.9 | 0.03 | 0.3 | 1.3 |
| 4 | Li et al. 2013 [24] | 53.5 | 31.5 | 5.5 | 2.0 | 2.4 | 0.8 |
| 5 | Zhao et al. 2010 [35] | 56.0 | 26.0 | 7.0 | 4.7 | 3.2 | 0.5 |
| 6 | Fomenko et al. 2016 [36] | 56.3 | 25.1 | 9.6 | 1.35 | 2.9 | 0.8 |
| 7 | Kolay and Bhusal 2014 [10] | 52.5 | 30.0 | 7.5 | 1.15 | 2.0 | |
| 8 | Wang et al. 2013 [13] | 62.3 | 19.9 | 5.9 | 1.5 | - | - |
| 9 | Blanco et al. 2000 [37] | 56.0 | 25.0 | 6.9 | 4.2 | 3.4 | 2.2 |
| 10 | Wang et al. 2012 [26] | 58.1 | 32.0 | 3.7 | 0.9 | 3.1 | 0.8 |
| 11 | Hanif et al. 2017 [39] | 73.1 | 16.7 | 2.0 | 1.06 | 3.9 | 2.4 |
| 12 | Xu et al. 2015 [9] | 61.0 | 32.1 | 3.1 | 0.71 | 1.7 | 3.1 |
| 13 | Kannan et al. 2016 [40] | 65.0 | 36 | 3.0 | 0.5 | - | - |
| 14 | Chen et al. 2016 [41] | 58.0 | 35 | 1.5 | 1.9 | 0.3 | 0.7 |
| 15 | Liu et al. 2017 [18] | 57.4 | 31.9 | 2.3 | 0.44 | 1.4 | 0.54 |
| 17 | Chen et al. 2019 [43] | 58.9 | 27.2 | 5.4 | 0.99 | 4.18 | |
| 18 | Patel et al. 2020 [44] | 56.2 | 33.4 | 2.8 | 1.46 | - | - |
| 19 | Kolay and Singh 2001 [1] | 52.3 | 30.1 | 7.5 | 1.15 | 1.98 | 0.02 |
| 20 | Xi et al. 2020 [45] | 59.8 | 32.3 | 2.2 | 0.24 | 0.52 | 0.33 |

| | | | | | | | |
|---|---|---|---|---|---|---|---|
| 21 | Patel et al. 2019 [46] | 56.24 | 33.4 | 2.8 | 1.46 | 1.97 | 0.02 |
| 22 | Liu et al. 2019 [28] | 57.4 | 31.9 | 2.3 | 0.44 | 1.4 | 0.54 |
| 23 | Yoriya et al. 2019 [4] | 56.0 | 25.0 | 13.0 | 30.0 | 4.1 | - |
| 24 | Beddu et al. 2018 [17] | 57.67 | 22.63 | 9.09 | 3.75 | 3.62 | |
| 25 | Zyrkowski et al. 2016 [47] | 60.45 | 28.3 | 5.43 | 14.81 | 3.62 | 4.83 |
| 26 | Yang et al. 2013 [23] | 57.83 | 23.89 | 3.43 | 6.73 | - | - |
| 28 | Chávez-Valdez et al. 2011 [48] | 61.58 | 24.76 | 2.0 | 2.64 | 0.44 | 0.22 |
| 29 | M. Wang et al. 2011 [32] | 62.212 | 29.704 | 3.531 | 0.902 | 1.702 | - |
| 30 | Anshits et al. 2010 [49] | 64.64 | 20.85 | 4.05 | 2.24 | 3.19 | 0.93 |
| 31 | Sokol et al. 2000 [50] | 58.98 | 27.91 | 4.83 | 1.17 | 2.85 | 0.38 |

Table 4 indicates that the mineralogy of FAC is dominated by $SiO_2$ and $Al_2O_3$ followed by $Fe_2O_3$. These mineral components impart much-desired properties such as smooth surface, inertness, chemical stability, and lightweight to FAC.

The important manifestation of $SiO_2$ and $Al_2O_3$ concentration in FAC is its inertness and smooth surface structure. This characteristic also provides thermal stability above 1000 $^0$C.

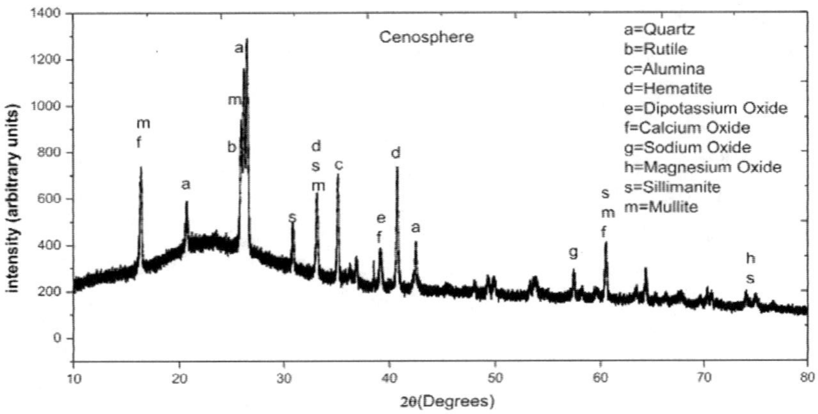


**Figure 3.** XRD diffraction Spectra of FAC [5].

Author studied the mineralogical characteristics of FAC originated from coal fly ash of Talcher Super Thermal Power Station, Kanhia, Odisha [5]. The coal in this thermal power station originates from the Kaniha coal block of Mahanadi Coalfields Ltd. The XRD image of the FAC sample is presented in figure 3 and the details of constituents are presented in Table 5 this indicates that the mineralogical composition of FAC is dominated by $SiO_2$ and $Al_2O_3$ similar to that reported by some authors [30, 31].

**Table 5.** Chemical composition (wt. %) of the FAC sample compound [5].

| Compound | $SiO_2$ | $Al_2O_3$ | $Fe_2O_3$ | Na2O | K2O | MgO | $TiO_2$ | $P_2O_5$ | LOI |
|---|---|---|---|---|---|---|---|---|---|
| Percentage | 48.62 | 27.79 | 3.75 | 0.86 | 4.23 | 0.46 | 2.12 | 7.99 | 4.18 |

Similarly, many other authors studied the mineralogical composition of FAC and concluded that $SiO_2$ and $Al_2O_3$ are the principal components of FAC. Table 4 presents a summary of the mineralogical composition of FAC reported by various authors.

*3.2 Physical Characteristics of FAC*

Many authors [3, 6-8, 16, 51] reported that FACs are smooth and spherical in shape and have a diameter of 1 μm to 500 μm [4, 5, 10, 11, 14, 17, 19, 20, 48]. The researchers used scanned electron microscopy for ascertaining the physical properties of FAC.

The majority of researchers however reported that the size of FAC ranges from 45 μm to 600 μm [4, 52]. The shell thickness of FAC reported by the authors ranges from 1μm to 2 μm [52]. The bulk density is attributed as the principal character of FAC which makes FAC commercially useful [10, 11, 19 -21, 42, 47]. The bulk density of FAC varies from 0.25 $g/cm^3$ to 0.7 $g/cm^3$ which indicates that FAC is lighter than water.

Blanco et al. 2000 [37] however contradicted the observation of earlier researchers on the particle size of FAC. They reported the minimum particle size of 0.2 mm based on granulometric analysis of FAC. This contradictory observation is possibly due to the source of coal. The variation in morphological properties of FAC is attributed to the origin of coal from which the FAC is extracted. Table 6 presents a summary of the morphological characteristics of FAC.

**Table 6.** Morphological characteristics of FAC.

| Sr. No. | References | Size (μm) | Shell Thickness (μm) | Mean Diameter (μm) | Wall Thickness (μm) |
|---|---|---|---|---|---|
| 1 | Yoriya et al. 2020 [16] | 45 - 220 | ~0.7 - 21.6 | 25.9 - 229.2 | 0.7 - 21.6 |
| 2 | Petrus et al. 2020 [51] | 2 - 200* | 2 - 30* | 1 - 10 | NA |
| 3 | Yoriya et al. 2019 [4] | 5 - 500 | 2 - 30* | < 100 | 2–30 |
| 4 | Sarkar et al. 2007 [33] | 20.72 - | 2 - 30* | 112.56 | 2.072 - |
| 5 | Beddu et al. 2018 [17] | 6.91 - 149.6 | 2 - 30* | 6.91 - 149.6 | 0.691 - |
| 6 | Ranjbar and Kuenzel 2017 [19] | 5 - 500* | 1–18 | 20 - 300 | 1–18 |
| 7 | Liu et al. 2017 [42] | 10 - 200 | 2 - 30* | 10 - 200 | 0.1 - 20 |
| 8 | Zyrkowski et al. 2016 [16] | 5 - 500* | 10 - 100 | 20 - 80 | < 25 |
| 9 | Fomenko et al. 2015 [11] | 10 - 200 | 5–35 | 47 - 225 | 2.4 - 12.7 |
| 10 | Xu et al. 2015 [9] | 5 - 500* | 2 - 30* | 94 | 0.5 - 50 |
| 11 | Kolay and Bhusal 2014 [10] | 2 - 250 | 10–20 | NA | 20 |
| 12 | Joseph et al. 2013 [14] | 5 - 500* | 1–20 | 0.8 - 80 | 1–20 |
| 13 | Li et al. 2012 [25] | 63 - 250 | 2–10 | 10–200 | 1–3 |
| 14 | Wang et al. 2012 [53] | 5 - 500* | 2 - 30 * | 10 -200 * | 0.5 - 50 |
| 15 | Chávez-Valdez et al. 2011 [48] | 17 | 2 - 30 * | 10 -200 * | 0.17 |
| 16 | Fomenko et al. 2011 [54] | 5 – 500* | 2.5 – 8.2 | 59 - 143 | 2 - 5 |
| 17 | Anshits et al. 2010 [49] | 120 - 325 | 1.5 - 16 | 50 - 350 | 2 - 16.2 |
| 18 | Ngu et al. 2007 [34] | 45 - 150 | 1 - 35 | 10 - 200* | 2 - 26 |
| 19 | Anshits et al. 2005 [55] | 500 | 3 - 5 | 10 - 200* | 50 |
| 20 | Vereshchagina et al. 2004 [56] | 10 - 500 | 2 - 30* | 10 - 200* | 1 - 50 |
| 21 | Vassilev et al. 2004 [52] | 20 - 600 | 2 - 30* | 10 - 200* | NA |
| 22 | Kolay and Singh 2001 [1] | 40 - 140 | 2 – 30* | 10 - 200 | 0.4 - 14 |

Note: * Approximate values are written. NA — Not Available.

The properties presented in Table 6 are evaluated by researchers using various techniques SEM, XRD, etc. A glimpse of the summary of properties of FAC indicates that the particle size of FAC falls in the range of nano size. The inertness of FAC coupled with its low density and nano-size makes it an ideal material for the replacement of fine aggregate from concrete. Another, the shell thickness, and the smooth surface of FAC make it an ideal candidate for use as a roller bearing in concrete improves the workability.

## 4 Application of FAC

One of the objectives of the study is to embark upon the application of FAC in replacing cement and aggregate from concrete. There have been several studies that have investigated the effect of the addition of FAC in cement concrete as a replacement of fine aggregate/cement [40, 57-59]. The major advantage of replacing fine aggregate/cement is the reduction in the unit weight of concrete without compromising the strength of concrete. Many research attempts are directed towards the replacement of fine aggregate/ cement in concrete by FAC mainly because of the reduction in the unit weight of concrete without compromising the strength.

Another important characteristic of FAC is its inert nature and nanoparticle size making it an ideal replacement of fine aggregate/cement. Therefore, a number of research reports are available on the effect of replacement of fine aggregate/cement from concrete. Few researchers have also tried other applications of FAC such as synthetic in foam and composite material.

This section deals with the application of FAC in the following areas;

4.1 Replacement of cement
4.2 Replacement of fine aggregate
4.3 Uses as a composite material

*4.1 Replacement of cement by FAC from Concrete:*

The FAC originates from the fly ash as described in the preceding paragraph. It is, therefore, obvious that the FAC possesses that properties like fly ash. The FAC possesses pozzolanic and cementitious properties of fly ash which are very useful for the replacement of cement from concrete. Other properties such as bulk density, particle size, and surface area help in accelerating the hydration process of cement. The hollow FAC particles carry water which helps in the internal curing of concrete thereby reducing shrinkage cracks in concrete [18].

Author utilized FAC in combination with silica fumes as a replacement of cement in mortar and reported that the mechanical strength of mortar reduced with an increase in replacement percentage of FAC [40, 58]. However, the authors have not reported the optimum percentage of replacement of FAC where the mechanical strength will not fall below the desired value. While some authors [38, 40] reported that the loss in strength due to the addition of FAC can be compensated by using silica fume. One of the important draws back of the addition of FAC to cement mortar for plaster is that it increases the porosity which may increase seepage.

Author investigate and reported that though the addition of FAC may improve the strength of cement paste initially, over the period the strength decreases [39]. The increased pozzolanic activity due to the addition of FAC plays an important role in the reduction of strength over the period. They have reported that 10% replacement of FAC by weight is an optimum replacement of FAC. The increase in porosity is reported by the authors due to the addition of FAC. The higher percentage of FAC addition adversely affected the durability of concrete.

Here author reported that the FAC is an ideal material for the replacement of cement to some extent [60]. However, the increased proportion of FAC reported the loss of strength of concrete. Author [37] experimented with a filling of gaps using microfilters in concrete using FAC. They reported that when the optimum gaps are filled with FAC paste, the thermal conductivity of the concrete sample improves with a simultaneous reduction in the mechanical strength of concrete. They further reported that the acoustic behaviour of concrete with FAC similar to that without FAC.

This work suggested the addition of Silica fume to improve the performance of concrete and reported compressive strength, tensile strength, flexural strength, fracture mechanism, and toughness of concrete resulting from the addition of FAC. Some investigations [61] concluded that the use of the FAC as a nanofiller increased the porosity because of their spherical nature [38]. Also, it seems that the addition of the FAC in cement paste reduces the density of concrete and makes it lightweight concrete. The lightweight concrete made up of FAC will help for improving the thermal conductivity and be used as a noise barrier [37].

*4.2 Replacement of fine aggregate by FAC from Concrete*

Many authors extensively reported the use of FAC as a partial replacement of fine aggregate. This has two objectives first as a nanofiller and second as a replacement of fine aggregate. The particle size distribution of fine aggregate and FAC is presented in figure 4.

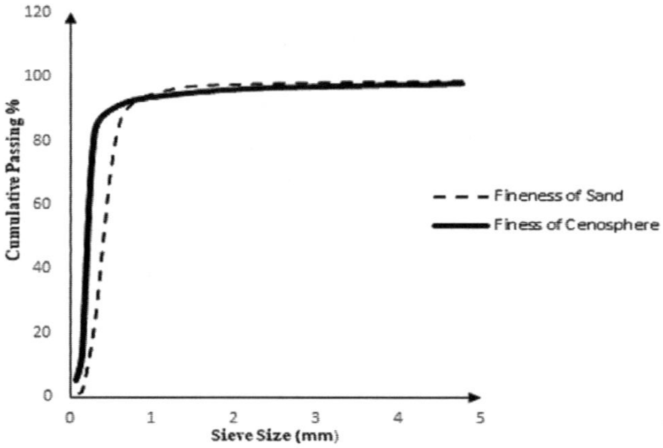


**Figure 4.** Fineness modulus of fine aggregate and FAC.

The particle size distribution of FAC and fine aggregate indicate that both have similar particle size distribution hence can be used as a replacement of each other. The effect of replacement of fine aggregate by FAC on concrete is studied by many researchers and have reported that the replacement affects the following properties of concrete.

1. Mechanical strength.
2. Workability.
3. Bulk density.
4. Surface Finish.
5. Porosity.
6. Thermal Conductivity.

Some investigations conducted by authors [46, 58] reported that the mechanical strength of concrete remains at the desired level even after the addition of FAC. However, the volume of pores in concrete has been found to increase due to the addition of FAC. This may adversely affect the permeability of concrete. Research [62] reported that the addition of FAC in any form in concrete adversely affects the hydration hence the addition of FAC to the tune of 5 to 10% is only advisable in concrete. Author [63] reported that though the addition of FCA as replacement of prime aggregate initially increased the mechanical strength, the further addition of FAC reduces the mechanical strength due to poor inter-particle bonding between cement and FAC. They suggested only a low volume of FAC replacement in concrete.

Investigation by author [58] reported various replacement proportions such as 33%, 67%, and 100 % of fine aggregate by FAC. Results of concrete produced after the replacement of fine aggregate by FAC indicate that the compressive strength decreased as the percentage replacement of FAC was increased [58]. Results of a similar experiment are reported by some other authors Satpathy et al. 2019 wherein they combined the use of FAC with Sintered Fly Ash (SFA) as a replacement of fine aggregate. Satpathy et al. 2019 reported that the optimum proportion of FAC and SFA is 50% and 75% respectively for achieving desired mechanical properties [57]. Author [64] reported that the addition of micro-size FAC in lightweight concrete reduced the ductility and thermal conductivity of concrete.

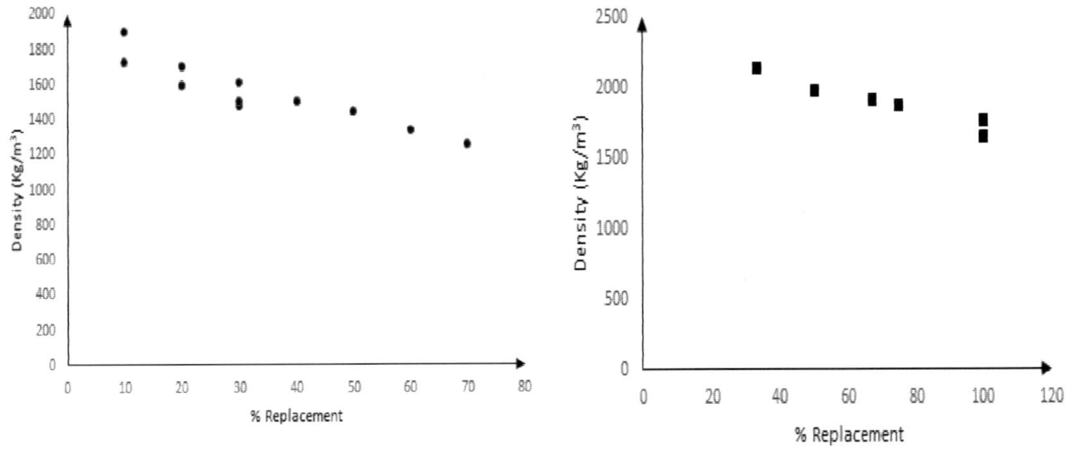


(a) (b)

**Figure 5.** Variation in density with % replacement of (a) cement (b) fine aggregate with FAC.

Figure 5 and figure 6 presents the variation in compressive strength with percentage replacement of fine aggregate and cement by FAC reported by various authors. It is seen from the figure 5 that the compressive strength and density reduce with an increase in the percentage of replacement. This is reported by almost all researchers. However, no author has reported the globally optimize percentage replacement of fine aggregate/ cement with FAC which will give desired mechanical strength and bulk density. However, a higher proportion of addition of FAC decreases the strength and density on one hand and increases the porosity (seepage) on the other hand. This needs to be investigated in detail.

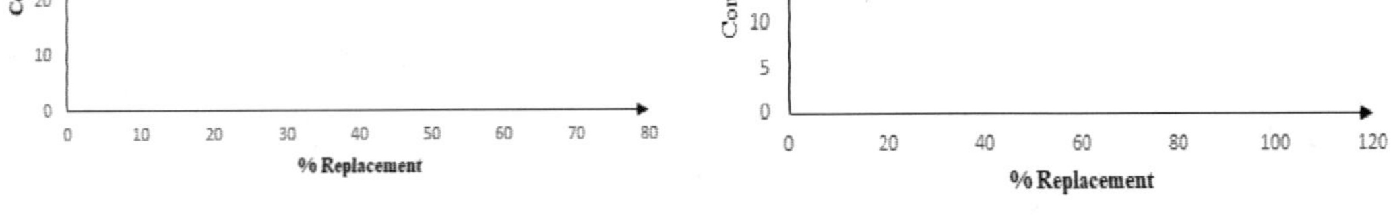


(a) (b)

**Figure 6.** Variation in compressive strength with % replacement of (a) cement (b) fine aggregate with FAC.

The FAC can also be used in many other forms in composite materials. The properties of FAC such as particle size, spherical shape, inertness, etc make it an ideal material to be used as a composite material. Various authors [65, 66] reported the use of FAC with polyethylene and zinc. Author [32] reported geo-polymeric use as metakaolin (MK)-based slurry. Few authors reported the research on silane treatment of FAC to be incorporated in thermoplastic high-density polyethylene. Some literature studied the effect of the FAC coated with nickel and ZSM-5 to form magnetic ZSM-5/Ni/Fly [66] ash hollow spheres and characterized by SEM, XRD, nitrogen adsorption, and vibrating sample magnetometer.

The summary of research efforts reporting the replacement of cement and fine aggregate by FAC is presented in Table 7.

**Table 7.** Summary of Results of Replacement of Fine Aggregates and Cement by FAC.

| Sr. No. | Reference | % Replacement | W/C Ratio | Density Kg/m$^3$ | Strength (N/mm$^2$) | | |
|---|---|---|---|---|---|---|---|
| | | | | | Compressive | Flexural | Split Tensile |
| 1 | de Souza et al. 2019 [58] | 33 | 0.2 | 2134 | 40.5 | - | - |
| | | 67 | | 1912 | 46.1 | | |
| | | 100 | | 1650 | 45.8 | | |
| 2 | Chen et al. 2019 [43] | 0 | 0.32 | 1870 | 58.87 | 10.80 | 31.49 |
| | | 10 | | 1724 | , 56.35 | 16.00 | 32.69 |
| | | 20 | | 1592 | 53.86 | 13.61 | 33.83 |
| | | 30 | | 1477 | 50.06 | 11.59 | 33.89 |
| 3 | Satpathy et al. 2019 [57] | 0 | NA | 2200 | 34.87 | 4.62 | 3.58 |
| | | 50 | | 1980 | 18.75 | 3.38 | 2.68 |
| | | 75 | | 1867 | 20.47 | 3.73 | 2.86 |
| | | 100 | | 1769 | 32.04 | 4.27 | 3.48 |
| 4 | Hanif et al. 2017a [39] | 30 | 0.3 | 1612 | 95.02 | 9.29 | 3.507 |
| | | 40 | | 1506 | 48.48 | 8.61 | 2.912 |
| | | 50 | | 1444 | 43.50 | 7.37 | 2.761 |
| | | 60 | | 1342 | 34.90 | 5.87 | 1.912 |
| | | 70 | | 1260 | 30.38 | 5.38 | 1.656 |
| 5 | Hanif et al. 2017 [67] | 10 | 0.3 | 1900 | 75 | - | - |
| | | 20 | | 1700 | 60 | | |
| | | 30 | | 1500 | 30 | | |
| 6 | Kannan et al. 2016 [40] | 5 | - | - | 55 | - | - |
| | | 10 | | | 54 | | |
| | | 15 | | | 52 | | |
| | | 20 | | | 48 | | |
| | | 25 | | | 47 | | |
| | | 30 | | | 46 | | |
| 7 | Patel et al. 2019 [46] | 50 | - | - | 22.66 | - | - |
| | | 75 | | | 24.44 | | |
| | | 100 | | | 17.78 | | |

Results show that the sample contains a hollow structure with a magnetic property [66]. Author [68] investigated the effect of the addition of FAC to concrete on acoustic properties of concrete. They reported that due to the reduction in density of concrete the acoustic properties of concrete improved considerably. Author [69] reported that the fracture toughness of concrete increases by 200% due to the addition of FAC.

**Table 8.** Presents the summary of use of FAC as a composite material.

| Sr. No. | Author | % Addition of FAC | Observation Reported |
|---|---|---|---|
| 1 | Venkatesha et al.2020 [70] | 0.5, 1, 1.5 ,2 | Flexural strength and internal laminar shear strength increased with 1.5% FAC composite and constant 2% bamboo and glass hybrid epoxy. |
| 2 | Umashankar et al. 2018 [71] | 20, 30, 40, 50 | Optimum mix obtained at 30%, at this mix tensile strength, flexural strength, and impact strength recorded maximum. |
| 3 | Ashoka et al. 2018 [72] | 3, 6, 9 | Modulus of elasticity, yield stress, and hardness increases with an increase in FAC but a decrease in the elongation |
| 4 | Rugele et al. 2017 [73] | 10, 30, 50, 60 | the composite material shows an increase in strength with the increase in firing temperature, and a decrease of mechanical reliability with a decrease in density, which is typical for porous materials. |
| 5 | Gupta and Kua 2020 [74] | 10, 20, 30, 40 | Design matrix mix of rice husk biochar, silica fume, and FAC 10 – 30% best mix as a replacement for cement. |
| 6 | Huang 2019 [43] | 0.1, 0.2, 0.3 | Lightweight toughness cement-based composites containing 20% FAC and 1% fibre volume have the best toughness and ductility. |

Table 8 indicates that the FAC can also be used in composite material successfully due to its unique characteristics such as lightweight, smooth surface finish, nanoparticle size, etc. Its inert nature contributes to its use as a composite material.

*4.3 Miscellaneous*

The FAC is utilized in many other ways such as phase change material [42], internal curing agent [18], nanofillers [75], etc. However, the distinguished properties of FAC make it the most suitable material for the replacement of fine aggregate and cement.

## 5. Results and Discussion

A closer look at the literature, reveals some gaps and shortcomings listed below. There is still a great deal of work to be done in the application of FAC in cement concrete.

1. While extensively reviewing the literature, it was found that no author evaluated the effect of the addition of FAC on the workability of concrete. Various tests to evaluate the workability of concrete, such as compaction factor, slump cone, and Vee — Bee can be used to access the effect of the addition of FAC on the workability of concrete.
2. The authors tried various percentages of replacement of FAC and have reported its effect on various properties of concrete. But there is no agreement on the optimum percentage of the addition of FAC to achieve optimum properties of concrete.
3. The wear resistance of concrete after the addition of the FAC has been found very rarely mentioned [65].
4. The dynamic compressive strength of concrete and the fracture toughness are important properties of concrete required for the surface durability of rigid pavement. These properties of concrete are not probed by the researchers. The permeability is an important characteristic of concrete. The researchers have not probed the effect of the addition of the FAC on the permeability of concrete.

## 6. Conclusions

The work described above is carried out to document various characteristics of the FAC reported by authors all over the world. It briefly accounts for the application of FAC such as a replacement material of fine aggregate and cement. This report concisely documents the contribution of authors world over in the study of FAC and its various applications.

Researchers have reported various characteristics of the FAC such as size, pore size, shell thickness, density, and chemical configuration of the FAC. The researchers have used XRD, XRF, SEM, and EDS to study various aspects of FAC. Following conclusions can be drawn from the work reported by the researcher's world over.

### *6.1 Characterization of FAC*

Characteristics in brief:

(i) The particle size of the FAC was found to vary from 10 μm to 500 μm.
(ii) The average shell thickness of the FAC is reported as 10 to 20 μm.
(iii) The unit weight of the FAC was found to vary from 0.65 to 0.75 $g/cm^3$.
(iv) The chemical composition of the FAC was reported as the sum of $Al_2O_3$, $SiO_2$, $Fe_2O_3$, CaO, MgO, $MgCl_2$, etc.
(v) The variation in characteristics of the FAC is mainly due to the origin of the FAC that is the origin of coal.

Though the researchers have reported different characteristics of FAC, there is wide variation in reporting the characteristics. Though this is attributed to the origin of coal fly ash, the use of different methods for characterization is also one of the reasons for the variation in reported characteristics. Hence it is required to decide a specific range for the characteristics which will encompass all the known sources of coal fly ash the world over.

### *6.2 Application of FAC*

The FAC is found to have a wide range of applications in the replacement of fine aggregate and cement. The mechanical strength and density of the concrete are found to decrease with an increase in the percentage of FAC addition. However, the optimum percentage of FAC addition needs to be ascertained for achieving the desired mechanical strength and bulk density.

## References

[1] Kolay P K and Singh D N 2001 *Cem. Concr. Res.* **31** 539
[2] Danish A and Mosaberpanah M A 2020 *J. Mater Sci.* **55** 4539
[3] Wrona J, Zukowski W, Bradlo D and Czupryński P 2020 *Energies* **13** 3576
[4] Yoriya S, Intana T and Tepsri P 2019 *Appl. Sci.* **9** 3792
[5] Sen S 2014 Physical properties of cenosphere, PhD diss. NIT Raulkela, 2014.
[6] Petrus H, Hirajima T, Oosako Y, Nonaka M, Sasaki K and Ando T 2011 *Int. J. Miner. Process.* **98** 15
[7] Acar I and Atalay M U 2016 *Fuel.* **180** 97
[8] Hirajima T, Petrus HTBM, Oosako Y, Nonaka M, Sasaki K and Ando T 2010 *Int. J. Miner Process.* **95** 18
[9] Xu B, Ma H and Hu C 2015 *Mater. and Struc.* **1** 8
[10] Kolay P K and Bhusal S 2014 *Fuel.* **117**118
[11] Fomenko E V et al. 2015 *Energy and Fuels.* **29** 5390
[12] Żyrkowski M J 2014 *Report of UG work* **1** 74
[13] Wang J Y, Chia K S, Liew J Y R and Zhang M H 2013 *Cem. Concr. Compos.* **43** 39
[14] Joseph K V, Francis F, Chacko J, Das P and Hebbar G 2013 *Int. J. Eng. Res. Technol.* **2** 18
[15] Li Y and Wu H 2012 *Energy and Fuels.* **26** 130
[16] Yoriya S and Tepsri P 2020 *Appl. Sci.* **10** 1

[17] Beddu S, Zainoodin M, Kamal N M, Mohamad D, Nabihah S and Nazri F M *Green Design and Manuf. Adv. and Emer. App. AIP Conf. Proc.* **2030** 020259-1
[18] Liu F, Wang J, Qian X and Hollingsworth J 2017 *Cem. Concr. Res.* **95** 39
[19] Ranjbar N and Kuenzel C 2017 *Fuel.* **207** 1
[20] Zyrkowski M, Neto R C, Santos L F andWitkowski K 2016 *Fuel.* **174** 49
[21] Luong D, Lehmhus D, Gupta N, Weise J and Bayoumi M 2016 *Materials (Basel).* **9** 115
[22] Chandel V and Bhatia O S 2016 *Int. J. Eng. Trends Technol.* **29** 133
[23] Yang B, Yang Y F and Gai G S 2013 *Adv. Mater. Res.* **826** 215
[24] Li Y, Gao X and Wu H 2013 *Energy Fuels.* **27** 8118151.
[25] Li Y 2012 *Curtin University* **1** 140
[26] Wang J Y, Zhang M H, Li W, Chia K S and Liew R J Y 2012 *Cem. Concr. Res.* **42** 721
[27] Wang J, Dong S, Wang D, Yu X, Han B and Ou J 2019 *J. Mater. Civ. Eng.* **31** 04019030
[28] Liu Z, Zhao K, Tang Y and Hu C 2019 *Adv. Mater. Sci. Eng.* **2019** 1
[29] Petrus H T B M, Hirajima T, Oosako Y, Nonaka M, Sasaki K and Ando T 2011 *Int. J. Miner. Process.* **98** 15
[30] Agrawal U S and Wanjari S P 2017 *Mater. Today Proc.* **4** 9797
[31] Adesina A 2020 *Dev Built Environ.* **4** 100029
[32] Wang M, Jia D, He P and Zhou Y 2011 *Ceram. Int.* **37** 1661
[33] Sarkar A, Rano R, Mishra K K and Mazumder 2015 *Taylor Fr.* **30** 271
[34] Ngu L N, Wu H and Zhang D K 2007 *Energy and Fuels.* **21** 3437
[35] Zhao Y, Zhang J, Tian C, Li H, Shao X and Zheng C 2010 *Energy and Fuels.* **24** 834
[36] Fomenko E V et al. 2016 *Solid Fuel Chem.* **50** 238
[37] Blanco F, Garciéa P, Mateos and P Ayala J 2000 *Cem. Concr. Res.* **30** 1715
[38] Hanif A, Parthasarathy P, Ma H, Fan T and Li Z 2017 *Cem. Concr. Compos.* **81** 35
[39] Hanif A, Lu Z, Diao S, Zeng X and Li Z 2017 *Constr. Build Mater.* **140** 139
[40] Kannan S K, Andal and Shanmugasundaram M 2016 *Adv. Mater. Sci. Eng.* **2016** 1
[41] Li C and Qiao X 2016 *Chem. Eng. J.* **302** 388
[42] Liu F, Wang J and Qian X 2017 *Cem. Concr. Compos.* **80** 317
[43] Chen W and Huang Z 2019 *Materials (Basel)* **12** 1
[44] Patel S K, Satpathy H P, Nayak A N and Mohanty C R 2020 *J. Inst. Eng. Ser. A.* **101** 179
[45] Xi B et al. 2020 *J. Clean Prod.* **262** 121274
[46] Patel S K, Majhi R K, Satpathy H P and Nayak A N 2019 *Constr. Build Mater.* **226** 579
[47] Zyrkowski M, Neto R C, Santos L F and Witkowski K 2016 *Fuel.* **174** 49
[48] Chávez-Valdez A, Arizmendi-Morquecho A, Vargas G, Almanza J M and Alvarez-Quintana J. 2011 *Acta Mater.* **59** 2556
[49] Anshits N N, Mikhailova O A, Salanov A N and Anshits AG 2010 *Fuel.* **89** 1849
[50] Sokol E and Volkova N 2000 F*uel Proc. Tech.* **67** 35
[51] Petrus H T B M et al. 2020 *J. Environ. Chem. Eng.* **8** 104116
[52] Vassilev S V, Menendez R, Diaz-somoano M and Martinez-tarazona M R 2004 *Fuel* **83** 585
[53] Wang B, Li Q, Wang W, Li Y and Zhai J 2011 *Appl. Surf. Sci.* **257** 3473
[54] Fomenko E V, Anshits N N, Pankova M V, Solovyov L A and Anshits A G 2011 *World Coal Ash Conf–May.* **2011** 9
[55] Anshits N N, Vereshchagina T A, Bayukov O A, Salanov A N and Anshits A G 2005 *Glass Phy. Chem.* **31** 306
[56] Vereshchagina T A et al. 2004 *Glass Phy. Chem.* **30** 247
[57] Satpathy H P, Patel S K and Nayak A N 2019 *Constr. Build Mater.* **202** 636
[58] De Souza F B, Montedo O R K and Grassi R L E G P A 2019 *Rev. Matéria* **24** 1
[59] Dhivya P, Minnalkodi G and Dhanalakshmi G 2018 *IRJET* **5** 2010
[60] Shukla S, Seal S, Akesson J, Oder R, Carter R and Rahman Z 2001 *App. Surf. Sci.* **181** 35
[61] Barbare N, Shukla A and Bose A 2003 *Cem. Concr. Res.* **33** 1681
[62] Pundiene I, Prantskevichene I, Kligis M and Kairite A G 2019 *Refract. Ind. Ceram.* **59** 482

[63] Shukla A, Bose A, Wayne Lee K and McBride S P 2001 *J. Mater. Scie.* **37** 4217
[64] Zhou H and Brooks A L 2012 *Constr. Build Mater.* **198** 512
[65] Chand N, Sharma P and Fahim M 2010 *Mater. Sci. Eng. A.* **527** 5873
[66] Tao H, Yao J, Zhang L and Xu N 2009 *Mater. Lett.* **63** 203
[67] Hanif A, Lu Z, Diao S, Zeng X and Li Z 2017 *Constr. Build Mater.* **140** 139
[68] Tiwari V, Shukla A and Bose A 2004 *App. Acou.* **65** 263
[69] Cardoso R J and Shukla A 2002 *J. Mater. Sci.* **37** 603
[70] Venkatesha B K and Saravanan R 2020 *Int. J. Veh. Struct. Syst.* **12** 447
[71] Umashankar LV, Sasikumar T, Raghu Vikram M V, Reddy P T and Reddy S M 2018 *Int. J. Civ. Eng. Technol.* **9** 246
[72] Ashoka E, Sharanaprabhu C M, Krishnaraja G K and Kudari SK 2018 *AIP Conf Proc.* **2018** 1943
[73] Rugele K, Lehmhus D, Hussainova I, Peculevica J, Lisnanskis M and Shishkin A 2017 *Materials (Basel).* **10** 1
[74] Gupta S and Kua H W 2020 *Constr Build Mater.* **253** 119083
[75] Hanif A, Lu Z and Li Z 2017 *Constr Build Mater.* **144** 37

# Fuzzy modeling of surface roughness during wet turning of AISI D3 tool steel

**Pawan Kumar[1] and P K Karsh[2]**

[1] Department of Mechanical Engineering, National Institute of Technology Kurukshetra, Thanesar 136119, India
[2] Department of Mechanical Engineering, Parul Institute of Engineering & Technology, Parul University

*Corresponding author's e-mail address: pradeepkarsh@gmail.com

**Abstract.** In this study, an attempt has been made to develop a Mamdani fuzzy inference system-based predictive model for predicting surface roughness during wet turning of AISI D3 tool steel. The influence of machining parameters on surface roughness is also investigated. The experimental runs were performed on a lathe using Taguchi's $L_9$ orthogonal array to know the effect of cutting parameters on machining surface roughness. The performance of the fuzzy model was also compared with the regression model. The fuzzy model's prediction efficiency (92.10%) was found better than the regression model (70.33%). Therefore, the resultant model so developed is found adequate for surface roughness prediction purposes within the parametric range.

## 1. Introduction

Tool steel is mainly used as working parts of a tool that should endure high cyclic load during its service and should be wear-resistant. Since the tool steel is primarily used at a higher temperature, its selection as tool material also must include high toughness and wear resistance. Plain carbon steel cannot satisfy all tooling problems; that's why alloying elements are added to plain carbon steel to enhance the mechanical properties. Achieving a better surface quality, tool life, and dimensional accuracy are the crucial concerns in turning hardened steels [1]. The cutting tool is the main critical element in hard turning, and an increase in wear resistance has characterized the development of newer cutting tool materials. Some work has been done on various hardened steels with CBN and PCBN cutting tools. According to recent research, few studies have been reported on the wet turning of AISI D3 steel. Ceramic tools are widely used in the manufacturing industry to machine various hard materials such as alloy steels, die steels, high-speed steels, bearing steels, white cast iron, and graphite cast iron [2-6]. Some studies have been carried out on machining parameters such as surface roughness, tool wear, productivity, and cutting force by employing ANOVA and other modeling and optimization techniques in hard turning [7-9]. Bouchelaghem et al. [10]applied response surface methodology (RSM) to determine the effect of machining variables on the output response using CBN inserts. In contrast, Varaprasad et al.[11] developed a model based on RSM based central composite design to predict the ceramic flank wear during machining of AISI D3 steel. Kribes et al.[12] carried out statistical analysis to find the effect of feed rate, cutting speed, and depth of cut on surface roughness during hard turning of 42CrMo4steel. Neseli et al.[13]used RSM to optimize the tool geometry of hard turning of AISI 1040 with P25 tool for minimum surface roughness.

In this paper, the fuzzy-based model is employed to model machining parameters of the wet turning of AISI D3 tool steel. The combined effects of feed, cutting speed, and depth of cut are determined on surface roughness. Analysis of variance (ANOVA) analysis is also performed to quantify the effect of machining parameters. First, a linear regression model is presented because of its simplicity in predicting the response. Furthermore, the artificial intelligence-based predicted model is proposed using fuzzy logic to attain a remarkable improvement of surface roughness prediction. In this context, very few research studies have presented such a prediction model during the machining of D3 tool steel.

## 2. Experimental Procedure

Taguchi's $L_9$ OA is utilized for experimentation considering three input cutting parameters viz. cutting speed (s), feed (f), and depth of cut (d). Turning of D3 tool steel was performed under wet conditions. The levels of input parameters considered in this study are given in Table1.

**Table 1.** Input parameter levels.

| Parameters | Units | Parameter levels | | |
|---|---|---|---|---|
| **Cutting speed (s)** | rpm | 192 | 325 | 420 |
| **Feed (f)** | mm/rev | 0.05 | 0.1 | 0.2 |
| **Depth of cut (d)** | mm | 0.2 | 0.6 | 1 |

## 3. Machining process

In the present study, D3 steel was chosen as workpiece material. As per the experimental plan (i.e., $L_9$ orthogonal array), experiments were carried out on NH22 lathe (*Make: HMT*) with CCMT 060208 (*Make: SANDVIK–COROMANT*) inserts, as shown in figure 1. These inserts are having a 0.8 mm nose radius. The METACOOL (*SAE 40 grade*) cutting fluid is also used to assist the machining.

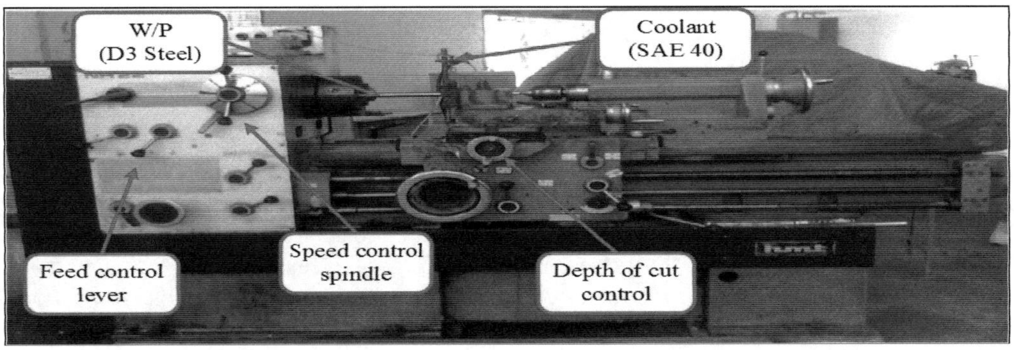


**Figure1.** Lathe Machine and its components.

The machined surface's surface roughness is measured using the SJ301 surface roughness tester (Make: Mitutoyo). These results constituted an experimental dataset. This dataset is utilized for regression analysis and analysis of the proposed fuzzy inference system (FIS) model. The correlation between surface roughness and machining parameter was investigated using the model analysis results. For the data analysis, a statistical significance level of 0.05 is selected.

**Table 2.** Experimental results.

| Exp. Trial Runs | Cutting parameter | | | Response parameter Surface Roughness (μm) | | | Residuals | |
|---|---|---|---|---|---|---|---|---|
| | Cutting speed s (rpm) | Feed f (mm/rev) | Depth of cut d (mm) | Experimental | Regression Predicted | Fuzzy Predicted | Regression | Fuzzy |
| **1** | 420 | 0.20 | 1.0 | 4.638 | 4.417 | 4.228 | 0.220 | 0.409 |
| **2** | 420 | 0.10 | 0.6 | 2.798 | 2.990 | 2.330 | -0.192 | 0.467 |
| **3** | 420 | 0.05 | 0.2 | 2.376 | 2.025 | 2.293 | 0.351 | 0.083 |
| **4** | 325 | 0.20 | 0.6 | 3.910 | 3.731 | 3.261 | 0.179 | 0.649 |
| **5** | 325 | 0.10 | 0.2 | 2.451 | 2.304 | 2.330 | 0.147 | 0.120 |
| **6** | 325 | 0.05 | 1.0 | 1.883 | 2.851 | 2.300 | -0.968 | - |
| **7** | 192 | 0.20 | 0.2 | 2.374 | 2.972 | 2.293 | -0.598 | 0.081 |
| **8** | 192 | 0.10 | 1.0 | 3.704 | 3.057 | 3.261 | 0.647 | 0.443 |
| **9** | 192 | 0.05 | 0.6 | 2.311 | 2.091 | 2.293 | 0.219 | 0.018 |
| | | | | MAPE | | | 15.61 | 9.88 |

## 4. Modeling preliminaries

The regression technique correlates the controllable and response variables to found the relationship between these variables. A mathematical expression is formulated between these variables as described by equation 1.

$$y = \beta_0 + \beta_1 x \quad (1)$$

Where y is the response parameter, x is the predictor variable and $\beta_0$ and $\beta_1$ denotes coefficients. Regression models are primitive, yet they solved numerous problems. But sometimes, they lack to capture to nonlinearity present in the system. With the advancement in technology, several new and advanced methods are emerging to solve this problem. A fuzzy logic method is an artificial intelligence based method that can model the system's nonlinearity.

## 5. Development of Prediction Models

From the ANOVA study of the experimentally generated dataset, it is found that the surface roughness is directly proportional to feed (s) and depth of cut (d). Figure 2 is depicting the increase in surface roughness with increasing levels of feed and depth of cut. This increase in surface roughness values is due to increased chip cross-sectional area, which increases with the increasing feed (f) and depth of cut (d). This increase in the cross-sectional area raises friction in the cutting zone, resulting in low surface quality. The effect of feed can also be easily understood by its theoretical definition. The cutting speed also influenced the surface roughness at its higher level to some extent. It can be attributed to an increase in tool wear with increasing cutting speed.

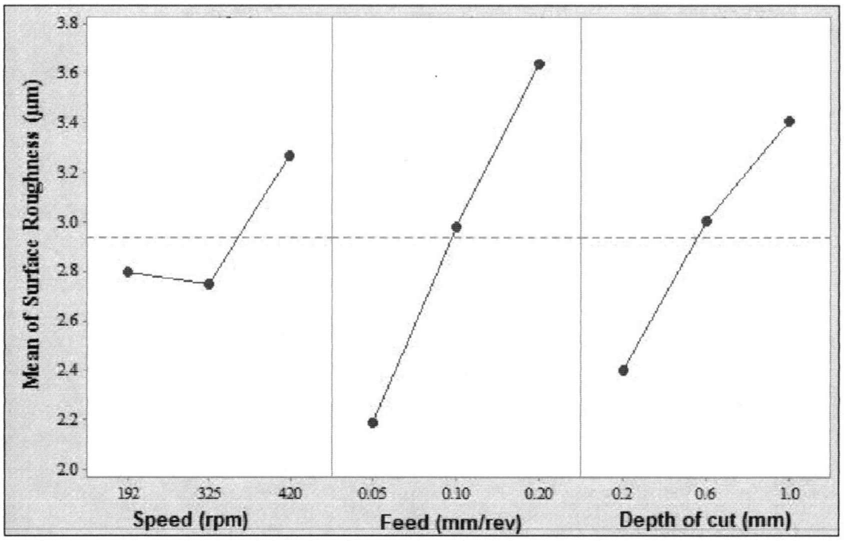


**Figure 2.** Main effect plot for surface roughness.

From the statistical analysis, it was also found that there is a strong relationship between response parameter with feed (f) and depth of cut (d). Still, it will be interesting to know the contribution of machining parameters to the response parameter. Therefore ANOVA analysis is performed using statistical software (MINITAB). The results of ANOVA analysis are shown in Table 3.

**Table 3.** Results of analysis of variance.

| Source | dof | Sum of Squares | Mean Square | F Value | p-value Prob> F | Contribution (%) |
|---|---|---|---|---|---|---|
| **Cutting Speed (s)** | 1 | 0.2912 | 0.2912 | 0.72 | 0.435 | 04.27 |
| **Feed (f)** | 1 | 2.9799 | 2.9799 | 7.36 | 0.042 | 43.70 |
| **Depth of Cut (d)** | 1 | 1.5241 | 1.5241 | 3.77 | 0.11 | 22.35 |
| **Residual** | 5 | 2.0232 | 0.4046 | | | 29.68 |
| **Total** | 8 | 6.8184 | | | | |

Also, an attempt has been made to develop a predictive model using machining dataset by regression analysis. Such models seek to predict the response variables for the predefined parametric space without conducting experiments. Hence, these developed models help for desired parameter selection with the least utilization of resources. The developed regression model to predict the surface roughness is shown in equation (2).

$$SR\ (\mu m) = 0.505 + 0.00192 \times s + 9.23 \times f + 1.26 \times d \tag{2}$$

An artificial intelligence-based fuzzy model is further developed to explore the system's nonlinearity understudy against the linear regression model. The fuzzy model uses fuzzy rules to define the relationship

IOP Conf. Series: Materials Science and Engineering 1120 (2021) 012026 doi:10.1088/1757-899X/1120/1/012026

between input and output process variables. These fuzzy rules use linguistic variables and consist of general statements [14]. The general structure of IF-THEN rules as follows:

Rule 1: If $s=a_1$, $f=b_1$, $d=c_1$ then $SR=e_1$; else

Rule 2: If $s=a_2$, $f=b_2$, $d=c_2$ then $SR=e_2$; else

.

.

.

.

Rule n: If $s=a_n$, $f=b_n$, $d=c_n$, then $SR=e_n$

The various input parameters were weighed to formulate fuzzy rules according to their respective association to the degree of membership. The input and output variables are fuzzified and represented by means of membership functions (MF), as shown in figure 3. After that Mamdani based fuzzy inference system (FIS) was developed using various IF-THEN rules. The output of FIS system was eventually defuzified using the centroid of area method. The fuzzy model prediction results along with the experimental results and the regression model prediction are shown in Table 2.

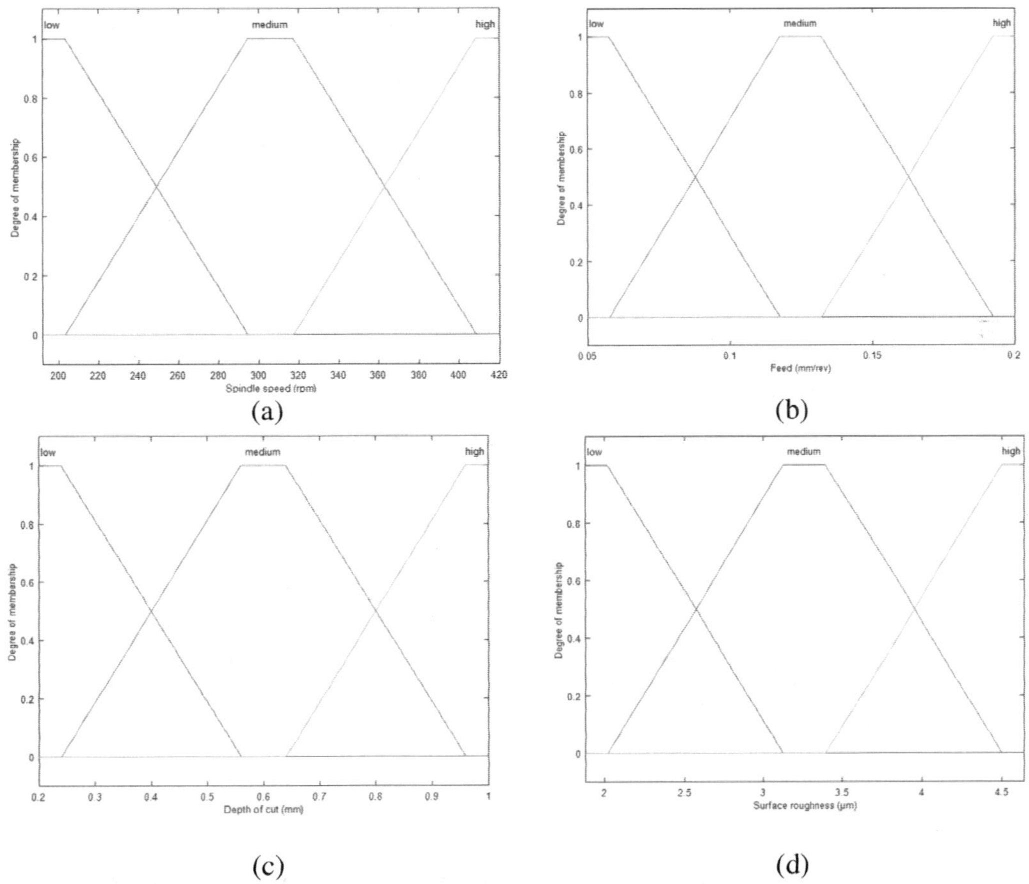


(a) (b) (c) (d)

**Figure 3.** Fuzzy modeling: (a-c) machining parameters and (d) response parameter.

The prediction efficiency of both models is also compared, as depicted in figure 4. It can be observed that the fuzzy model predicted values are along the trend line (below line). The fuzzy model has 92.10% prediction efficiency for the present machining process to predict the surface roughness. In contrast, the prediction efficiency of the regression model was found 70.33%.

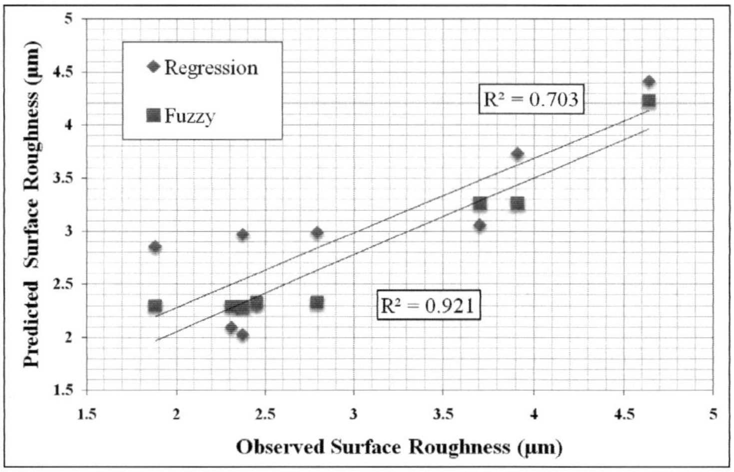


**Figure 4.** Performance plot for developed models.

Figure 5 is showing the plots of residuals for the predictive models. As the fuzzy model has less variation than the regression model, it is better for the prediction purpose.

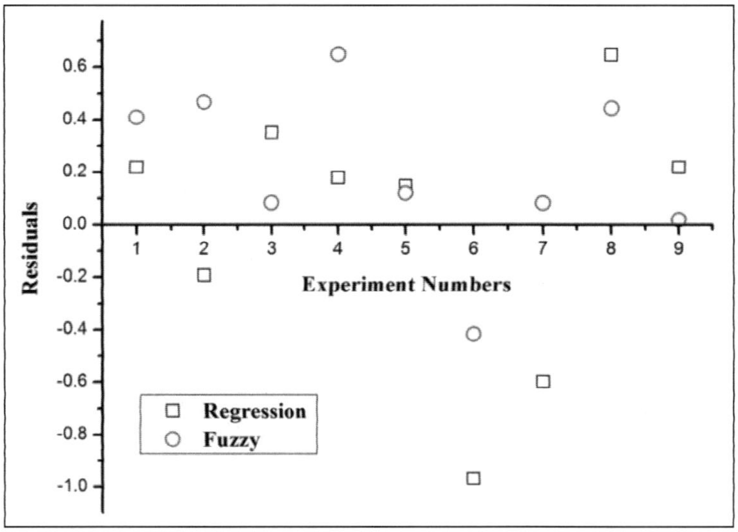


**Figure 5.** Residuals plots.

The model predicted values are also compared with experimental outcomes for easy comparison of both models. Figure 6 is depicting the comparison of both the models to predict the response parameter. The fuzzy model predicted surface roughness close to the observed values. The fuzzy model is found to have a mean absolute error (MAPE) of 9.88% only, which is permissible under the statistical limit and representing the robustness of the predictive model.

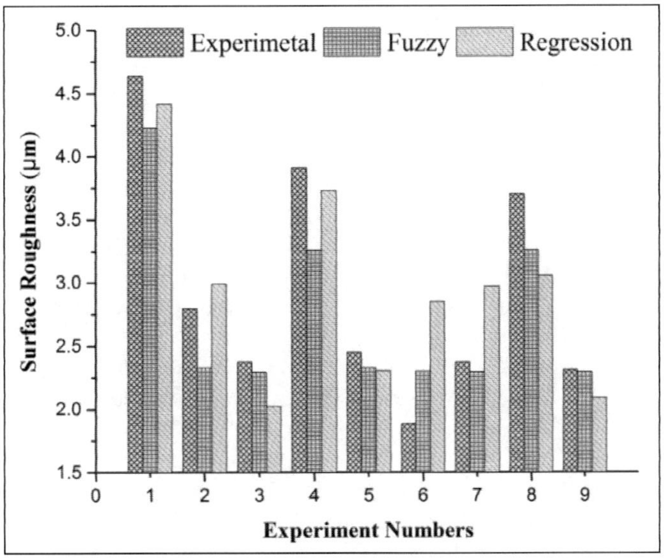


**Figure 6.** Actual Vs predicted results.

As the present study's fuzzy model is found superior to the regression model, the response plot is also developed for the whole design space using this model. Figure 7 is depicting the response plot for easy parameter selection for machining of D3 tool steel.

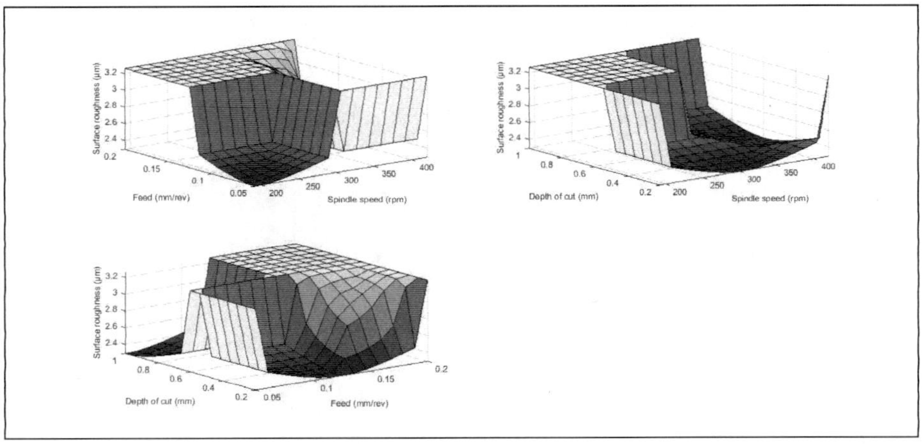


**Figure 7.** Response plots for surface roughness.

The prediction of machining characteristics helps in machining parameter selection for the quality improvement of the product. Therefore, an attempt is made to improve the prognosis of the machined part's surface quality in this study through the model study.

## 6. Conclusion

The experimental work and data analysis found that feed (f) and depth of cut (d) have a substantial influence on the surface roughness during the wet turning of D3 tool steel. They cumulatively contribute more than half (66.05%) to the surface roughness. To obtain the low surface roughness, lower level of feed (f), and depth of cut (d)) is recommended, along with a middle level of cutting speed (s) for the present study. The accuracy and adequacy of developed models are also checked, and the fuzzy model is found adequate for prediction purposes with an accuracy of 92.10%.

## References

[1] Mohammadi A and Zarepour H 2010 *Advanced Materials Research* **83-86** 303
[2] Lalwani D, Mehta N and Jain P 2008 *Journal of materials processing technology* **206** 167
[3] Suresh R, Basavarajappa S and Samuel G 2012 *Measurement* **45** 1872
[4] Fnides B, Yallese M A, Mabrouki T and Rigal J F 2011 *Sadhana* **36** 109
[5] Azizi M W, Belhadi S, Yallese MA, Mabrouki T and Rigal J F 2012 *Journal of mechanical science and technology* **26** 4105
[6] Hessainia Z, Belbah A, Yallese M A, Mabrouki T and Rigal J F 2013 *Measurement* **46** 1671
[7] Bouzid L, Yallese MA, Chaoui K, Mabrouki T and Boulanouar L 2015 Mathematical modeling for turning on AISI 420 stainless steel using surface response methodology. *Proceedings of the Institution of Mechanical Engineers, Part B: Journal of Engineering Manufacture* **229** 45
[8] Aouici H, Yallese MA, Belbah A, Ameur M F and Elbah M 2013 *Sadhana* **38** 429
[9] Fnides B and Yallese M 2008*Mechanics* **70** 73
[10] Bouchelaghem H, Yallese M A, Mabrouki T, Amirat A and Rigal J F 2010 *Machining Science and Technology* **14** 471
[11] Bh V, Ch S Rand Vinay P 2014 *Procedia Engineering* **97** 338
[12] Nabil K, Zahia H, Yalles M A and Ouelaa N 2012 *Mechanics* **18** 605
[13] Neşeli S, Yaldız S and Türkeş E 2011 *Measurement* **44** 580
[14] Zadeh L A 1979 *Advances in fuzzy set theory and applications* **11** 3

IOP Conf. Series: Materials Science and Engineering 1120 (2021) 012027 doi:10.1088/1757-899X/1120/1/012027

# Acoustical characterization of nickel aluminide ($Ni_3Al$) at low temperatures

**Krishna Kumar Pandey[1*], Vikas Lahariya[2], Balgovind Tiwari[3] and Abhilasha Singh Mathuriya[4]**

[1] Department of Physics, School of Basic Sciences and Research, Sharda University, Greater Noida-201310, India
[2] Amity School of Applied Sciences, Amity University Haryana, Gurgaon, Haryana-122413, India
[3] Dept. of Physics, IIIT-RKValley, RGUKT, A.P., India
[4] Department of Life Sciences, School of Basic Sciences and Research, Sharda University, Greater Noida-201310, India

*Corresponding author's e-mail address: krishnakumar.pandey2@sharda.ac.in

**Abstract.** Present research work deals with the acoustical characterization of Nickel Aluminide ($Ni_3Al$) at low temperature region from 10K to 50K. Particularly, the Ultrasonic waves have been used as a Non-destructive testing (NDT) tool to characterize this $Ni_3Al$. For this purpose, in our technique, the two parameters such as the nearest-neighbour distance and the hardness parameter is taken as two input parameters. Ultrasonic attenuation due to the cause of e-p (electron-phonon) interaction has been evaluated from 10K to 50K, for longitudinal as well as shear waves along <100> crystallographic direction and the elastic constants (second order) for this intermetallic alloy of nickel and aluminium is also evaluated in low temperature region and the obtained results have been discussed.

## 1. Introduction

$Ni_3Al$ is an alloy consisting of Nickel and Aluminium both and it has the properties similar to ceramic and metal both. The different properties of $Ni_3Al$ ($L1_2$ phase) such as structural, mechanical and elastic properties have been investigated and discussed in recent years. In the literature study several data related with the mechanical properties are available for this alloy, but especially the data related with the ultrasonic characterization at low temperature region is not available. Therefore the present investigation is much required in which for characterizing this alloy acoustically, we have used the ultrasonic waves and the ultrasonic attenuation studies have been done in low temperature region i.e. due to the electron-phonon interaction.

There are several uses of different alloys which are Nickel based such as they are used as in aircraft engines, in turbine disks. The overview of different applications of $Ni_3Al$ based intermetallic alloys is discussed in detail by Jozwik et al [1]. Also in the ultrasonic attenuation studies in intermetallics is done [2].

IOP Conf. Series: Materials Science and Engineering 1120 (2021) 012027 doi:10.1088/1757-899X/1120/1/012027

## 2. Theory

A review of these [3] studies shows that the ultrasonic attenuation varies from substance to substance in crystallographic directions and also according to the temperature regions studied. By taking the two input parameters such as the nearest-neighbour distance and hardness parameter, for the $Ni_3Al$ Alloy having FCC structure (As per the already available data of X-ray diffraction analysis, $Ni_3Al$ retains FCC structure) and by using the detailed concept of electrostatic and short-range repulsive potentials, the different elastic contants (Second Order) at 0K viz. $C_{ij}^0$ have been obtained by following the Brugger's definition [4]. By using the above concepts of potentials and taking interactions upto next nearest neighbours, the expressions for Second and third order elastic constants at any temperature, T are given as:

$$C_{ij}(T) = C_{ij}^{0} + C_{ij}^{vib.} \quad (1)$$

& $$C_{ijk}(T) = C_{ijk}^{0} + C_{ijk}^{vib.} \quad (2)$$

The expressions for $C_{ij}$'s and $C_{ijk}$'s have already been disscussed [5, 6]. As per the basic concepts it is already well known that at low temperature in metals, the mean free path of a conduction electron becomes comparable to the wavelength of an acoustical phonon. The coupling of the conduction electron with a propagating ultrasonic wave causes dissipation of the energy of the wave, and thus a viscous loss occurs. The attenuation caused by the energy loss due to the compressional and shear viscosities of the lattice [7], at low temperatures are

$$\left(\frac{\alpha}{f^2}\right)_{Longitudinal} = \frac{2\pi^2}{dv_l^3}\left(\frac{4}{3}\eta_e + \chi_e\right) \quad (3)$$

and $$\left(\frac{\alpha}{f^2}\right)_{Shear} = \frac{2\pi^2}{dv_s^3}\eta_e \quad (4)$$

In the above expressions, "d" is the density of the crystal, "f" is the ultrasonic frequency, $\eta_e$ is the transverse viscosity, $\chi_e$ is the compressional viscosity and this $\chi_e$ is zero as explained by Bommel [8]. The detailed behaviour of acoustical Phonons in Metals in low temperature region is investigated by S K Kor et al [9,10]. Also at any temperature, the resulting elastic constant is obtained by an addition of the vibrational energy contribution to the second order elastic constants SOEC at absolute zero and by using the anharmonic theory for lattice dynamics [11-13]. The expression for $v_l$ and $v_s$ (longitudinal and shear wave velocities) are as follows:

$$v_l = \sqrt{\frac{C_{11}}{d}} \quad (5)$$

and $$v_s = \sqrt{\frac{C_{44}}{d}} \quad (6)$$

The viscosity of an electron gas is given by

$$\eta_e = \frac{Nme\bar{l}\bar{v}}{3} \quad (7)$$

Where N is the electron density, m is its mass, $\bar{l}$ and $\bar{v}$ are mean free path and mean velocity of electron.

Where from the theory of electron gas [14], by substituting the values of $\bar{l}$ and $\bar{v}$ in equation **7**, $\eta_e$ becomes :

$$\eta_e = \frac{9x10^{11}\hbar^2}{5e^2R}\left(3\pi^2N\right)^{2/3} \quad (8)$$

Where R is electrical resistivity and N is free electron density.

## 3. Results and Discussions

The obtained results of the elastic constants (second order) for $Ni_3Al$ alloy in the low temperature region i.e. from 10K to 50K is presented in table 1. According to which our evaluated value at 10K is $C_{11}$ = 27.94227 x $10^{11}$ dynes/$cm^2$. Since no specific results are available for such an alloy in the said temperature region, So we have compared our result with the available results for other intermetallic alloys [2] and we have found that the orders of the results are the same. The "ultrasonic attenuation coefficient" due to e-p interaction for both the modes of propagation i.e. the "longitudinal as well as shear" from 10K to 50K i.e. $\left(\frac{\alpha}{f^2}\right)$ is evaluated by the help of the equations (3) to (8) and the same is as presented in figure 1. From this figure we can see that as usual this ultrasonic coefficient decrease with temperature but in the case of such a specific alloy Ni3Al, this rate of decrease is very much small as compared to the other intermetallic alloys [2]. This nature of decrease in attenuation coefficient can be attributed to the inherent nature of this alloy [1]. The temperature variation of ultrasonic wave velocities (longitudinal and shear waves) is as presented in figure 2. This result of wave velocity is as expected [9]. So, we can conclude that the approach which is used in this research paper is correct and most easy and helpful to the different researchers.

**Table 1.** Second Order elastic constants ($10^{11}$dynes/$cm^2$) for $Ni_3Al$ from 10K to 50K.

| Temp (K) | $C_{11}$ | $C_{12}$ | $C_{44}$ |
|---|---|---|---|
| 10 | 27.94227 | 4.709194 | 4.736072 |
| 15 | 27.94226 | 4.709173 | 4.736073 |
| 20 | 27.94222 | 4.708981 | 4.736074 |
| 25 | 27.94219 | 4.708376 | 4.736078 |
| 30 | 27.94236 | 4.707228 | 4.736088 |
| 35 | 27.94292 | 4.705556 | 4.736105 |
| 40 | 27.94402 | 4.703449 | 4.736129 |
| 45 | 27.94547 | 4.701007 | 4.736160 |
| 50 | 27.94797 | 4.698315 | 4.736197 |

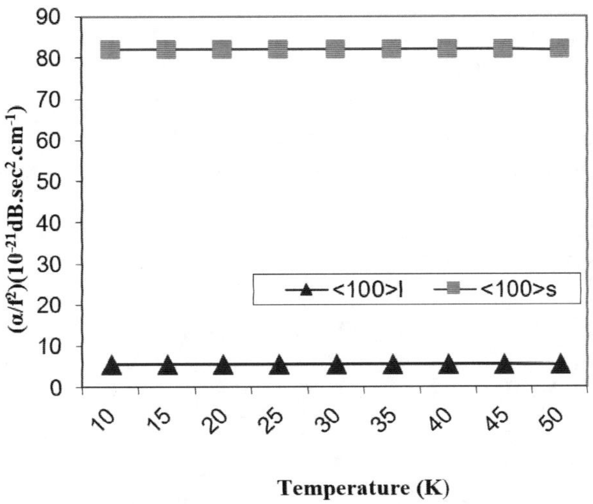


**Figure 1.** Temperature variation of ($\alpha/f^2$) due to e-p interaction in Nickel Aluminide crystal for longitudinal and shear waves along <100> direction.

IOP Conf. Series: Materials Science and Engineering 1120 (2021) 012027 doi:10.1088/1757-899X/1120/1/012027

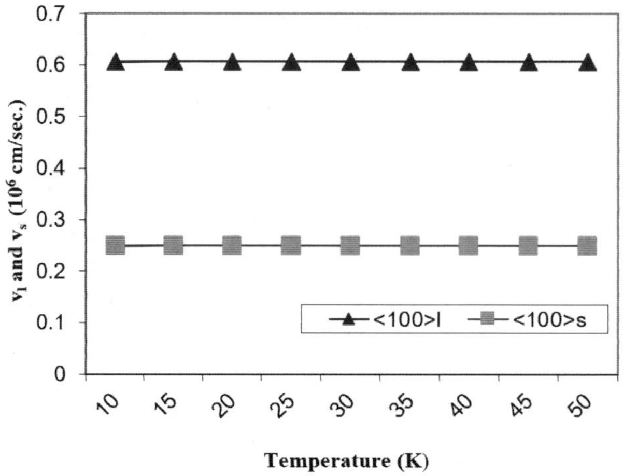


**Figure 2.** Temperature variation of Ultrasonic wave velocities (Longitudinal and Shear waves), $v_l$ and $v_s$ due to e-p interaction in Nickel Aluminide along <100> direction.

## 4. Conclusion

As per the detailed results and discussions as shown above, we can conclude that, in this present research paper, we have efficiently applied the concepts of two types of potentials such as "Born-Mayer and repulsive" and we have used very simple method for the present characterization technique by taking the two input parameters only. The detailed behaviour of $Ni_3Al$ ($L1_2$ phase) such as its response to the ultrasonic wave when it passes through this material at low temperature as discussed above will definitely be very useful for different researchers who are working in the area of NDT Materials characterization for the materials having FCC structures and the observed results will definitely be very useful for many applications in industry.

## References

[1] Pawel Jozwik *et al* 2015 *Materials* **8** 2537
[2] Yadav *et al* 2001 *Intermetallics* **9** 189
[3] W P Mason 1965 Physical acoustics, vol 3-Part B. New York, London Academic Press 237-284
[4] K Brugger 1964 *Phys. Rev.* A **133** 1611
[5] S K Kor and R K Singh 1993 *Acustica* **79** 292
[6] S K Kor and R K Singh 1993 *Acta Physica Polonica A* **83** 751
[7] W P Mason 1950 Piezoelectric crystals and their applications to ultrasonics. New York, Van Nostrand 478
[8] H E Bommel 1954 *Physics Rev.* **96** 220
[9] S K Kor *et al* J. *Phys. Soc.* Jpn. **56** 2428
[10] Kailash and S K Kor *et al* *Physica* B **145** 209
[11] G Leibfried and H Hahn 1958 *Z. Phys.* **150** 497
[12] G Leibfried and W Ludwig 1961 *Solid State Physics* **12** (Academic, New York)
[13] Y Hiki 1981 Higher Order Elastic Constants *Annual Rev. Mat. Science Japan* **11** 51
[14] C Kittel 1981 Introduction to solid state physics, John Wiley Press, New York, 7th edition

IOP Conf. Series: Materials Science and Engineering 1120 (2021) 012028 doi:10.1088/1757-899X/1120/1/012028

# Comparative study of mineral and soya bean oil at faulty temperature for transformer application

**Rishabh Roy*** **and Srikant Prasad**

Department of Electrical & Electronics Engineering, O.P Jindal University, Chhattisgarh, India

*Corresponding author's e-mail address: rishabhroy.980@gmail.com (R. Roy)

**Abstract.** Mineral oil is one of the most important insulating materials used for electro insulating & cooling of transformers. For monitoring & diagnosing the incipient fault in power transformers, dissolved gas analysis (DGA) technique is used. DGA have some drawbacks such as it requires regular instrument calibration, standard carrier gas and so on. To overcome these drawbacks, Fourier transform infrared spectroscopy (FTIR) & UV- Visible spectroscopy are used to determine dissolved gas present & ageing of oil. Once the gases are quantified from the FTIR result they can be interpreted using various diagnostic methods mentioned in IEEE Std.C57.104-2008. Also, absorption spectroscopy is considered for condition assessment tool to inspect the oil. By adopting Ultraviolet absorption technique, transformer health assessment is developed: UV-Visible spectroscopy is used. As in search of solution to the harmful & ecological problems, due to non-biodegradability & toxicity by mineral oil, the soya bean seed oil was carried out for the production of transformer insulating oil. In this proposed work, we carried out results from both FTIR & UV-Vis spectroscopy for both mineral oil & soya bean oil & compared both oil by preparing samples heated at 150 $^0$C for 24hrs & 48hrs.

## 1. Introduction

The transformer is very costly, indispensable, and most important equipment of any electric power system. Almost all of the large power transformers in power delivery systems around the world are still being insulated with mineral-based insulating oil [1-2]. Power transformers play an important role in the field of power transmission and distribution. But they are always under the influence of electrical, thermal, mechanical and environmental stresses. Due to these stresses, insulating oil degrades, leading the operation of the transformer to fail over a period of time. These stresses cause generation of gases from decomposition of insulating oil [3-5]. For many years the method of analysis of gases dissolved in the oil has been used as a tool in transformer diagnostics in order to detect incipient faults, to supervise suspect transformers, to test a hypothesis or explanation for the probable reasons of failures or disturbances which have already occurred and to ensure that new transformers are healthy. Dissolved gas analysis (DGA) technique is used for monitoring and diagnosing the incipient faults in power transformers. However, DGA has some drawbacks like it cannot localize the fault, no mistakes are okay while taking samples from the transformer, regular instrument calibration, cleaning of instrument after every test, maintenance cost is high, requires standard carrier gases and so on. An optical diagnostic method like Fourier transform infrared (FTIR) spectroscopy and UV-Visible spectroscopy are considered to overcome these drawbacks as they show greater ability to diagnose transformer health

IOP Conf. Series: Materials Science and Engineering 1120 (2021) 012028 doi:10.1088/1757-899X/1120/1/012028

conditions [6]. FTIR spectrometer has capability to identify the C-H and C-O chemical bond [6-8]. The released gases are Hydrogen ($H_2$), Methane ($CH_4$), Ethane ($C_2H_6$), Ethylene ($C_2H_4$), Acetylene ($C_2H_2$), and Carbon monoxide (CO), Carbon dioxide ($CO_2$) if cellulose also involves [7-9]. Also UV-Visible spectroscopy analysis method is proved and very much useful to be qualitative measurement of gases dissolved in transformer oil. Comparative results show that absorption spectra obtained by wavelet transmission are consistent to those with their real background and are independent of the spectrometer. Through UV-Visible Spectroscopy ageing of oil can be determined [10-12]. Oil constitutes a major component of power transformers. Conventionally, mineral or synthetic oils have been used. In an attempt to overcome some of the drawbacks with mineral oils, vegetable oil has been considered as an alternative. Used in the industry for several decades it has proved to be a viable alternative to mineral oil and has several advantages over mineral oil. As mineral oil is non- biodegradable, researchers find alternative oil for this purpose. Cooking oil can be used instead of mineral oil. Use of soya bean oil and palm kernel oil can be used as alternative to mineral oil in transformer insulation. Crude samples of these oils and their blend in varied proportions were tested for dielectric strength, pour point, flash point, kinematic viscosity, density and moisture content. The results showed that soya bean oil and palm kernel oil have good properties to act as insulating and cooling liquid in a transformer. These properties could be further improved when the oils are refined and purified [13-16]. In terms of economic costs and environmental considerations, soya bean oil and palm kernel oil appear to be viable alternatives to transformer oils. In this work, we are showing and comparing results of FTIR & UV-Visible spectroscopy for mineral and soya bean oils heated at 150 °C (faulty temperature) in the laboratory.

## 2. Spectroscopy method for oil analysis

Infrared (IR) spectroscopy is a very useful method for detecting the characteristic bonds of many functional groups through their absorption of infrared light. If you shine infrared light on a molecule, it is possible that the molecule absorbs energy from light. Absorbed energy can cause a bond to stretch or bend. We call that a stretching or bending vibration. These vibrations occur only at specific frequencies, which correspond to the frequency of IR light. When the frequency of IR light matches the frequency of a particular vibrational mode, the IR light is absorbed, and you can tell which frequencies are absorbed by looking at your infrared spectrum. Different kinds of bonds vibrate at different frequencies, so they absorb different frequencies of IR light, so it is possible to determine the functional groups present.
IR spectra can be divided into two main regions:
1) Diagnostic region – generally has fewer peaks and provides the clearest information. This region contains all signals that arise from all bonds in a molecule.
2) Fingerprint region – contains signals resulting from the vibrational excitation of most single bonds (stretching and bending).
Since the fingerprint region generally contains many signals and is more difficult to analyze, we can ignore it. Such spectra will be the same in the diagnostic region, but in the fingerprint region won't. Thus this region is called a fingerprint because each compound has a unique pattern of signals in this region, much the way each person has a unique fingerprint.

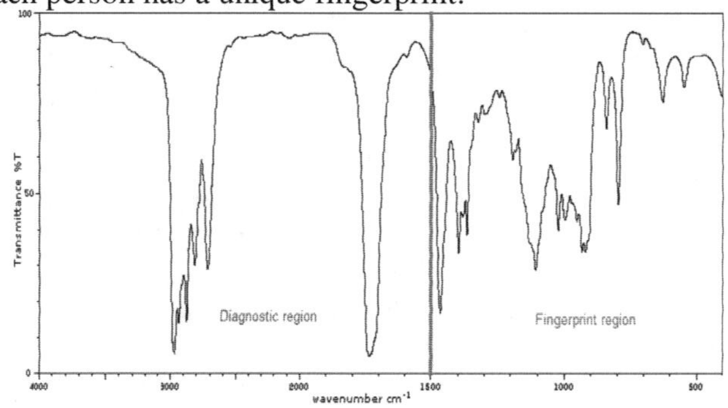


**Figure 1.** IR diagram showing Diagnostic and Fingerprint region.

IOP Conf. Series: Materials Science and Engineering 1120 (2021) 012028 doi:10.1088/1757-899X/1120/1/012028

## 3. Ultraviolet–Visible spectroscopy

Ultraviolet-visible spectroscopy or ultraviolet-visible spectro-photometry (UV-Vis or UV/Vis) refers to absorption spectroscopy in the ultraviolet-visible spectral region. This means it uses light in the visible and adjacent (near-UV and near-infrared (NIR)) ranges. UV/Vis spectrophotometer is used in the quantitative determination of concentrations of the absorber in the solutions of transition metal ions and highly conjugated organic compounds.

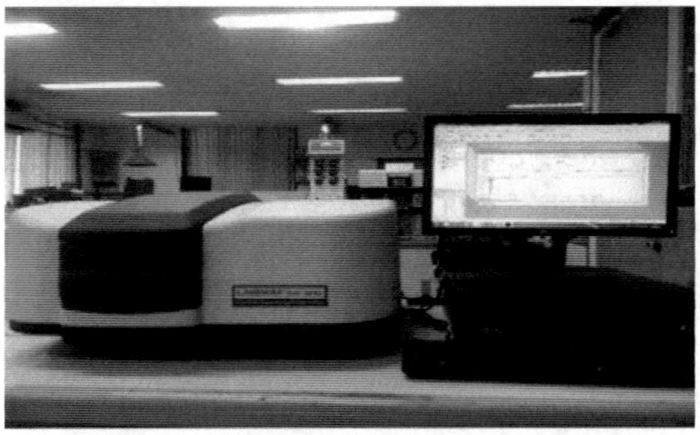

**Figure 2.** UV-Visible spectroscopy model- labindia UV 3092.

The UV-Visible spectroscopy analysis has been done on different transformer oils like in-service aged oil, laboratory aged and fresh transformer oils. The method is very much useful and proved to be qualitative measurement of gases dissolved in transformer mineral oil. Comparative results show that absorption spectra obtained by wavelet transmission are consistent to those with their real background and are independent on spectrometers.

## 4. Fourier transform infrared spectroscopy

FTIR spectroscopy converts raw data into the actual spectrum. It is a non-destructive technique which is used to acquire an infrared spectrum of absorption of a solid, liquid and gas by collecting spectral data in a wide spectral range. When IR radiation is passed through a collected sample, some of it is absorbed by the sample (Absorption) and some of it is transmitted (Transmission) creating a molecular fingerprint of the sample. FTIR spectroscopy provides information by identifying unknown materials. It also determines quality of a sample and amount of components. In fact FTIR identifies a fingerprint of a sample with absorption peaks that correspond to the frequency of vibrations between the bonds of molecular structure of the sample. In view of much technology advancements, Infrared is found to be an excellent tool for qualitative and quantitative analysis as it has greater optical throughput. This is developed mainly to overcome the limitation with dispersive instruments. It produces a unique type of signal which encodes all the infrared frequencies into it.

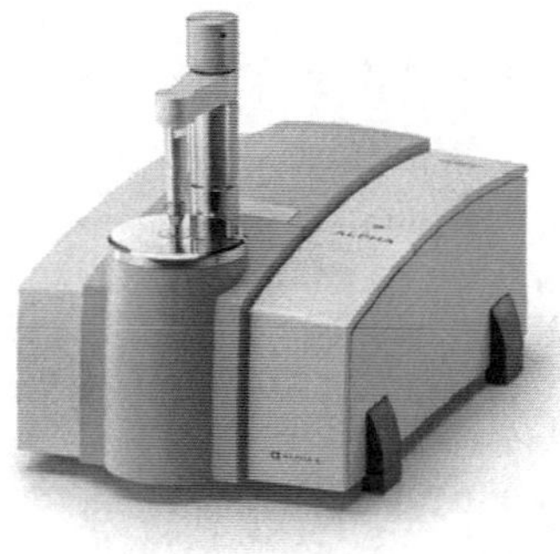

**Figure 3.** Bruker alpha-2 spectrometer.

IOP Conf. Series: Materials Science and Engineering 1120 (2021) 012028 doi:10.1088/1757-899X/1120/1/012028

Basic principal - Molecular bonds vibrate at various frequencies depending on the elements and the type of bonds. For any given bond, there are several specific frequencies at which it can vibrate. According to quantum mechanics, these frequencies correspond to the ground state (lowest frequency) and several excited states (higher frequencies). One way to cause the frequency of a molecular vibration to increase is to excite the bond by having it absorb light energy. For any given transition between two states the light energy (determined by the wavelength) must exactly equal the difference in the energy between the ground state and the first excited state.

## 5. Absorption spectroscopy

Many atoms emit or absorb visible light. In order to obtain a fine line spectrum, the atoms must be in a gas phase. This means that the substance has to be vaporized. The spectrum is studied in absorption or emission. Visible absorption spectroscopy is often combined with UV absorption spectroscopy in UV-Vis spectroscopy. According to Beer-Lambert relationship, the change in light intensity due to absorption is determined by the number of absorbing types. The Beer-Lambert relationship states that there is a logarithmic dependence between the transmission, of light through a substance and the product of the absorption coefficient of the substance, α, and the distance the light travels through the material i.e. the path length 'l'.

**Table 1.** The major properties of soya bean and castor bio-transformer oil compared with ASTM limits of conventional transformer oil.

| Properties | Soya bean Bio-transformer oil | Transformer oil |
|---|---|---|
| Flash point ($^0$C) | 250 | 140 |
| Pour point | -16 | -7 |
| Viscosity(cst) at 27 $^0$C | 7.29 | 9.3 |
| Specific gravity at 27 $^0$C | 0.88 | 0.89 |
| Density (g ml-1) | 0.9404 | 0.89 |
| Dielectric strength (KV) | 21 | 24 |

$$A= -\log T= \log I_0/I= a, c, b \quad (1)$$

Where A is the absorbance , T is the transmittance defined as I/Io , I is the incident light intensity & Io is the transmitted light intensity , c is the concentration of the absorbing types (gm/l), b is the path length traversed by the light, a is the absorbtivity of the absorbing types.

## 6. Experimental Setup

This experiment is carried out by UV- Visible spectroscopy and FTIR spectroscopy by taking two different types of oil namely – Mineral oil and Soya bean oil respectively and heated them at 150 $^0$C (faulty temperature for transformer oil) for 24hrs and 48 hrs. Also, we try to mimic the internal condition of a transformer by adding copper pieces and insulation paper in them during heating. Here 4 beakers are taken and named them Beaker A, B, C, & D. Where,

Beaker A- consists of 500ml of fresh mineral oil heated at 150 $^0$C for 24hrs and 48hrs.

Beaker B- consists of 500ml of fresh mineral oil + 10 gram of insulation paper + 30 gram of pure copper pieces heated at 150 $^0$C for 24hrs and 48hrs.

**Figure 4.** Beaker A consists of 500ml of fresh mineral oil (right) & beaker B consists of 500ml of fresh mineral oil + 10 gram of insulation paper + 30 gram of pure copper pieces (left).

Beaker C- consists of 500ml of fresh soya bean oil heated at 150 ºC for 24hrs and 48hrs. Beaker D- consists of 500ml of fresh soya bean oil + 10 gram of insulation paper + 30 gram of pure copper pieces heated at 150 ºC for 24hrs and 48hrs.
Beaker D- consists of 500ml of fresh soya bean oil + 10 gram of insulation paper + 30 gram of pure copper pieces heated at 150 ºC for 24hrs and 48hrs.

**Figure 5.** Beaker C consists of 500ml of fresh soya bean oil (right) & beaker D consists of 500ml of fresh soya bean oil + 10 gram of insulation paper + 30 gram of pure copper pieces (left).

From these beakers we have taken out 15ml of oil from each of them after 24hrs and 48 hrs of heating.

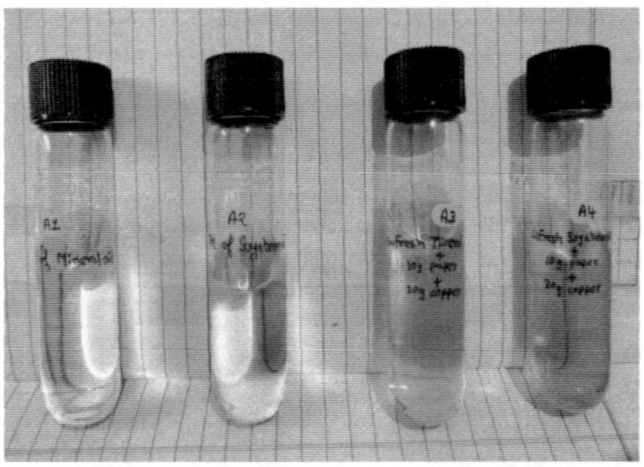

**Figure 6.** Sample A1= fresh mineral oil heated at 150 ºC for 24hrs.Sample A2= fresh soya bean oil heated at 150 ºC for 24hrs.Sample A3= fresh mineral oil + 10gram of paper+30gram of copper pieces heated at 150 ºC for 24hrs.Sample A4= fresh soya bean oil+ 10 gram of paper + 30 gram of copper

IOP Conf. Series: Materials Science and Engineering 1120 (2021) 012028 doi:10.1088/1757-899X/1120/1/012028

pieces heated at 150 °C for 24hrs.

**Figure 7.** Sample B1= fresh mineral oil heated at 150 °C for 48hrs.Sample B2= fresh soya bean oil heated at 150 °C for 48hrs.Sample B3= fresh mineral oil + 10gram of paper+30gram of copper pieces heated at 150 °C for 48hrs.Sample B4= fresh soya bean oil+ 10 gram of paper + 30 gram of copper pieces heated at 150 °C for 48hrs.

## 7. Result and its analysis

For analysis of results we follow Infrared spectroscopy table to find chemical bond in each peak. We observe only alkanes, alkynes, alkenes (CH, CH-O, C-C etc).

*7.1 For FTIR Spectroscopy*

**Sample A1-**

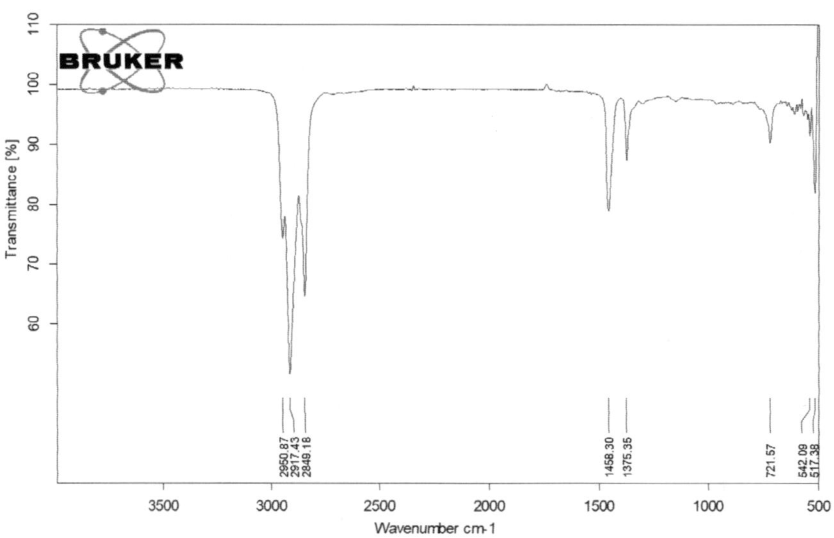


**Figure 8**. Result of Fresh mineral oil heated at 150 °C for 24 hrs.

Peaks observed and its analysis =

At range 3000-2500 cm-1, 2950.87, 2917.43, 2849.18 = medium intensity, C-H stretching, alkane. At range 1600-1300 cm-1, 1458.30, 1375.35 = medium intensity, C-H bending, alkane, methyl group. At range 1000-650 cm-1, 721.57 = strong intensity, C=C bending, alkene, disubstituted (cis).

**Sample A2-**

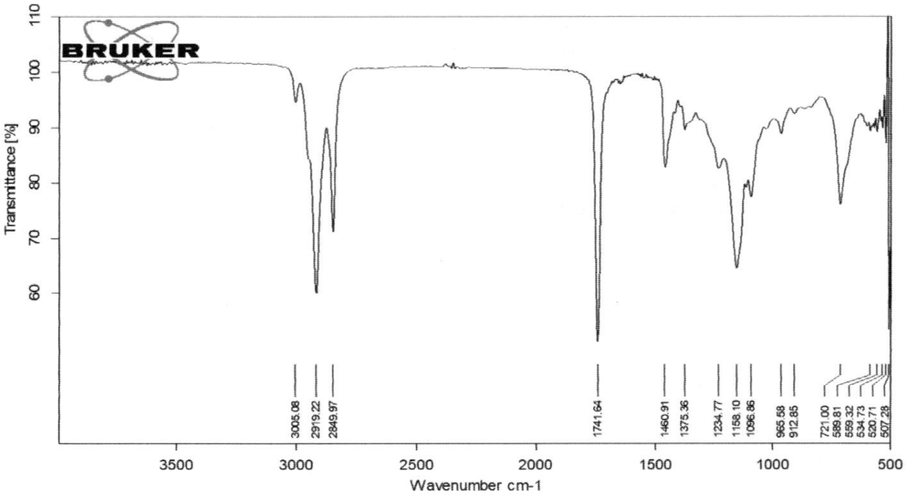


**Figure 9**. Result of Fresh Soya bean oil heated at 150 °C for 24 hrs.

Peak observed and its analysis=
At range 3000-2500 cm-1, 3005.08 = medium intensity, C-H stretching, alkene & 2919.22, 2849.97 = medium intensity, C-H stretching, alkane. At range 2000-1650 cm-1, 1741.64 = weak intensity, C-H bending, aromatic compound, overtone. At range 1600-1300 cm-1, 1460.91 = medium intensity, C-H bending, alkane, methylene group & 1375.36 = medium intensity, C-H bending, alkane, methyl group. At range 1400-1000 cm-1, 1234.77 = strong intensity, C-O stretching, vinyl ether, 1158.10 = strong intensity, C-O stretching, tertiary alcohol & 1096.86 = strong intensity, C-O stretching, secondary alcohol. At range 1000-650 cm-1, 965.58 = strong intensity, C=C bending, alkene, disubstituted (trans), 912.85 = strong intensity, C=C bending, alkene, monosubstituted, 721.00 = strong intensity, C=C bending, alkene, di-substituted (cis).

**Sample A3-**

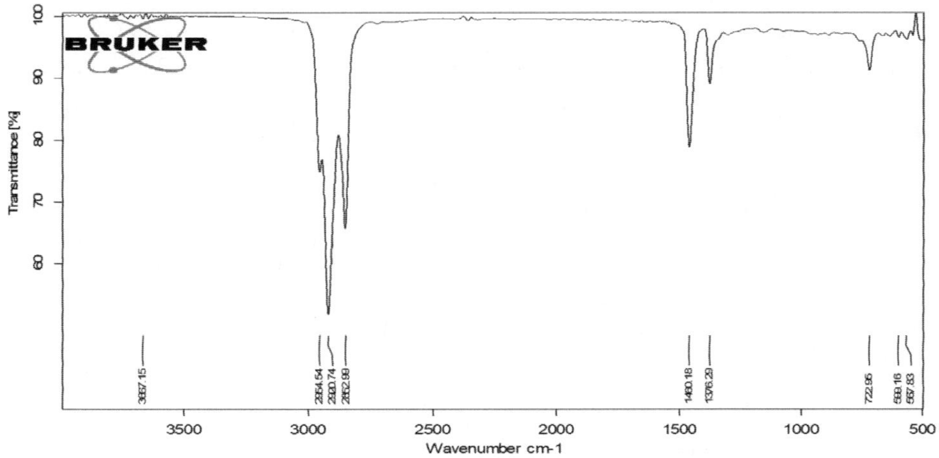


**Figure10**. Result of fresh mineral oil + 10gram of paper+30gram of copper pieces heated at 150 °C for 24hrs.

Peak observed and its analysis=
At range 4000-3000 cm-1, 3667.15 = strong intensity, O-H stretching, alcohol, free. At range 3000-2500 cm-1, 2954.54, 2920.74, 2852.99 = medium intensity, C-H stretching, alkane. At range 1600-1300 cm-1, 1460.18 = medium intensity, C-H bending, alkane, methylene group & 1376.29 = medium intensity, C-H bending, alkane, methyl group. At range 1000-650 cm-1, 722.95 = strong intensity, C=C bending, alkene, disubstituted.

**Sample A4-**

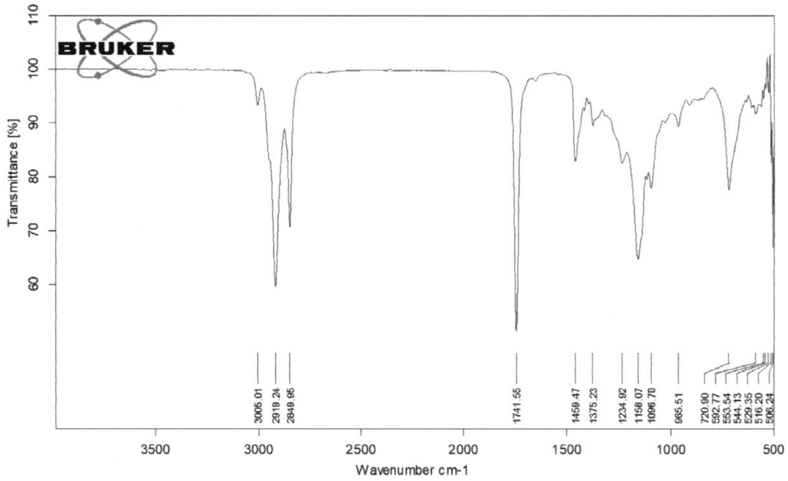


**Figure 11**. Result of fresh soya bean oil+ 10 gram of paper + 30 gram of copper pieces heated at 150 ºC for 24hrs.

Peak observed and analysis =
At range 3000-2500 cm-1, 3005.01 = medium intensity, C-H stretching, alkene, 2919.24, 2849.95 = medium intensity, C-H stretching, alkane. At range 2000-1650 cm-1, 1741.55 = weak intensity, C-H bending, aromatic compound, overtone. At range 1600-1300 cm-1, 1459.47 = medium intensity, C-H bending, alkane, methylene group & 1375.23 = medium intensity, C-H bending, alkane, methyl group. At range 1400-1000 cm-1, 1234.92 = strong intensity, C-O stretching, alkyl aryl ether, 1158.07 = strong intensity, C-O stretching, tertiary alcohol & 1096.70 = strong intensity, C-O stretching, secondary alcohol. At range 1000-650 cm-1, 965.51 = strong intensity, C=C bending, alkene, monosubstituted & 720.90 = strong intensity, C=C bending, alkene, disubstituted (cis).

**Sample B1-**

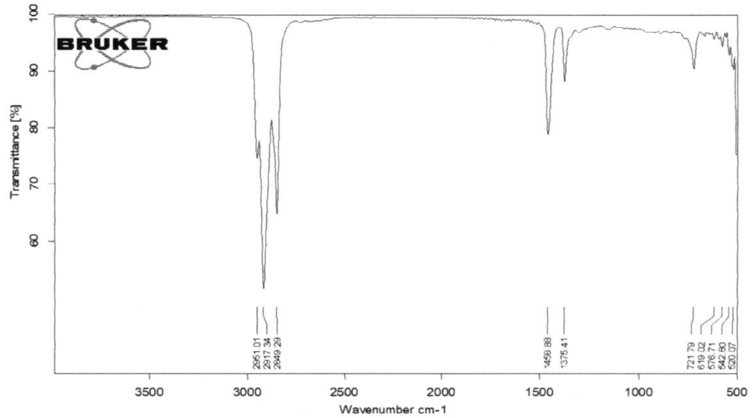


**Figure 12**. Result of fresh mineral oil heated at 150 ºC for 48hrs.

Peak observed and analysis =
At range 3000-2500 cm-1, 2951.01, 2917.34, 2849.29 = medium intensity, C-H stretching, alkane. At range 1600-1300 cm-1, 1458.88 = medium intensity, C-H bending, alkane, methylene group & 1375.41 = medium intensity, C-H bending, alkane, methyl group. At range 1000-650 cm-1, 721.79 = strong intensity, C=C bending, alkene, disubstituted (cis).

**Sample B2**-

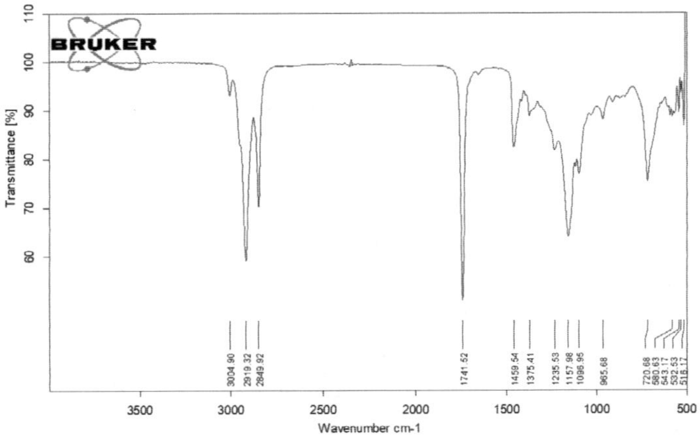


**Figure 13**. Result of fresh soya bean oil heated at 150 $^oC$ for 48hrs.

Peak observed and analysis =
At range 3000-2500 cm-1, 3004.90 = medium intensity, C-H stretching, alkene & 2919.32, 2849.92 = medium intensity, C-H stretching, alkane. At range 2000-1650 cm-1, 1741.52 = C-H bending, aromatic compound, overtone. At range 1600-1300 cm-1, 1459.54 = medium intensity, C-H bending, alkane, methlylene group & 1375.98 = medium intensity, C-H bending, alkane, methyl group. At range 1400-1000 cm-1, 1235.53 = strong intensity, C-O stretching, alkyl aryl ether, 1157.98 = strong intensity, C-O stretching, tertiary alcohol & 1096.95 = strong intensity, C-O stretching, secondary alcohol. At range 1000-650 cm-1, 965.68 = strong intensity, C=C bending, alkene, disubstituted (trans) & 720.68 = strong intensity, C=C bending, alkene, disubstituted (cis).

**Sample B3**-

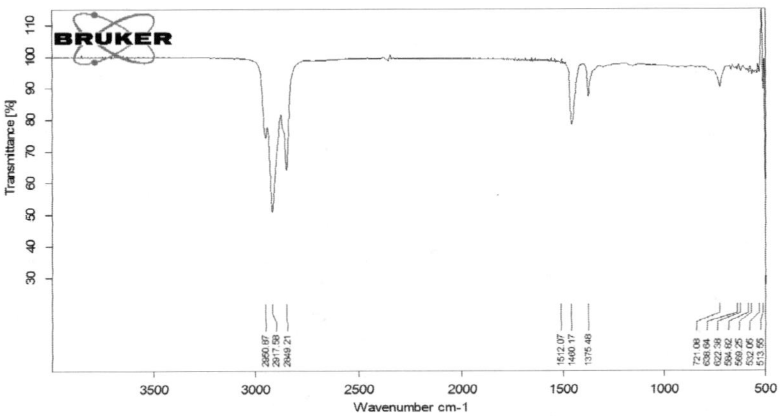


**Figure 14**. Result of fresh mineral oil + 10gram of paper+30gram of copper pieces heated at 150 $^oC$ for 48hrs.

At range 3000-2500 cm-1, 2950.87, 2917.58, 2849.21 = medium intensity, C-H stretching, alkane. At range 1600-1300 cm-1, 1512.07 = strong intensity, N-O stretching, nitro compound, 1460.17 = medium intensity, C-H bending, alkane, methylene group & 1375.48 = medium intensity, C-H bending, alkane, methyl group. At range 1000-650 cm-1, 721.08 = strong intensity, C=C bending, alkene, disubstituted (cis).

**Sample B4-**

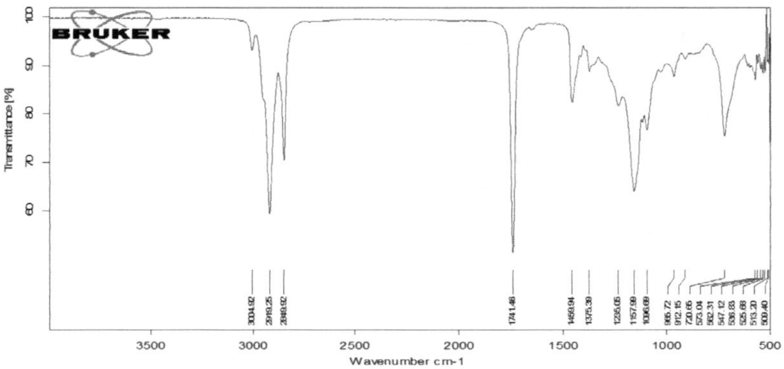


**Figure 15**. Result of fresh soya bean oil + 10 gram of paper + 30 gram of copper pieces heated at 150 °C for 48hrs.

Peak observed and analysis =
At range 3000-2500 cm-1, 3004.92 = medium intensity, C-H stretching, alkene, 2919.25 = medium intensity, C-H stretching, alkane & 2849.92 = medium intensity, C-H stretching, alkane. At range 2000-1650 cm-1, 1741.46 = weak intensity, C-H bending, aromatic compound, overtone. At range 1600-1300 cm-1, 1459.94 = medium intensity, C-H bending, alkane, methylene group & 1375.39 = medium intensity, C-H bending, alkane, methyl group. At range 1400-1000 cm-1, 1235.05 = strong intensity, C-O stretching, alkyl aryl ether, 1157.99 = strong intensity, C-O stretching, tertiary alcohol & 1096.69 = strong intensity, C-O stretching, secondary alcohol. At range 1000-650 cm-1, 965.72 = strong, C=C bending, alkene, monosubstituted, 912.15 = strong, C=C bending, alkene, monosubstituted & 720.65 = strong, C=C bending, alkene, disubstituted (cis).

*7.2 For UV-Visible spectroscopy*

**Sample A1**

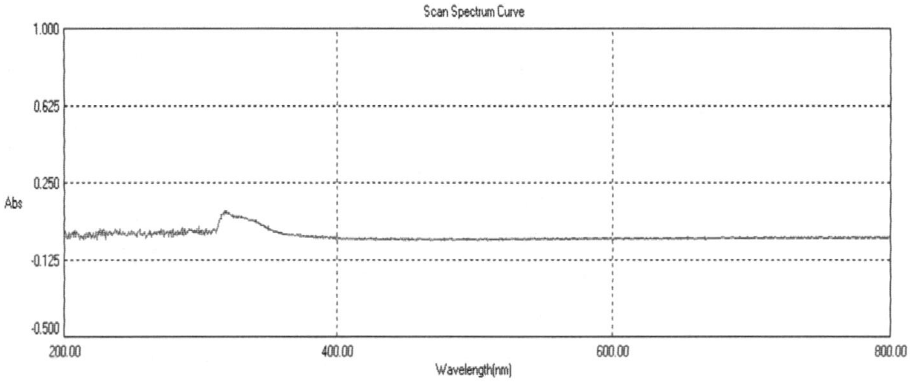


**Figure 16**. Result of fresh mineral oil heated at 150 °C for 24hrs

Peak observed at 318 nm.

Sample A2

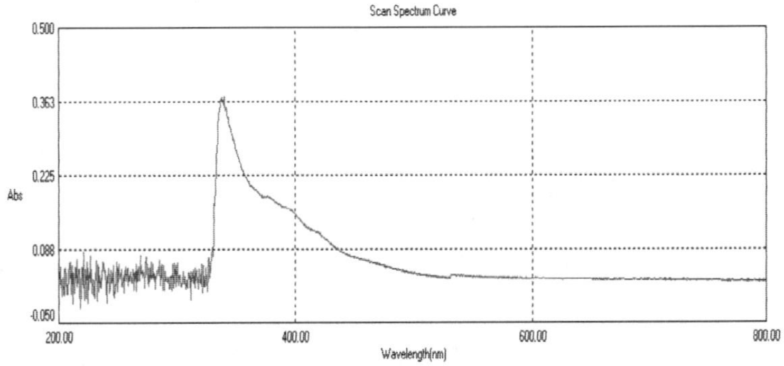


**Figure 17**. Result of fresh mineral oil heated at 150 ºC for 24hrs.

Peak observed at 339 nm.

Sample A3-

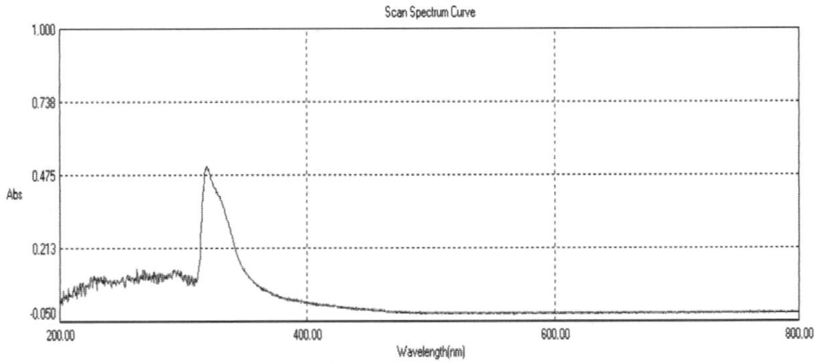


**Figure 18**. Result of fresh mineral oil + 10gram of paper+30gram of copper pieces heated at 150 ºC for 24hrs.

Peak observed at 319.5 nm.

Sample A4-

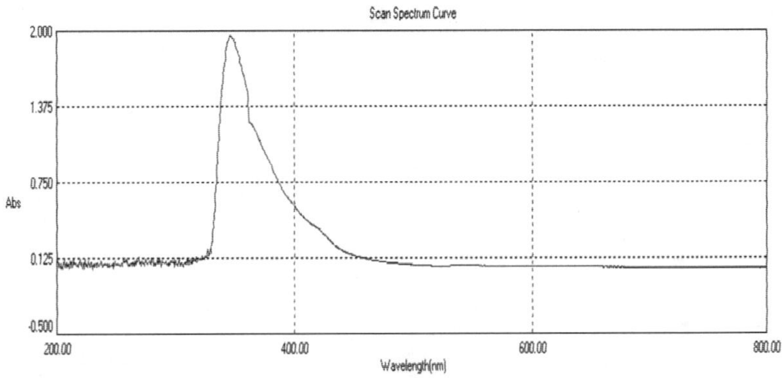


**Figure 19**. Report of fresh soya bean oil+ 10 gram of paper + 30 gram of copper pieces heated at 150 ºC for 24hrs.

Peak observed at 347.5 nm.

Sample B1-

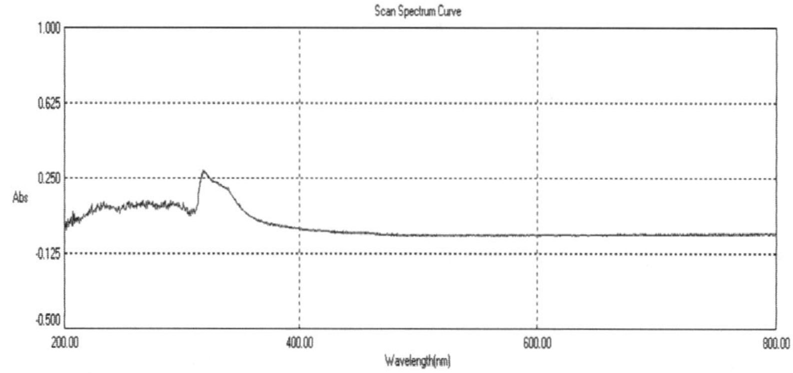


**Figure 20**. Result of fresh mineral oil heated at 150 °C for 48hrs.

Peak observed at 319.5 nm

Sample B2-

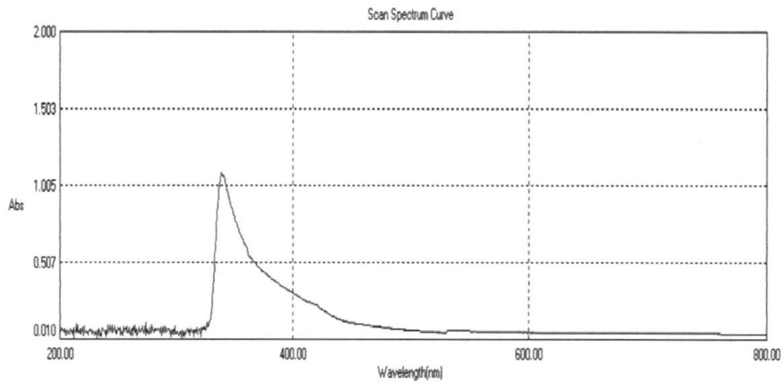


**Figure 21**. Report of fresh soya bean oil heated at 150 °C for 48hrs.

Peak observed at 341 nm.

Sample B3-

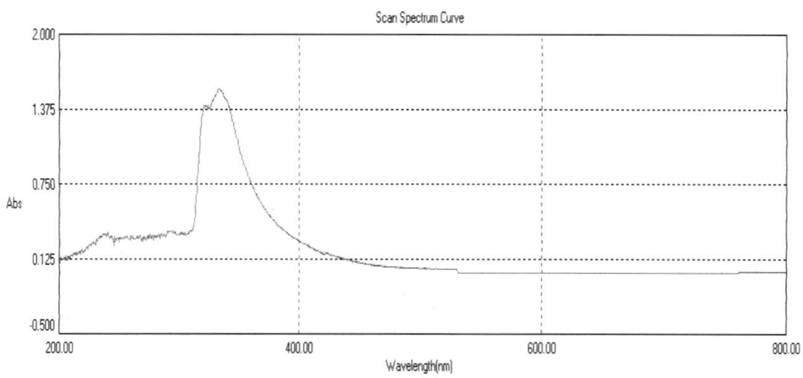


**Figure 22**. Report of fresh mineral oil + 10gram of paper+30gram of copper pieces heated at 150 °C for 48hrs.

Peak observed at 332.5 nm.

Sample B4-

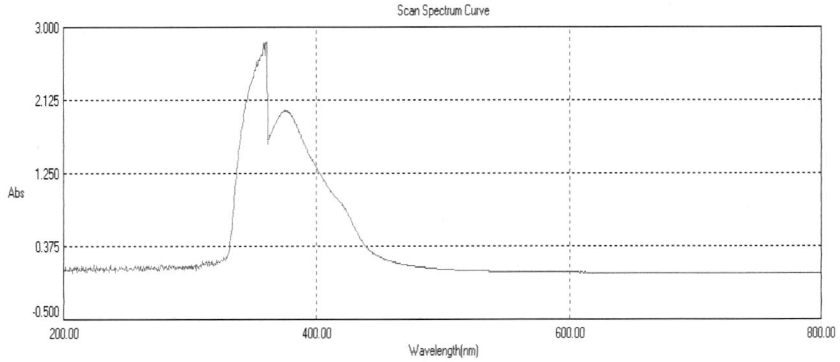


**Figure 23**. Result of fresh soya bean oil+ 10 gram of paper + 30 gram of copper pieces heated at 150 ºC for 48hrs.

Peak observed at 361nm.

**Comparison graph =**

1) All mineral oil samples

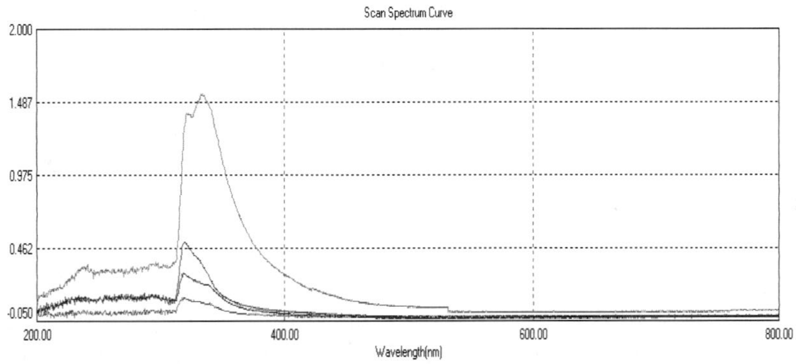


**Figure 24**. Comparison graph of all mineral oil samples.

2) All soya bean oil samples

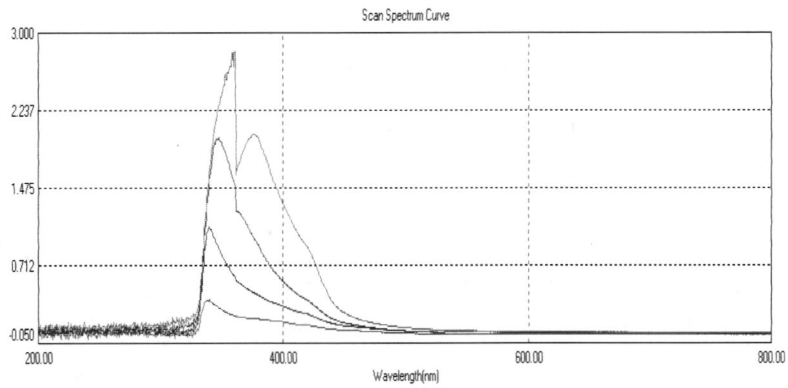


**Figure 25**. Comparison graph of all soya bean oil samples.

IOP Conf. Series: Materials Science and Engineering 1120 (2021) 012028 doi:10.1088/1757-899X/1120/1/012028

## 8. Conclusion

FTIR Spectroscopy Gases formed in insulating oils in a transformer due thermal and electrical stresses are H2 (Hydrogen), CH4 (Methane), C2H2 (Acetylene), C2H4 (Ethylene),C2H6 (Ethane), CO (Carbon monoxide), CO2 (Carbon dioxide).

As we know FTIR spectroscopy shows excellent diagnostic capability to identify the C-H and C-O chemical bond [6]. Due to various stresses, gases formed shows incipient faults, to find out the gases quantitatively as well as qualitatively IR spectroscopy is used.

From the result analysis we observed:-

**Table 2.** Peak observed by FTIR at different range by different oil.

| Different oils | Peak observed($cm^{-1}$) at range | Heating time |
|---|---|---|
| Mineral oil | 4000-3000, 3000-2500, 1600-1300, 1000-650 | 24hrs |
| Mineral oil | 3000-2500, 1600-1300, 1000-650 | 48hrs |
| Soya bean oil | 3000-2500, 2000-1650, 1600-1300, 1400-1000, 1000-650 | 24hrs |
| Soya bean oil | 3000-2500, 2000-1650, 1600-1300, 1400,1000, 1000-650 | 48hrs |

Once gases are quantified, the quantified gases can be further interpreted using various no of diagnostic method in IEEE Std. C57.104-2008 [6]. From the analysis of obtained results we can clearly see the transmittance of IR at different wavelength by sample oils. That shows amount of gases releases in both soya bean and mineral oil samples which were heated at different time period, the degradation of oil takes place. From the analysis FTIR results of different samples of oil we see that soya bean oil transmit/absorb more shows higher level of gases release.

### UV-Visible spectroscopy

Main aim for any absorption spectroscopy is to obtained and measure how well a sample absorbs light at each wavelength. Many monochromatic beam of light of different wavelength passes through samples repeatedly to observe how light is absorbed by sample in a UV-Visible spectrometer. In this work, UV-Vis spectrometer is used to determine the decomposition and ageing of two different kind of oil in different condition and heated at different time period. Here we can see that absorption take place in between 300nm to 370nm wavelength. Below table shows different oil's absorption in different time period. In the region 300 to 320 nm wavelength which means decomposition of oil takes place. For mix mineral oil samples for 24hrs and 48 hrs is 319.5 nm and 332.5 nm which means absorb value increase than the mineral oil samples, and also as time increases the decomposition of oil is more.

**Table 3.** Observed results in UV- Visible spectroscopy for different samples.

| Sample Oil | Peak observed at wavelength(nm) for 24hrs | Peak observed at wavelength(nm) for 48hrs |
|---|---|---|
| Mineral oil 500ml heated at 150 °C. | 318 | 319.5 |
| Mineral oil 500 ml + 10 gram of paper + 30 gram of copper pieces. | 319.5 | 332.5 |
| Soya bean oil 500 ml heated at 150°C. | 339 | 341 |
| Soya bean + 10 gram of paper + 30 gram of copper pieces | 347 | 361 |

As compared to mineral oil samples, soya bean oil shows more absorption which means decomposition of oil is more than mineral oil. From the above results we clearly see that for 24 hrs and 48 hrs heated fresh soya bean oil shows peaks at 340 nm and 339 nm. Where for mix soya bean oil for 24hrs and 48hrs shows absorption peak at 347 nm and 361 nm, by which we can say that decomposition of these sample gets higher due to increment of time period. From the comparison graph of fig no- 24 and fig no 25 we clearly see the differences. As UV-Visible spectroscopy gives the clear results, it can be used as a tool for investigating the Transformer oil for its deterioration and ageing due to electrical and thermal stresses [9].

Hence, Bio transformer oil is made for actual use of soya bean oil in transformer by adding solvent to refined soya bean oil [12].

**References**

[1] Palmer J A, Wang X, Shoureshi R A, Mander A, Torgerson D and Rich C 2000 Effect of aging on the spectral response of transformer oil *Conference Record of the 2000 IEEE International Symposium on Electrical Insulation.* p 460-464

[2] Suleiman A A, Mohamad N A B, Alghamdi A S and Aizam M 2012 Improving Accuracy of DGA Interpretation of Oil-filled Power Transformers Needed for Effective Condition Monitoring *2012 IEEE International Conference on Condition Monitoring and Diagnosis.* A 374

[3] Gray I A R 2005 Transformer, A GUIDE TO TRANSFORMER OIL ANALYSIS.

[4] Skelly D 2012 Photo-acoustic spectroscopy for dissolved gas analysis: Benefits and Experience *2012 IEEE International Conference on Condition Monitoring and Diagnosis.* pp. 29-43

[5] IEEE Std. C57.104 IEEE 2008 Guide for the Interpretation of Gases Generated in Oil -Immersed Transformers.

[6] Hussain K and Karmakar S 2015 Incipient Fault Diagnosis in Oil-filled Power Transformers using Fourier Transform Infrared Spectroscopy, *International Conference on High Voltage Engineering and Technology (ICHVET-2015)* **9654** p 29

[7] Tang X, Wang W, Zhang X, Wang E and Li X 2018 Energies **11** 3192

[8] Meena R R, Chaki S, Khimani A J and Deshpande M P 2018 Petroleum and Coal ***60*** 872

[9] Hussain K and Subrata Karmakar 2014 Condition Assessment of Transformer Oil using UV-Visible Spectroscopy, In 2014 Eighteenth National Power Systems Conference (NPSC), p 1-5. IEEE

[10] Thakur N and Rahi O P 2011 *International Journal of Electronics Signals and Systems.* **1** 134

[11] Karthik R, Raja T S R and Shunmugam S S 2014 *Acta Scientiarum Technology.* **36** 245

[12] Zhang W, Li N, Feng Y, Su S, Li T and Liang B 2015 *Food Chem. 2015.* **185** 326

[13] Egbuna C N 2016 *International Journal of Engineering Sciences and Research Technology.* **5**

IOP Conf. Series: Materials Science and Engineering 1120 (2021) 012028 doi:10.1088/1757-899X/1120/1/012028

105
[14] Ursula Biermann and Jürgen O. Metzger 2013 Application of Vegetable Oil-Based Fluids as Transformer Oil, Oleo chemicals under Changing Global Conditions *International Journal of Engineering Research & Technology (IJERT).* **2** 1
[15] Rycroft M 2014 Energize **4** 37
[16] Bandara K, Ekanayake C, Saha T and Ma H 2016 *Energies* **9** 258
[17] M. A. Usman, O. O. Olanipekun and U. T. Henshaw 2012 *Journal of Emerging Trends in Engineering and Applied Sciences* **3** 33

IOP Conf. Series: Materials Science and Engineering 1120 (2021) 012029 doi:10.1088/1757-899X/1120/1/012029

# Growth and photoemission studies of titanium carbide coating by laser carburizing

**Sheetal Soni[1,2*], D M Phase[3] and Ratnesh Gupta[1]**

[1] School of Instrumentation, Devi Ahilya Vishwavidyalaya, Takshshila campus Khandwa road, Indore 452001, India.
[2] Department of Physics, National Institute of Technology Raipur 492010, India.
[3] UGC-DAE CSR, Indore, Khandwa Road, Indore 452 017, India.

*Corresponding author's e-mail address: ssoni2312@gmail.com

**Abstract:** Carburizing of metals is well known technique to modify the tribological properties like hardness and corrosion resistance of metal surfaces. Besides this the electronic structures of nitrides and carbides of titanium metal have great importance from both fundamental and technological point of view. Resonance photoemission electron spectroscopy is a well-known technique to reveal the electronic structure of these alloys. In the present work titanium carbide (TiC) coating has been synthesized by laser carburizing on commercial available pure titanium sheet in a controlled methane atmosphere. X-ray diffraction studies reveal the formation of TiC phase after laser treatment. The hardness of TiC was performed by nano-indentation and its electronic structure has been characterized by x-ray and valence band photoelectron spectroscopy. The measured hardness is six times higher than the pure Ti substrate. Resonance photoemission spectrum of TiC gives maxima at 46 eV, which is a characteristic feature of Ti *3d* states in the valence band of TiC coating.

## 1. Introduction

TiC belongs to the family of refractory metals and hence exhibits similar properties. Due to its extraordinary chemical, thermal and physical properties it has been used in cutting tools, space technology, microelectronics, aerospace industry and biomaterial since last two decades. These fascinating properties of TiC make it very motivating to study theoretically as well as experimentally. Aforementioned properties of TiC mainly are governed by on its growth, electronic structure and bonding characteristics [1-4]. Carburizing of titanium metal and its compound to form a coating of TiC implant in a metallic substrate has been extensively studied with an aim to enhance its tribological properties. Plenty of work has been carried out for investigating the various properties of TiC coatings prepared from different methods i.e. gas carburizing, chemical vapour deposition (CVD), physical vapour deposition (PVD), evaporation, ion plating and sputtering [4-8] the literatures involving the fabrication of TiC coating in $CH_4$ atmosphere prepared by laser carburizing and resonant photoemission study of Ti *3d* states in TiC are still scarce. Resonant photo emission spectrum (RPES) of the Ti *3d* character in Ti, TiN, TiCN and $TiO_2$ has been observed previously [9-11], i.e. above the photon energy of Ti *3p* absorption edge. Although various techniques are employed for the fabrication of TiC coatings on different substrate, laser processing in gaseous atmosphere is relative new technique to modify near surface region of metals and alloys; however the formation of TiCN on pure titanium sheet via laser

IOP Conf. Series: Materials Science and Engineering 1120 (2021) 012029 doi:10.1088/1757-899X/1120/1/012029

carbonitriding had already been reported earlier [12-13]. In this work TiC coating was fabricated on pure titanium sheet by laser carburizing.

## 2. Experimental Methods

### *2.1 Formation of TiC coating by laser carburizing*
Pure commercial sheet of titanium was polished by SiC grinding paper and diamond paste. We have used pulsed Kr-F excimer laser of wavelength 248 nm to irradiate polished titanium sheet at laser energy density of 6.8 J/cm$^2$ in the controlled atmosphere of $CH_4$ gas (2 bar pressure). The reacting chamber has been evacuated to a pressure of $10^{-4}$ mbar and we have flushed the chamber with the pressure of $CH_4$ gas three times before achieving the 2 bar pressure of $CH_4$ for laser irradiation.

### *2.2 Characterization on coated film*
As prepared laser irradiated TiC coating was characterized with different characterization techniques to reveal its hardness, microstructure, chemical analysis and bonding characteristics. The micro-hardness of the coated film was investigated by nano-indentation hardness measurements. Phase analysis of the coating was characterized by XRD with monochromatic Cu Kα (λ= 1.5406 Å) radiation source. The core level photoelectron spectrum was studied by x-ray photoelectron spectroscopy (XPS) using Al Kα excitation source of energy 1486.6 eV. Before the performance of XPS study, Ar+ ion etchings performed at 3 keV in the preparation chamber to eliminate the atmospheric contamination on the specimen. To perform valence band photo electron spectroscopy (VBS), we have used BL-1 beam line of INDUS-I, RRCAT, Indore with the incident energy ranging from 30 eV-60 eV. XPS and VBS measurements have been done one after another in the same chamber without breaking the vacuum.

## 3. Result and Discussion
### *3.1 Structural characterization by XRD*
The diffraction pattern of prepared TiC coating have shown in figure 1. Some of the diffraction peak of the specimen belongs to FCC TiC (JCPDS PDF#32-1383) compound and rest of them corresponds to α-hcp Ti phase due to the higher penetration depth of Cu Kα in the material used in θ-2θ geometry. There is no peak belongs to any phase of titanium oxide.

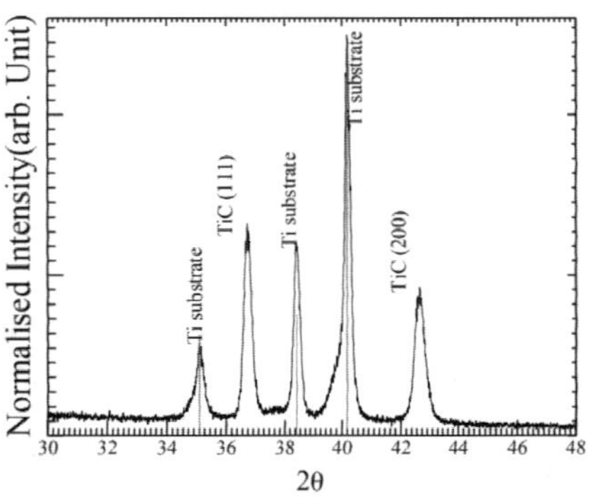

**Figure 1.** XRD patterns of TiC coatings at methane ambient pressure of 2 bar.

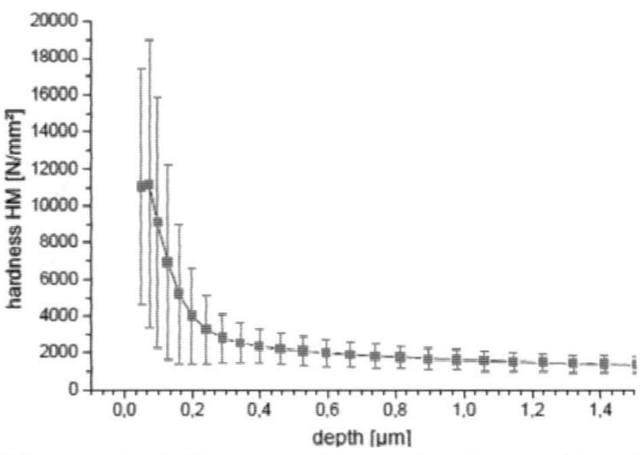

**Figure 2.** Micro-hardness depth profile of TiC coatings at methane ambient pressure of 2 bar.

### *3.2 Hardness measurement*
In the nano-indentation technique a diamond indenter used to indents the coating surface on the application of force F. The force F increases from 0 to F, then return to 0 continuously. The obtained hardness HU (N/mm$^2$) is well-defined as the ratio of the applied force F and the area A(h). Figure 2

shows the composite hardness profile as a function of depth (indentation). From the fitting of the experimental data clearly suggest that the hardness of the TiC compound formed on the Ti substrate has higher value.

*3.3 Chemical analysis by XPS:*
Core level XPS spectrum of the constituents present in the coating were carried out to understand the chemical state and bonding states of the coated TiC film via laser carburizing. Core level spectra of Ti *2p* region (figure 3(a)) clearly showed spin-orbit doublet Ti $2p_{3/2}$ and Ti $2p_{1/2}$ of Ti *2p* for all the chemical bonds between carbon and titanium, hence further confirmed the formation of Ti-C phase. We have fitted Ti *2p* spectra with three sets of peaks at binding energies corresponds to Ti $2p_{3/2}$ and Ti $2p_{1/2}$ doublet, peaks at 456.2eV and 462.2 eV attributed to the Ti-C bonding [2]. Satellite peaks corresponding to TiC appeared at 458.7 eV and 464.7 eV for the Ti $2p_{3/2}$ and Ti $2p_{1/2}$ doublet, respectively. The difference between the doublet and corresponding satellite peak is approximately 2.1 eV and matches with the literature [14]. Production of satellite peaks may be related with the presence of different phases defined in the literature. The shake-up satellite is due to the sudden change in columbic potential as the emitted photoelectron passes through the valence band. The loss in energy is equivalent to the difference in binding energy of main peak and shake-up peak [14-15].

Figure 3(b) shows the typical XPS core level spectra of C *1s*region. The binding energy at 281.6 eV corresponds to C-Ti bond again indicating formation of Ti-C phase. However, intense peak at binding energy of 284.0 eV appeared, that has been attributed to presence of carbon in carbon matrix agreed well with the early reports [2]. Quantitative analysis of the component could not be carried out due to the excessive presence of the carbon in carbon matrix of the coated film.

*3.4 Resonant photoemission of Ti 3d*
It is well known that *3d* electron states show strong resonant in the valence band to Ti and its interstitial compound between the photon energy range 40 eV to 50 eV. Figure 4 shows the energy distribution curves of valance band of TiC in the photon energy between 30 eV to 60 eV. One can clearly see that when we tuned the photon energy above 40 eV, the intensity variation of peak near Fermi edge changes rapidly. It shows maxima at 46 eV, this intensity increment is correlated with the result of resonance photoemission of Ti *3d* character in the valence band of titanium and its compound [9-10].

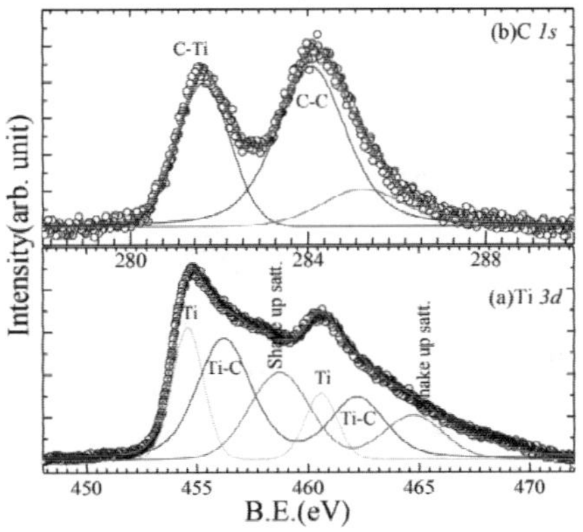


**Figure 3.** (a) Ti 2p and (b) C 1s, Core level Spectra of TiCcoating.

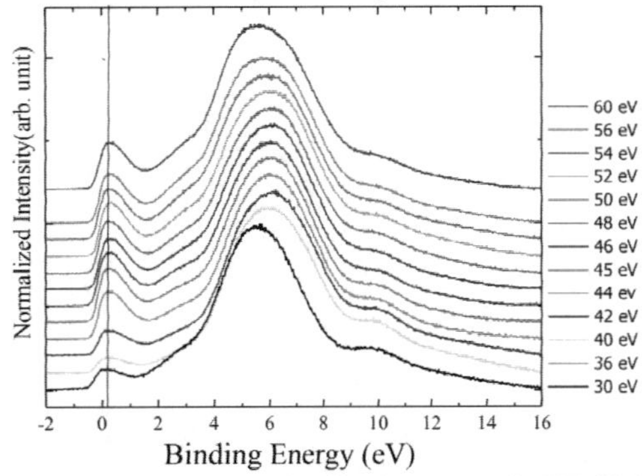


**Figure 4.** Energy distribution curve's(EDC's)ranging from 30 eV ≤ hυ≤ 60 eV for valence band of TiC.

## 4 Conclusion
Non-stoichiometric oxide free FCC phase of TiC has been formed during the laser carburizing on titanium sheet. The hardness of the coating is quite high in comparison to Ti substrate. XPS further confirmed the growth of TiC. RPES study reveals the resonant photoemission of Ti *3d* states at 46 eV

photon energy involved in the valance band of TiC. The hardness of the coatings depends on the strong hybridization of Ti *3d* with C *2p*.

## Acknowledgements

This work was sponsored by UGC-DAE CSR, Indore. Authors are thankful to Prof. Peter Schaaf, Technical University, Ilmenau, Germany for the nano-indentation measurements. One of the author thanks to Mr. A. Wadekar for his help during PES measurements.

## References

[1] Dong B-X, Qiu F, Li Q, Shu S-L, Yang H-Y and Jiang Q-C 2019 *Nanomaterials* **9** 1152
[2] Gong Y, Tu R and Goto T 2013 *Mater. Res. Bull.* **48** 2766
[3] Guemmaz M, MosserA and Parlebas J C 2000*J. Electron Spectrosc.* **107** 91
[4] Kim H K, Kim H G, Lee B-S, Min S-H, Ha T K, Jung K-H, Lee C-W and Park H-K 2017 *mater. Trans.* **58** 592
[5] Izman S, Abdul-Kadir M R, Anwar M, Nazim E M, Rosliza R, Shah A and Hassan M A 2012 *Titanium Alloys-towards achieving enhanced properties of diversified applications* 214
[6] Teghil R, D'Alessio L, Bonis A D, Galasso A, Villani P and Santagata A 2006 *Thin Solid Films* **515** 1411
[7] Kusano E, Satoh A, Kitagawa M, Canto H and Kinbara A 1999 *Thin Solid Films* **343-344** 254
[8] Tang B, Liu H, Wang L, Wang X, Gan K, Yu Y, Wang Y, Sun T and Wang S 2004 *Thin Solid Films* **186** 320
[9] Fuentes G G, Prieto P, Morant C, Quiros C, Nunez R, Soriano L, Elizalde E, andSanz J M 2001 *Phys. Rev.* B **63** 075403
[10] Sanchez-Agudo M, Soriano L, Quiros C, Roca L, Perez-Dieste V and Sanz J M 2002 *Surf. Sci.* **507-510** 672
[11] Soni S, Nair K G M, Phase D M and Gupta R 2012 *AIP confproc* **1447** 677
[12] Gupta R, Soni S and Phase D M 2015 *Appl. Phys. A* **118** 191
[13] Soni S, Pandey A, Sinha A K and Gupta R 2015 *AIP confproc* **1665** 080055
[14] Kotani A and Ogasawara H 1992 *J. Electron Spectrosc. Relat. Phenom.* **60** 257
[15] Jolly W L 1972 *Electron Spectroscopy* (North-Holland, Amsterdam)

IOP Conf. Series: Materials Science and Engineering 1120 (2021) 012030 doi:10.1088/1757-899X/1120/1/012030

# Thermal and electrical properties of $In_xAl_{1-x}N$ alloy

**Jay Kumar Mehra* and Bijay Kumar Sahoo**

Department of Physics, National Institute of Technology Raipur, G.E Road, Raipur - 492010 (C.G) India

*Corresponding author's e-mail address: jkmehra.phd2019.phy@nitrr.ac.in

**Abstract.** $In_xAl_{1-x}N$ is a promising semiconducting material due to its wide range of potential applications in optoelectronics, bio sensing and in thermoelectric power generation. It has a tunable band gap (6 to 0.7eV), high electron saturation velocity, mobility (290 $cm^2/Vs$) and non-toxic nature. Due to this, the material's electrical and thermal properties require further study in detail to explore more potential applications. In this work, we have investigated theoretically the thermal and electrical properties of $In_xAl_{1-x}N$ alloy to predict thermoelectric (TE) figure of merit and performance to identify the most promising composition and carrier densities for optimum power generation at room temperature and above.

## 1. Introduction

III- nitride semiconductors are very useful materials for high-power and high temperature opto-electronics and thermoelectric (TE) devices. In addition to their wide band gap, nitride materials also demonstrate excellent electronic and thermal properties. III-nitride based materials are increasingly used in a wide variety of application such as laser diodes, LEDs and solar cells with a number of commercial products already available [1,2]. Thermoelectric effect is a phenomenon in which heat energy is converted into electrical energy and vise-versa [3]. In order to improve efficiencies, new materials with high TE figure of merit (ZT) must be developed. The efficiency of a TE device is directly related to ZT. Figure of merit is given by

$$ZT = \frac{S^2 \sigma T}{k_{lattice} + k_{electron}} \quad (1)$$

Here, σ is electrical conductivity, S is Seebeck coefficient, *k* is thermal conductivity due to the lattice vibration ($k_p$) and electron conduction ($k_e$), and T is the absolute temperature. The ZT value of the best commercial TE material such as III- nitride semiconductors (GaN, AlN and InN) has attracted attention as potential system for optical and electrical devices. Liu et al investigated ZT value for bulk GaN at T=300K, is about 0.0017 while it can reach 0.2 in the thermally resistive $Al_{0.4}Ga_{0.6}N$ alloy at 1000K [4]. Taylor and Francies investigated ZT for InN/AlN heterojunction is 0.014 at 8K and then decreases up to 40K. Recently, ternary nitrides (InAlN etc) are promising materials for high ZT. InGaN alloy have been shown to have high ZT at room temperature [5,6]. Sztein et al predicted ZT of 0.57 at 1200K for AlGaN. The TE properties of AlInN alloy experimentlly investigated by Tong et al [15]. Experimental result has been motivated to research to explore theoretically. In this work, we present the theory of thermal, electrical properties and ZT of $In_xAl_{1-x}N$ alloy.

## 2. Electrical and thermal Transport properties

*2.1 Electron* transport model

The TE characteristics, electrical conductivity (σ), Seebeck coefficient(S), and $k_e$ are derived from the solution of the Boltzmann transport equation.

$$\sigma = \frac{-2q^2}{3m}\int_{E=0}^{\infty}\frac{\partial f_0}{\partial E}D(E).E.\tau dE \quad (2)$$

$$S = \frac{1}{qT}\left(\frac{\int_{E=0}^{\infty}\frac{\partial f_0}{\partial E}D(E).E[E-\mu].\tau dE}{\int_{E=0}^{\infty}\frac{\partial f_0}{\partial E}D(E).E.\tau dE}\right) \quad (3)$$

$$k_e = \frac{2}{3mT}\left(\frac{\int_{E=0}^{\infty}\frac{\partial f_0}{\partial E}D(E).E^2.\tau dE}{\int_{E=0}^{\infty}\frac{\partial f_0}{\partial E}D(E).E.\tau dE} - \int_{E=0}^{\infty}\frac{\partial f_0}{\partial E}D(E).E^3.\tau dE\right) \quad (4)$$

Here, q is charge of electron, E is electron energy, D(E) is density of state, μ is chemical potential and $f_0(k)$ is the Fermi - Dirac distribution function which is expressed by

$$f_0(k) = \frac{1}{\exp\left(\frac{E(k)-\mu}{k_\beta T}\right)+1} \quad (5)$$

The energy dependent electron relaxation time $\tau_{tot}(E)$ as per relaxation time approximation (RTA) is given by [14]

$$\tau_{tot}^{-1}(E) = \sum \tau_i^{-1} \quad (6)$$

Here, the index *i* describes different scattering term. The scattering mechanism is combination of piezoelectric scattering, impurity scattering, alloy scattering, polar optical phonon scattering, deformation potential scattering, and charge dislocation scattering.

*2.2 Phonon transport model*

The phonon component of thermal conductivity is first proposed by Callaway [7]. The Thermal conductivity of a material can be expressed by $k = k_{ph} + k_e$ here $k_{ph}$ is the thermal conductivity due to phonons and $k_e$ is thermal conductivity due to electrons. Florsecu et al. [8] experimentally found that $k_e = 10^{-3}\ k_{ph}$. So, $k_e$ is neglected and for a medium doped semiconductor $k = k_{ph}$ . The Callaway model for $k_{ph}$ InAlN can be written by

$$k = \frac{k_b^4 T^3}{2\hbar^3\pi^2 v}\int_0^{\frac{\theta_d}{T}}\tau_c(x)\frac{x^4 e^x}{(e^x-1)^2}dx \quad (7)$$

Here *v* is average group velocity, $\theta_d$ is Debye temperature and $\tau_C$ represent the combined scattering time due to both normal and resistive processes. The variable of *x* is connected to the phonon frequency(ω), is defined by

$$x = \frac{\hbar\omega}{k_b T} \tag{8}$$

The phonon normal scattering ($\tau_N^{-1}$) represent phonon - phonon scattering where momentum is conserved and in resistive scattering ($\tau_R^{-1}$), the momentum is not conserved [10-13]. The key resistive rates are three phonon Umklapp scattering ($\tau_U^{-1}$), dislocation scattering ($\tau_D^{-1}$), phonon-electron scattering ($\tau_{e-p}^{-1}$) and boundary scattering ($\tau_B^{-1}$). virtual atomic mass *M* describes by alloy components $M = y_A M_A + (1 - y_A) M_B$, $y_A$ is mass fraction of alloy constituent A; and $M_A$ and $M_B$ represent atomic masses of constituents A and B respectively. The normal scattering is given by

$$\tau_N^{-1} = \frac{k_b \gamma^3 V}{M\hbar^2 v^5} \varpi^2 T^3 \tag{9}$$

Here $\gamma$ is Gruneisen parameter, V volume per unit atom, phonon- phonon scattering events where momentum is not conserved called Umklapp scattering and given by [13].

$$\tau_U^{-1} = \frac{\hbar\gamma^2 T\omega^2}{Mv\theta_d} e^{(-\theta_d/3T)} \tag{10}$$

Point defect scattering is a very important scattering which having due to heavily doped atoms. It is also known as mass defect scattering following the approach of Liu et al. [14].

$$\tau_P^{-1} = \frac{V_0 \Gamma \omega^4}{4\pi v^3} \tag{11}$$

Here, $\Gamma$ is point defect scattering factor [14]. The phonon scattering rate at the core of the dislocation is proportional to the cube of the phonon frequency and is given by.

$$\tau_D^{-1} = \eta N_D \frac{V_0^{4/3}}{v^2} \omega^3 \tag{12}$$

Here, $N_D$ is density dislocation lines, and $\eta$ is weight factor to account for the mutual orientation of the direction of the temperature gradient and the dislocation line. The average value of $\eta = 0.55$, at low doping levels, electron-phonon scattering is given by,

$$\tau_{e-p}^{-1} = \frac{n_e \varepsilon_l^2 \omega}{\rho v^2 k_b T} \sqrt{\frac{\pi m^* v^2}{2k_b T}} \exp\left(-\frac{m^* v^2}{2k_b T}\right) \tag{13}$$

Here $\eta_e$ is electrons concentration, $\epsilon_1$ is deformation potential, and $m^*$ is electron effective mass. The boundary scattering is,

$$\tau_B^{-1} = \frac{v}{L} \tag{14}$$

Here L is the dimension of the sample. The combined phonon relaxation time is given by.

$$\tau_t = \left[\frac{1}{\tau_N} + \frac{1}{\tau_U} + \frac{1}{\tau_P} + \frac{1}{\tau_D} + \frac{1}{\tau_B} + \frac{1}{\tau_{p-e}}\right]^{-1} \tag{15}$$

IOP Conf. Series: Materials Science and Engineering 1120 (2021) 012030 doi:10.1088/1757-899X/1120/1/012030

## 3. Results and discussion

We have computed combined relaxation time for phonons in $In_xAl_{1-x}N$ at room temperature, which is shown in figure 1. It can be found that combine relaxation time ($\tau_t$) decreases with phonon frequency at room temperature. It has been observed that point defect scattering is the dominant mechanism due to which phonon relaxation time decreases. It is important to keep in mind that the relaxation time affects thermal conductivity. Material parameter values have been taken from ref. [9,10].

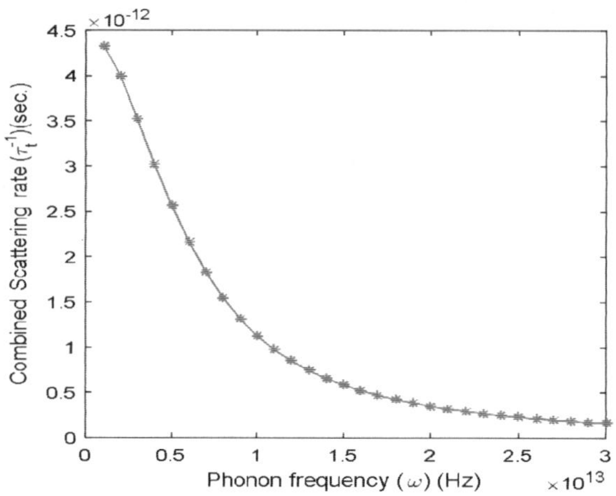


**Figure 1.** Combine relaxation time for $In_xAl_{1-x}N$ alloy at room temperature.

We have computed thermal conductivity of InAlN at 300K. Figure 2shows the $k$ of InAlN alloy. It can be seen that when Al-content varies from $x = 0.1$ to 0.5, thermal conductivity ($k$) of InAlN decreases initially and attains minima at $x$=0.5 and when Al content varies from $x$=0.5 to 0.9, $k$ increases and finally catches thin film limit. The simulated $k$ is compared with available experimental results. It can be observed that simulated $k$ is close to experimental results.

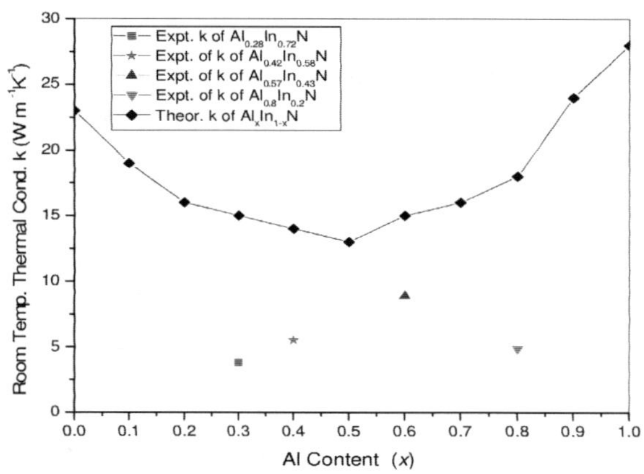


**Figure 2.** Thermal conductivity for $In_xAl_{1-x}N$ alloys with various Al-contents.

The thermoelectric property of InAlN is experimentally investigated by Tong et al and found that n-type $In_{0.17}Al_{0.83}N$ film ($n$= 5.1x10$^{18}$ cm$^{-3}$) has thermal conductivity $k$ equal to 4.87 W/(m-k), electrical conductivity $\sigma$ equal to 2.38x10$^{4}$/(Ω-m), and Seebeck coefficient S equal to -6.012x10$^{-4}$ (V/K). The power factor ($P$) for the $n$- $In_{0.17}Al_{0.83}N$ was measured as 0.864x10$^{-2}$ W/(mK$^{2}$) and the TE figure of merit ($ZT$) value of 0.532 was measured at room temperature [$T$=300K]. The result shows InAlN alloy exhibits excellent material for thermoelectric application.

## 4 Conclusions

Thermal and electrical properties $In_xAl_{1-x}N$ alloy are studied. Thermal conductivity is calculated by Callaway model for phonon transport. III-V nitride semiconductors are one of the good choices due to medium range of carrier concentration (5.1x10$^{18}$ cm$^{-3}$), high carrier mobility, high Seebeck coefficient and low thermal conductivity, which is relevant for improving thermodynamic efficiency. The accuracy of theoretical result is confirmed through comparison to experimental data.

## References

[1] Sztein A, Ohta H, Bowers J E, DenBaars S P and Nakamura S J 2011 *Appl. Phys.* **110** 123709
[2] Tritt T M and Subramanian M A 2006 MRS Bull. **31** 188
[3] Vineis Christopher J 2010 *Advanced Materials* **22** 3970
[4] LiuW L and Balandin A A 2005 *J. Appl. Phys.* **97** 123705
[5] Zhang J and Tansu N 2013 Optical and Polarization Properties with Staggered AlGaN Quantum Wells for Mid- and Deep-Ultraviolet Lasers and Light Emitting Diodes, Proc. of the SPIE Photonics West 2013, San Francisco, CA, January
[6] Zhu P F, Weigel P O, Liu G Y, Zhang J, Weldon A L, Muangnaphor T, Gilchrist J F and Tansu N2013 Optimization of Deposition Conditions for Silica / Polystyrene Microlens and Nanolens Arrays for Light Extraction Enhancement in GaN Light-Emitting Diodes, *Proc. of the SPIE Photonics West 2013*, San Francisco, CA, January
[7] Callaway J 1959 Phys. Rev. **113** 1046
[8] Florescu DI, Asnin V M and Pollak Fred H 2000 *J. Appl. Phys.* **88** 3295
[9] Liu W L and Balandin A A 2005 *J. Appl. Phys.* **97** 073710
[10] Sztein A, Haberstroh John, Bowers John E, Den Baars Steven P and Nakamura Shuji 2013 *J. Appl. Phys.* **113** 183707
[11] Liu W and Balandin A A 2005 *J. Appl. Phys.* **97** 073710
[12] Liu W and Balandin A A 2004 *Appl. Phys. Lett.* **85** 5230
[13] Morelli D, Heremans J and Slack G 2002 Phys. Rev. B **66** 195304
[14] Jimenez-Rioboo R J, Cusco R, Oliva R, Domenech-Amador N, Prieto C, Ibanez J, Boney C, Bensaoula A and Artus L 2012 *Appl. Phys.* **101** 062103
[15] Tong H, Zhang J, Liu G, Herbsommer J A, Huang G S and Tansu N 2010 *Appl. Phys. Lett.* **97** 112105

# Polarization electric field effect on cross-plane thermal conductivity of GaN/$In_xGa_{1-x}N$/GaN superlattice ($x \leq 0.3$)

**Anju Pansari and Bijay Kumar Sahoo**

Department of Physics, National Institute of Technology, Raipur – 492010

*Corresponding author's e-mail address: anjupansari20@gmail.com

**Abstract**. GaN/$In_xGa_{1-x}N$/GaN superlattice (SL) has electric field at interfaces termed as interfacial polarization electric (IPE) field. In this work, effect of this electric field on cross-plane thermal conductivity ($k_{cp}$) of the SL is investigated for Indium content $x \leq 0.3$. IPE field revises phonon velocity which enhances interfacial scattering and thermal boundary resistance (TBR). This is due to unequal changes in specific heat and phonon velocity which leads to decrease in phonon transmission and more mismatches of acoustic properties of material. This reduces $k_{cp}$. Room temperature $k_{cp}$ in presence (absence) of IPE field for GaN (10nm)/$In_xGa_{1-x}N$ (5nm) SL are 4.652 (5.720) and 4.282 (5.221) $Wm^{-1}K^{-1}$ respectively, for x=0.1 and 0.3. This work demonstrates that electric field of nitride SL can be utilized to reduce *k* for optimum thermoelectric power production.

## 1. Introduction

Efficiency of a thermoelectric (TE) material is generally quantified by dimensionless figure of merit (ZT) which is directly proportional to electrical conductivity ($\sigma$) and Seebeck coefficient (*S*); but inversely proportional to *k* of the material [1]. The main paradigm to achieve material's high ZT is to enhance their power factor ($S^2\sigma$) and/or reduce their *k*. A small reduction in *k* can improve TE efficiency significantly. Various approaches have been undertaken to reduce *k* of SL. Doping, alloying, period width variation, lattice mismatch, nano-particle substitution, interfacial roughness and atomic disorder etc [2-5]. Many of these studies have proved that *k* of SL is lower than the corresponding *k* of bulk material. Thus, more approaches must be explored to reduce *k* of SL. In this work, polarization property of nitride SL has been exploited to reduce *k* of SL.

Recent study reported that GaN /$In_xGa_{1-x}N$/ GaN SL can be a promising contender for TE devices due to its lowest *k* within the III-V nitride alloys [6]. Its ZT can approach best TE material Bi and Te. This motivates to explore methods to enhance ZT of GaN /$In_xGa_{1-x}N$ SL reducing *k* and increasing $\sigma$, *S* of the SL further. This SL has a unique advantage called polarization mechanism which produces an interfacial polarization electric (IPE) field at each interface of the SL due to difference in spontaneous (SP) and lattice mismatch piezoelectric (PZ) polarizations of layers [7]. Sztein et al. [6] demonstrated that IPE field of GaN/AlN/AlGaN SL improves $\sigma$ and *S* resulting in high ZT. However, effect of IPE field on *k* of the SL has not been studied. In this article, we have explored effect of IPE field on *k* of GaN/$In_xGa_{1-x}N$/GaN SL for indium content $\leq$30%.

## 2. Theoretical Model

### *2.1 IPE field and its effect*

Figure1 shows structure of a GaN/$In_xGa_{1-x}N$/GaN SL. Total polarization in a layer is $P=P^{sp}+P^{pz}$. In this SL, GaN is barrier while $In_xGa_{1-x}N$ performs as well. The interfacial electric field due to total polarization in a well $E_w$ (or in barrier $E_b$) layer is [7]: $E_w = l_b(P_b - P_w)/(\varepsilon_b l_w + \varepsilon_w l_b)$; $E_b = l_w(P_w - P_b)/(\varepsilon_b l_w + \varepsilon_w l_b)$. Here $l_w$ ($l_b$), $P_w$($P_b$) and $\varepsilon_w$ ($\varepsilon_b$) are width, polarization and dielectric constant of well (barrier) material, respectively. The IPE field is ~ 1 MV/cm.

**Figure 1.** Structure of GaN/$In_xGa_{1-x}N$/GaN SL.

Nitrides are piezoelectric semiconductors; so, as per Heckmann principle, the coupling between the strain field and electric field will induce additional electric polarization. This will contribute to elastic constant of the alloy. The elastic constant of $In_xGa_{1-x}N$ in the presence of IPE field has been found as [8]

$$C_{44,\,p}^{InGaN} = C_{44}^{InGaN} + [(e_{15}{}^2 + e_{31}{}^2 + e_{33}{}^2 + p_{sp}{}^2)/\varepsilon_0 \varepsilon\ ]^{InGaN}, \qquad (1)$$

where $C_{44}^{InGaN} = [48\,x + 105\,(1-x) - 16\,x(1-x)]$ and $C_{33}^{InGaN} = [224x + 398\,(1-x) - 38\,x(1-x)]$ GPa are elastic constant of $In_xGa_{1-x}N$ in the absence of IPE field [8]. The transverse and longitudinal phonon velocities are determined by $v_T = [C_{44}^{InGaN}/\rho]^{1/2}$ and $v_L = [C_{33}^{InGaN}/\rho]^{1/2}$, respectively; where $\rho$ is material density. The average phonon velocity is obtained as $v = [\tfrac{1}{3}(1/v_{T,1} + 1/v_{T,2} + 1/v_L)]^{-1}$. Debye frequency is determined by the expression $\omega_d = v[3N/4\pi V_0]^{1/3}$. Here, $N$ is the number of atoms per unit volume $V_0$. Debye temperature $\theta_d$ is calculated by $\theta_d = h\omega_d / k_\beta$; where $h$ is Planck's constant and $k_\beta$ is Boltzmann's constant.

### *2.2 k of GaN/InGaN/GaN SLs*

The $k$ of a semiconductor is written as sum of phonon and electron contribution $k = k_{ph} + k_e$. It is demonstrated that $k_e = 10^{-3} k_{ph}$. Thus, electronic contribution is neglected and for a semiconductor $k = k_{ph}$. In solids, the thermal energy is carried by acoustic phonons. These phonons suffer scatterings due to other phonons, defects and interfaces. Here, normal scattering ($\tau_n^{-1}$), phonon- phonon Umklapp ($\tau_u^{-1}$), alloy or point-defect ($\tau_p^{-1}$), dislocation ($\tau_d^{-1}$), phonon-electron ($\tau_{e-p}^{-1}$) and boundary ($\tau_b^{-1}$) scattering rates are taken into calculation. The combined phonon scattering can be written as $\tau_c^{-1} = [\tau_n^{-1} + \tau_u^{-1} + \tau_p^{-1} + \tau_d^{-1} + \tau_{e-p}^{-1} + \tau_b^{-1}]$. Thermal conductivity $k$ of a layer of the SL is determined by Callaway expression ($y = \hbar\omega / k_\beta T$) [8]

$$k = k_{ph} = [\frac{k_{\beta}^{4} T^{3}}{2\pi^{2} \hbar^{3} v} \int_{0}^{\theta_D / T} \frac{\tau_c y^4 e^y}{(e^y - 1)^2} dy ] \quad (2)$$

In order to determine $k$ of a SL, effect of interfaces has to be included. Interfaces generates thermal boundary resistance (TBR) which is computed by acoustic mismatch model (AMM)and diffuse mismatch model (DMM). If the interface is specular, then phonon transmissivity from a layer $i$ to layer $j$, $\tau_{ij}^{S}$ is [9]

$$\tau_{ij}^{S} = (4Z_i Z_j)/(Z_i + Z_j)^2 \quad (3)$$

Where $Z_i = v_i \rho_i$ is acoustic impedance, $v_i$ the phonon group velocity and $\rho_i$ the material density of the $i^{th}$layer. When interface is diffuse, diffuse transmissivity $\tau_{ij}^{D}$ is

$$\tau_{ij}^{D} = (C_j v_j)/ (C_i v_i + C_j v_j) \quad (4)$$

Here $C_i$ is specific heat of $i^{th}$ layer. With ($\mu_i = \cos\theta_i$), the total transmission coefficient $\Gamma_{ij}$ from material $i$ to material $j$ is

$$\Gamma_{ij} = p\Gamma_{ij}^{S} + (1-p)\Gamma_{ij}^{D} = p * 2\int \tau_{ij}^{S} \mu_i \, d\mu_i + (1-p) * \int \tau_{ij}^{D} \mu_i \, d\mu_i \quad (5)$$

Using $\Gamma_{ij}$ determined from Eq.(5), we can estimate TBRs from material $i$ to material $j$ by

$$R_{ij} = [\frac{4\pi^2 \hbar^3 v_i^2}{k_{\beta}^4 \Gamma_{ij} T_i^3}] / [\int_0^{\theta_D / T_i} y^4 e^y / (e^y - 1)^2 \, dy ] \quad (6)$$

Normally, $k$ of SL is a tensor because of different materials and interfaces which generates inhomogeneity in thermal flow. Depending on the direction of imposed temperature gradients, $k$ is divided into two parts: thermal flow parallel to a layer plane is termed as in-plane thermal conductivity ($k_{ip}$); whereas thermal flow across the layers is cross-plane thermal conductivity ($k_{cp}$). The $k_{cp}$of GaN/$In_xGa_{1-x}N$/GaN SLs with two layers per period is estimated by

$$k_{cp} = \frac{l_1 + l_2}{\frac{l_1}{k_1} + \frac{l_2}{k_2} + 2R} \quad (7)$$

Here $l_1(l_2)$ and $k_1(k_2)$ are layer thickness and thermal conductivities of GaN(InGaN), respectively. $R$ represents the TBRs between GaN and InGaN.

## 3. Result and Discussion

### *3.1 k of GaN/ $In_{0.1}Ga_{0.9}N$ /GaN SL*

Figure 2 shows $k_{cp}$ as a function of temperature for $x$=0.1 in the absence and presence of IPE field. The $k_{cp}$ has linear dependence on temperature from 20 <$T$< 120 K, whereas from 120 <$T$< 250K $k_{cp}$ has nonlinear dependence on temperature and $k_{cp}$ is relatively independent of temperature for $T$> 250 K [3].

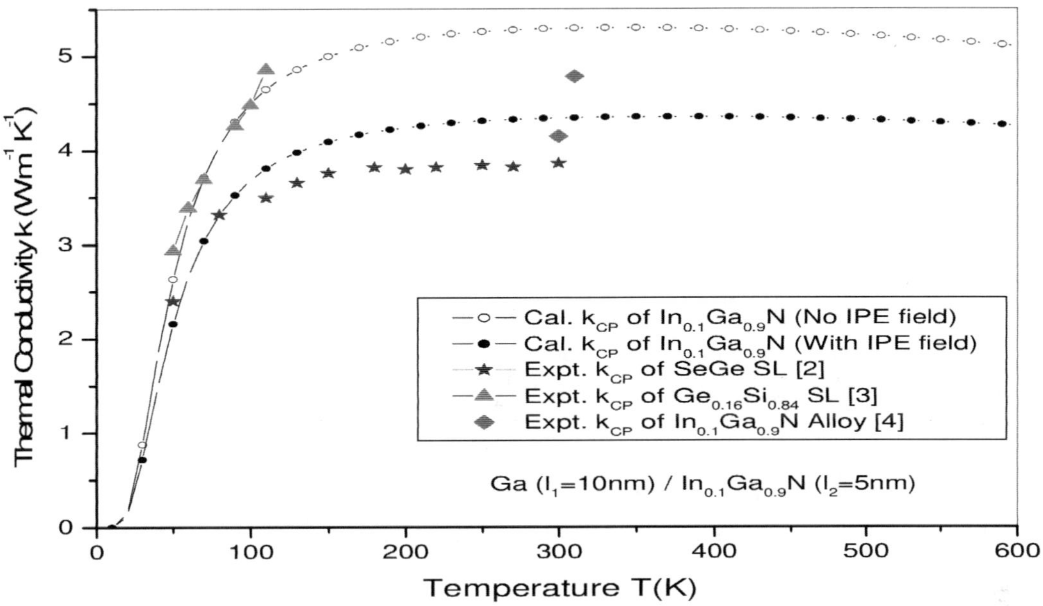


**Figure 2.** $k_{CP}$ of GaN/$In_{0.1}Ga_{0.9}N$/GaN SL as a function of temperature.

From figure 2, it can be found that $k_{cp}$ in presence of IPE field is lower than $k_{cp}$ in the absence of IPE field. The RT $k_{cp}$ of GaN (10nm)/$In_{0.1}Ga_{0.9}N$(5nm)/GaN SL in the absence and presence of IPE field is 5.720and 4.652$Wm^{-1}K^{-1}$respectively. This implies that IPE field reduces RT $k_{cp}$ and the reason for decrease of $k_{cp}$ is the increased TBR and decreased $k$ of layers due to IPE field. IPE field enhances phonon velocity and Debye temperature of the constituent material of the SL. Alloy effect raises scattering of high frequency phonons whereas boundary/interfaces enhance scattering of low frequency phonons. Low frequency phonons significantly contribute to $k$ of SL. IPE field increases the strength of boundary/interfaces scattering in the SL resulting in enhanced scattering of low and mid frequency phonons further. Phonon-mode mismatch between layers, which scales with the lattice mismatch at the interface is significantly modifying the $k$. Tong et al.[10] experimentally determined $k$ of $In_{0.1}Ga_{0.9}N$ alloy as 13$Wm^{-1}K^{-1}$with 233 nm thicknesses. Our predicted value is well below this alloy limit. Xu et al.[5] observed $k_{cp}$ of GaN/$In_{0.1}$ $Ga_{0.9}N$/GaN nanoporus SL as 4.2 $Wm^{-1}K^{-1}$which they attributed to additional phonon scattering from porosity of the sample. Zhang et al. [4] determined $k_{cp}$ of GaN/$In_{0.1}$ $Ga_{0.9}N$/ SL as 4.8 $Wm^{-1}K^{-1}$which is closure to our predicted value.

*3.2 k of GaN/$In_{0.3}Ga_{0.7}N$/GaN SL*

Figure 3 shows $k_{cp}$ of GaN/$In_{0.3}Ga_{0.7}N$/GaN SL as a function of temperature in the absence and presence of IPE field. Here also $k_{cp}$ curve can be divided into 3 regions: (i) linear dependence on temperature from 20 < T < 120 K which is attributed to coherent phonon transport through the layers and $T^3$ behaviour; (ii) nonlinear dependence on temperature from 120 < T < 250K which is due to the increased influence of incoherent effects and (iii) relatively independent of temperature for T > 250 K; but decreases very slowly with increasing temperature for T > 300 K [3-5].

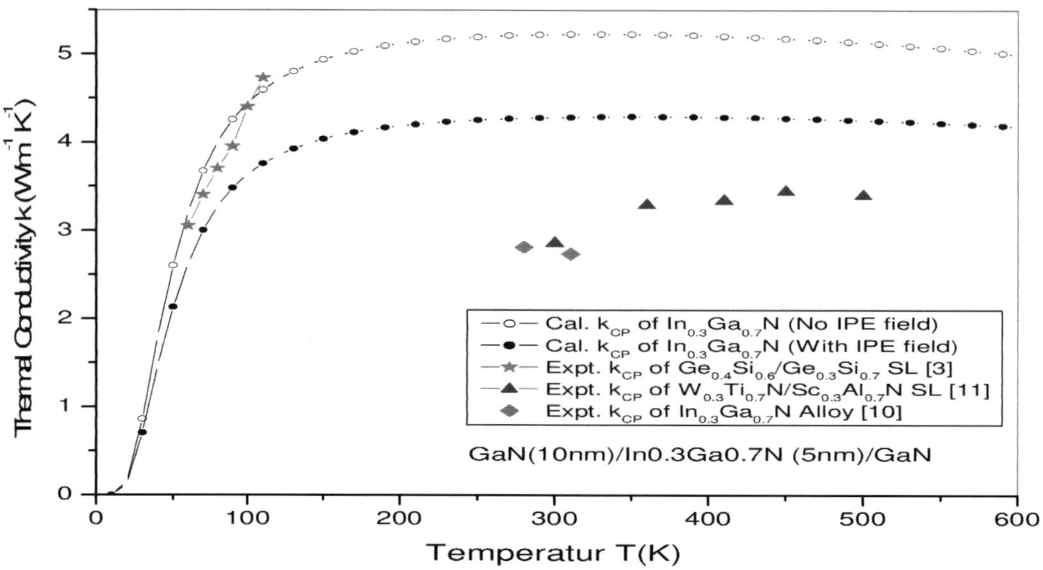


**Figure 3.** $k_{CP}$ of GaN/$In_{0.3}Ga_{0.7}N$/GaN SL as a function of temperature.

The nature of $k_{cp}$ in these regions can be explained in a way similar to GaN/$In_{0.1}$ $Ga_{0.9}N$/GaNSL under the fact that the low-frequency phonons are not affected in the temperature region T < 120K, whereas the high-frequency phonons are greatly affected for elevated temperature T > 300K due to anharmonic (Umklap) scattering which makes $k_{cp}$ to saturate with $T$. Our result shows that room temperature $k_{cp}$ of GaN/$In_{0.3}Ga_{0.7}N$/GaN SL in the absence and presence of IPE field is 5.221 and 4.282$Wm^{-1}K^{-1}$ respectively. This implies that IPE field reduces room temperature $k_{cp}$ and is due to enhanced TBR under influence of IPE field. From figure 3, it can be seen that a minute decrease of $k$ is found for T> 300K due to increase of Umklap scattering. At room temperature authors such as Huxtableet al. [3], Tong et al. [10] and Saha et al. [11] have experimentally determined $k_{cp}$ as 4.6, 2.9 and 3.3 respectively for $Si_{0.6}Ge_{0.4}/Si_{0.7}Ge_{0.3}$ SL, $In_{0.3}Ga_{0.7}N$ alloy and TiN/Al $_{0.7}Sc_{0.3}N$ SL. Low $k$ in the alloy samples reveals presence of high amount of dislocations, impurities and inhomogeneities.

## 4. Conclusion

In this work, effect of interfacial polarization electric (IPE) field on cross-plane thermal conductivity $k_{cp}$ of GaN/$In_xGa_{1-x}N$/GaN SL has been explored for Indium content $x \leq 0.3$. Our result reveals that TBR is enhanced at interfaces under the action of IPE field leading to decreased phonon transmission and more mismatches of acoustic properties. This caused reduction of $k_{cp}$ of the SL. TBR is found to be (2.10 to 5.30)$\times 10^{-9}$ $m^2KW^{-1}$ which is closer to the reported value of similar SLs. Room temperature (RT) $k_{cp}$ in the presence (absence) of IPE field for GaN (10nm)/$In_xGa_{1-x}N$ (5nm) SL are 4.652 (5.720) and 4.282 (5.221) $Wm^{-1}K^{-1}$ which demonstrate more than 20% reduction and are closely agreeing with experimental $k_{cp}$ of similar SLs. This work shows that desired value of $k$ can be achieved by tailoring electric field of nitride SL for maximum power production.

## Acknowledgments

Authors acknowledge with thank to National Institute of Technology Raipur, Govt. of India for financial support through Cumulative Professional Development Allowance (CPDA).

## References

[1] Nolas GS, Sharp J and Goldsmid HJ 2001 Thermoelectrics: Basic Principles and New Materials Developments (Springer, New York)
[2] Tasciuc T B, Liu W, Liu J, Zeng T, Song D W, Moore C D, Chen G, Wang K L and Goorsky M S 2000 Superlatt. and Microst. **28** 199
[3] Huxtable S T, Abramson A R, Tien C, Majumdar A, Bounty C La, Fan X, Zeng G, Bowers JE, Shakouri A and Croke ET 2002 *Appl. Phys. Lett.* **80** 1737
[4] Zhang J, Tong H, Liu G, Herbsommer J A, Huang G S, and Tansu N 2011 *J. Appl. Phys* **109** 053706
[5] Xu D, Wang Q, Wu X, Zhu J, Zhao H, Xiao B, Wang X and HAO Q 2018 *Front. Energy* **12** 136
[6] Sztein A, Bowers J E, DenBaars S P and Nakamura S 2014 *Appl. Phys. Lett.* **104** 042106
[7] Morkoc H 2013 Nitride Semiconductor Devices, *Wiley-VCH Verlag GmbH & Co., Weinheim.*
[8] Sahoo SK, Sahoo BK and Sahoo S 2013 *J. Appl. Phys.* **114** 163501
[9] Mei S and Knezevic I 2015 *J. Appl. Phys.* **118** 175101
[10] Tong T, Fu D, Levander A X, Schaff W J and Pantha B N 2013 *Appl. Phys. Lett.* **102** 121906
[11] Saha B, Koh Y R, Comparan J, Sadasivam S, Schroeder JL, Garbrecht M, Mohammed A, Birch J, Fisher T, Shakouri A and Sands TD 2016 *Phys. Rev. B* **93** 045311

IOP Conf. Series: Materials Science and Engineering 1120 (2021) 012032 doi:10.1088/1757-899X/1120/1/012032

# Full potential study of electronic and thermoelectric properties of GdNiSb compound

**Jagdish Kumar Satyam*  and  Sapan Mohan Saini**

Department of Physics, NIT Raipur, Raipur 492010, India

*Corresponding author's e-mail address: jagdishsatyam51@gmail.com

**Abstract.** In the present work we have investigated electronic and thermoelectric properties of GdNiSb compound. The calculations have been performed using local spin density approximation (LSDA) with Hubbard parameter (U) included for exchange correlation effect. The calculated results of a cubic half-Heusler phase of GdNiSb compound shows semiconducting behaviour due to a narrow band gap (0.279 eV) is opened. Thermoelectric properties such as electrical conductivity ($\sigma/\tau$), Seebeck coefficient (S) and figure of merit (ZT) was calculated. Values of thermoelectric parameters show that GdNiSb compound could be used as potential material for thermoelectric applications.

## 1. Introduction

The Heuslar alloys are new and unique class of intermetallic compounds. These compounds attracted attention due to having many different physical properties, such as magnetism, half-metallicity, spintronic and thermoelectric properties. They also have wide range of applications in electronic devices, such as magnetic sensor, random access memory (RAM), thermoelectric module and spintronic [1]. Recently in last decade ternary rare-earth based Heusler compound have been attracted interest [2-7], due to presence of strongly ordered and localized 4f unpaired electron in RE-element. In this context GdNiSb is ternary REME-type (where RE is rare-earth element, M is d-block transition metal, E is p-block element) half-Heusler compound is considered for investigation. The presence of f-electron in Gd- element, d-electron in Ni- element and p-electron in Sb-element of the compound are the decisive electronic states for the different physical properties [8, 9]. In this paper electronic and thermoelectric properties of GdNiSb were calculated using LSDA with exchange correlation effect implemented in full potential local orbital method in Wien2k package [10, 11]. The structure of GdNiSb is face centered cubic (fcc) MgAgAs-type having space group *F43m* (216). The Wyckoff positions in unit cell of Gd, Ni and Sb atoms are 4 (a) (0, 0, 0), 4(b) (¼, ¼, ¼ ) and 4 (d) (½, ½, ½), respectively. We have used the experimental lattice parameter $a_{exp.}$= 6.338 Å [12] to performed full potential study of electronic and thermoelectric properties of GdNiSb compound.

## 2. Computational method

In order to study the electronic and thermoelectric properties of GdNiSb compound, we performed first principle calculations using density functional theory (DFT). Coulomb corrected local spin density approximation (LSDA+U) [13, 14] has been used for exchange correlation potential as implemented in wien2k. Self consistent calculations were performed within $18\times 18\times 18$ $k$-mesh point. The cutoff energy was taken as -7.0 Ry with charge and energy converged at 0.0001 e and 0.0001 Ry respectively.

Due to presence of localized f- electron in Gd element we used the method of effective Hubbard parameter $U_{eff}$= U-J (where U is Hubbard potential and J is exchange) for more accurate calculations [15]. The value of $U_{eff}$ vary with different atoms due to correlation effect induced by the coulomb interaction of localized f-electron [16] and we used the value $U_{eff}$ = 6.0eV (where U= 6.7eV and J= 0.7eV) for Gd atom. These values have been calculated by Harman et al. [17] using by the super cell approach.

## 3. Results and Discussions

### *3.1 Electronic properties*

In order to study electronic properties of GdNiSb compound the total and partially density of states (DOS) were calculated and shown in Fig. 1.

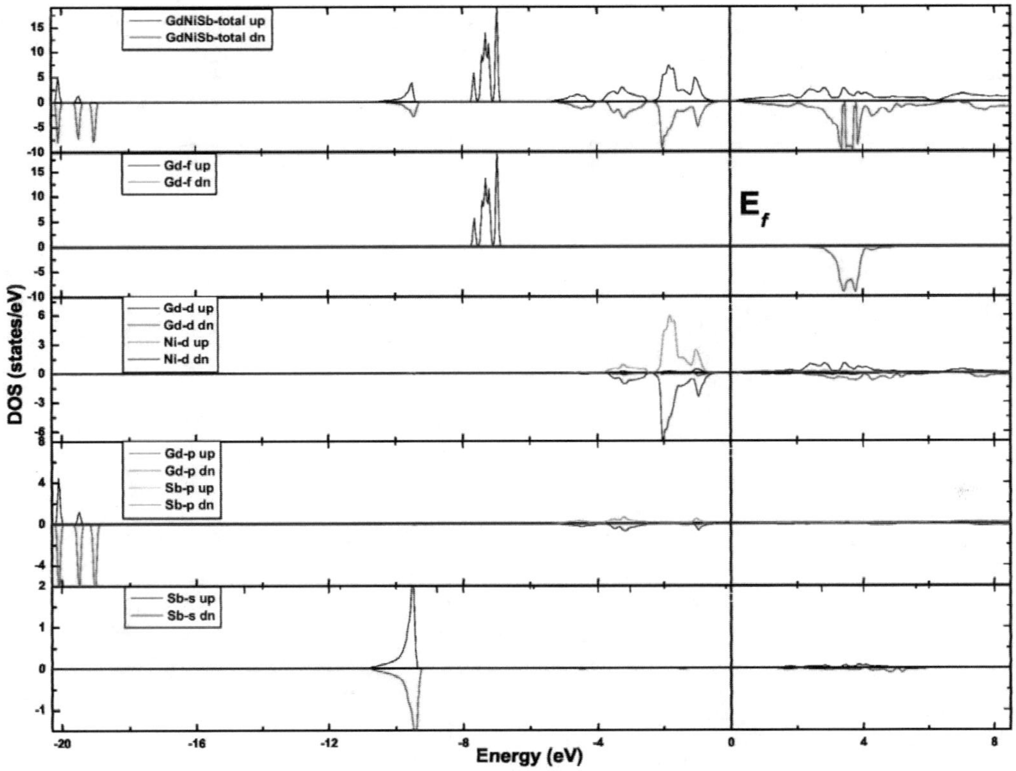


**Figure 1.** Total and partial DOS for both spin-up and spin-down states of GdNiSb by using LSDA+U approximation.

One can notice that almost zero DOS near Fermi level ($E_F$) is appeared in both spin (up and down) channel and small narrow energy gap (0.279 eV) is opened which reflect the semiconductor nature of GdNiSb compound. We found that Gd - *d,f* and Ni – *d* states mainly contribute in total DOS. While, the contribution of Sb- *p* states are negligible. In the energy range -7eV to 7eV Ni-*d* and Gd –*f* states are mainly dominated in total DOS. In the partial DOS distribution we see spin up Gd- *f* states lie around -7 eV and can be identified as the large peak in the graph. Spin down Gd-*f* states lie around 4 eV above $E_F$ and contribute in the conduction band. Both spin up and down Ni-*d* states lie around 2 eV below eV and contribute to the valance band. Thus Ni-*d* states and Gd-*f* states mainly contribute to electronic transitions from valence band to conduction band. The Gd-5*d* states hybridize with Ni -3*d* states at

IOP Conf. Series: Materials Science and Engineering 1120 (2021) 012032 doi:10.1088/1757-899X/1120/1/012032

~ 4 eV above the $E_F$ and contribute to the CB. Contribution of Gd - 5*d* states to the total DOS is small (<2 state/eV/primitive cell) this is mainly because they are quite empty in nature and are partly involved in metallic bonding.

### *3.2 Thermoelectric properties*

Our DOS calculations show the semiconducting nature of GdNiSb compound. We have computed the thermoelectric parameters such as electrical conductivity (σ), Seebeck coefficient (S) and figure of merit (ZT) (performance indicator of material) of GdNiSb compound by using the Boltzmann transport equation (BTE) within a simple rigid-body approximation and relaxation time approximation (RTA).

#### *3.2.1 Electrical conductivity ($\sigma/\tau$)*

The electrical conductivity ($\sigma/\tau$) is an important component of transport properties. The value of $\sigma/\tau$ must be high for desired thermoelectric materials. The electrical conductivity ($\sigma/\tau$) of GdNiSb compound has been investigated with respect to temperature for both spin configurations and is shown in the Figure 2.

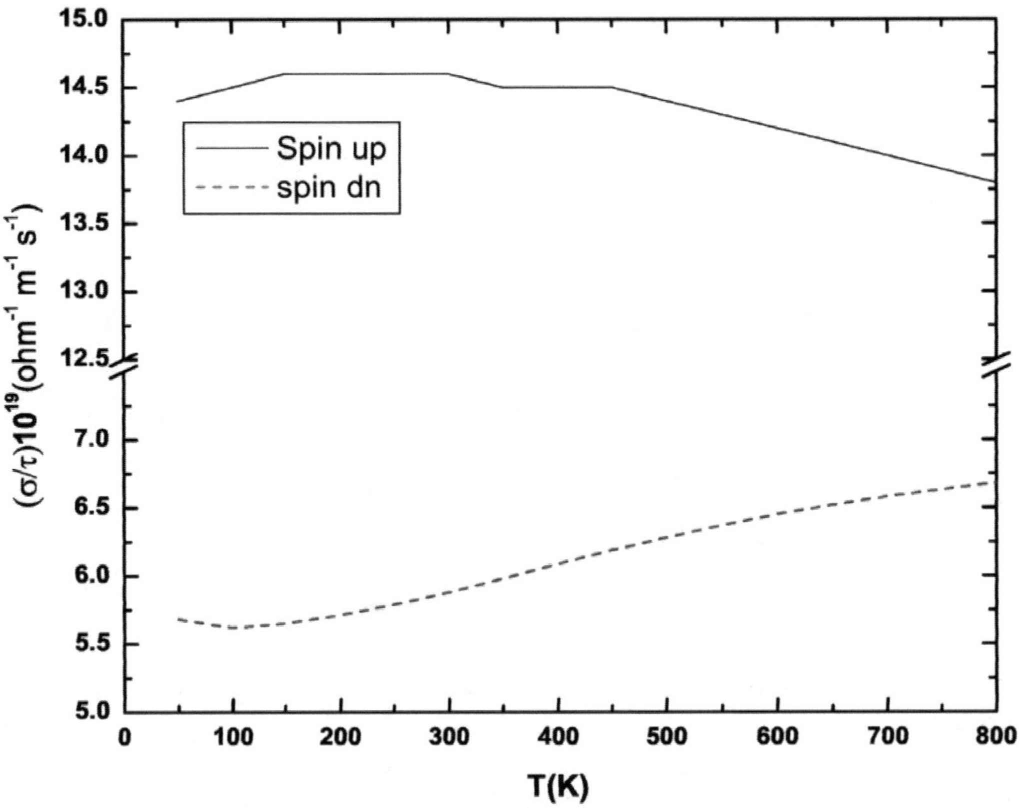


**Figure 2**. Variation of electrical conductivity ($\sigma/\tau$) verses temperature (T) for spin-up and spin-down.

The value of $\sigma/\tau$ for spin-down channel increases with temperature which shows it semiconductor behaviour. While, in the case of spin-up channel a small variation is observed with temperature. The observed value of electrical conductivity at room temperature is $14.6\text{x}10^{19}(\Omega\ \text{m s})^{-1}$ which decreases slightly and reaches to $13.8\times 10^{19}\ (\Omega\ \text{m s})^{-1}$ at 800 K in the spin-up configuration. On the other hand, for spin down configuration, its value increases from $5.88\ \text{x}10^{19}\ (\Omega\ \text{m s})^{-1}$ at room temperature to $6.68\times 10^{19}$

($\Omega$ m s)$^{-1}$ at 800 K. Overall the electrical conductivity of spin-up configuration is high, i.e. almost double as compare to spin-down configuration for all temperature range.

*3.2.2 Seebeck coefficient (S)*

The Seebeck coefficient (S) is important parameter of thermoelectric materials which is expressed by equation $S = \frac{8\pi^2 k_\beta^2}{3eh^2} m^* T[\frac{\pi}{3n}]^{2/3}$, where$k_\beta$, e, h, m*, n and T are Boltzmann constant, electric charge, Planck constant, effective mass, carrier concentration and absolute temperature respectively. From the equation, it is known that S depends directly on effective mass and absolute temperature and inversely proportional to carrier concentration. Figure3 shows the calculated values of Seebeck coefficient with different values of temperature for both spin up/down configurations.

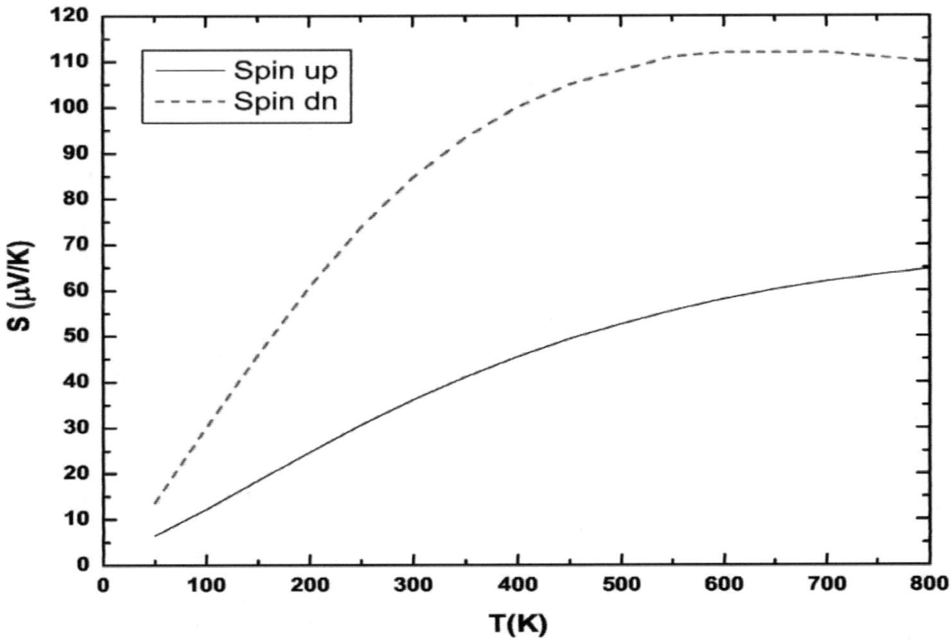


**Figure 3.** Seebeck coefficient (S) variation with respect to temperature (T) for both spins configurations.

The graph shows that the values of S increase non-linearly with temperature and it attain a maximum value of 110(μV/K) for spin-down channel and 65 (μV/K) for spin-up channel at 800 K temperature. The value of S remains positive for both the channels for whole temperature range, which indicates that holes are the majority charge carriers in both spin configurations.

Total Seebeck coefficient is calculated by two current modal and expressed by $S = \frac{\sigma'(\uparrow)S(\uparrow) + \sigma'(\downarrow)S(\downarrow)}{\sigma'(\uparrow) + \sigma'(\downarrow)}$, where $\sigma' = \frac{\sigma}{\tau}$. Figure4 shows the variation of total Seebeck coefficient with temperature. It shows positive values in entire temperature ranges indicate the leading role of p-type charge carriers. The total value of S increases with temperature and reaches highest80 (μV/K) at 800 K. The calculated value of total S at 380 K is 60 (μV/K). It is in good agreement with experimentally reported value 58 (μV/K) [12] at same temperature.

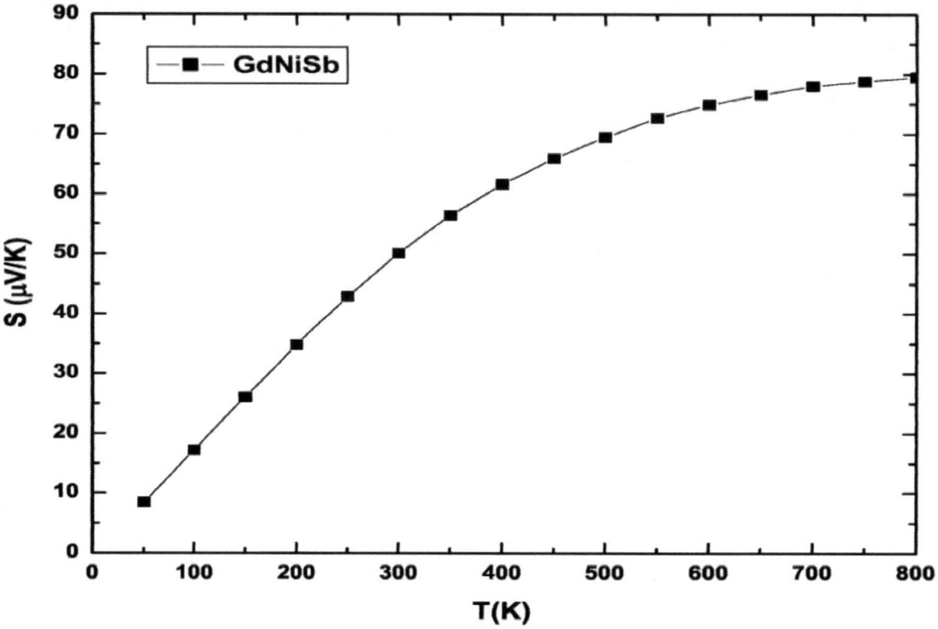


**Figure 4**. Variation of total Seebeck coefficient (S) verses temperature (T) for GdNiSb.

*3.2.3 Figure of merit (ZT)*

The Figure of merit is an important factor and performance indicator of any thermoelectric materials. The value of high ZT means materials that are more efficient for thermo electric phenomenon. The dimensionless ZT value calculated and by the equation $ZT = \frac{S^2\sigma T}{K}$ where S, σ, K and T are Seebeck coefficient, electrical conductivity, thermal conductivity and absolute temperature respectively. Figure 5 shows the variation of ZT with temperature of GdNiSb compound in both spin-up and spin-down configurations. The value of ZT remains high in spin down configuration in whole temperature range as compared to the spin up configuration. It reaches maximum of ~ 0.475 in spin-down channel at 800 K on the other hand in spin-up configuration its value is ~ 0.225 at same temperature.

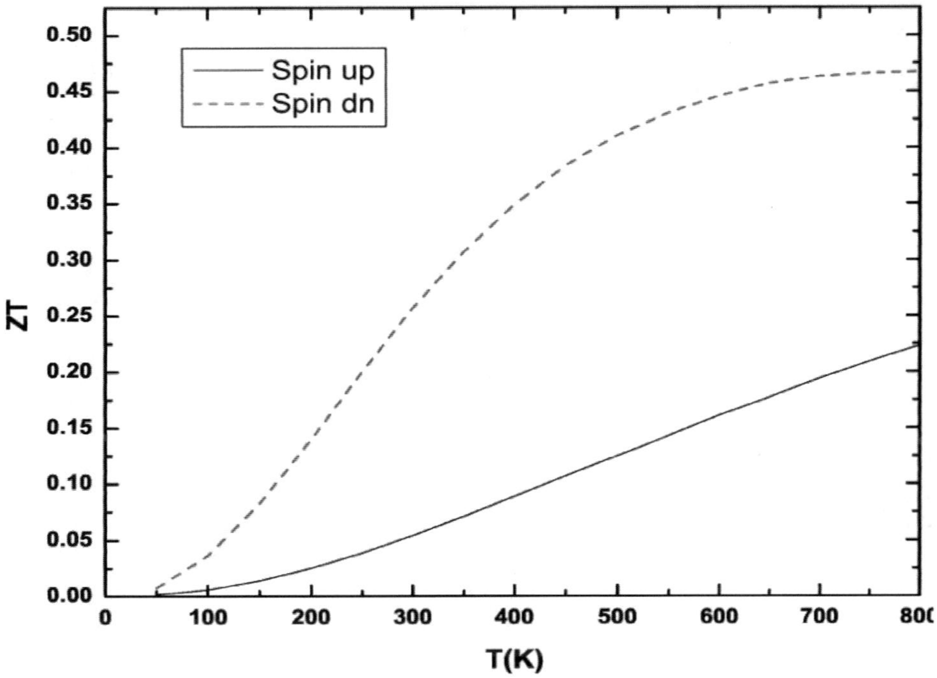


**Figure 5.** Figure of merit (ZT) verses temperature (T) for both spin configurations.

## 4. Conclusions

In this wok we have investigated the electronic and thermoelectric properties of rare-earth based GdNiSb compound using LSDA+U approximation. The calculated values of total and partially density of states (DOS)show the semiconductor behaviour of GdNiSb compound with narrow band gap (0.279eV). The valence band which spans from -5 eV to the $E_F$ is dominated by spin up/down Ni-3*d*. On the other hand, conduction band is mainly dominated by spin down Gd-4*f* bands and lie about 3.5 eV above the $E_F$. Also Thermoelctric parameters such as electrical conductivity, Seebeck coefficient and figure of merit were investigated. Values of all the calculated thermoelecrtic parameters indicate that GdNiSb compound could be a good and efficient candidate of thermoelectric material.

## References

[1] Felser C, Fecher G H and Balke B 2007 *Angew. Chem. Int. Ed.* **46** 668
[2] Daniel P, Javorský P, Prchal J, Šantavá E and Daniš S 2008 *Acta Phys. Pol. A* **113** 331
[3] Gondek L, Szytula A and Prokhnenko O 2008 *Acta Phys. Pol. A* **113** 1179
[4] Hermanowicz M, Jezierski A, Kaczkowski J and Kaczorowski D 2009 *Acta Phys. Pol. A* **115** 226
[5] Casper F, Felser C, Seshadri R, Sebastian C and Peter Pöttgen R 2008 *J. Phys. D, Appl. Phys.* **41** 035002
[6] Casper F, Kandpal H C, Fecher G H and Felser C 2007 *J. Phys. D, Appl. Phys.* **40** 3024
[7] Jirsa M, Rameš M and Muralidhar M 2008 *Acta Phys. Pol. A* **113** 223
[8] Liu X B and Altoanian Z 2010 *J. Appl. Phys.* **107** 173104
[9] Goremychkin E A, Natkaniec I, Muhle E and Chistyakov O D1989 *J. Mag. Magnet. Mat.* **81** 63
[10] Hohenberg P and Khon W 1964 *Phys. Rev. B* **136** 864
[11] Blaha P, Schwarz K, Madsen G H K, Kuasnicka D and Luitz J 2001 WIEN2KAn augmented plane wave+ local orbital program for calculating crystal properties, Technical Universitat, Wien, Austria.
[12] Skolozdra R V, Guzikk A, Goryne A M and Pierrec J 1996 *Acta Physica Polonica A* **92** 343.

IOP Conf. Series: Materials Science and Engineering 1120 (2021) 012032 doi:10.1088/1757-899X/1120/1/012032

[13] Schwarz K and Blaha P 2003 *Comput. Mater. Sci.* **28** 259.
[14] Schwarz K 2003 *J. Solid State Chem.* **176** 319.
[15] Anisimov V I, Soovyev I V, Korotin M A, Czyzyk M T and Sawatzky G A1993 *Phys. Rev. B* **48** 16929.
[16] Hubbard J 1963 Proc. R. Soc. (London) A **276** 238.
[17] Harman B N, Antropov V P, Liechtenstein A I and Anisimov V I 1995 *J. Phys. Chem. Sol.* **56** 1521

IOP Conf. Series: Materials Science and Engineering 1120 (2021) 012033 doi:10.1088/1757-899X/1120/1/012033

# Role of anharmonicity in the electronic heat capacity of superconductors

**Nitin P Singh[1], Preetam Singh Gour[1] and Sapan Mohan Saini[2*]**

[1] Department of Physics, Jaipur National University, Jaipur, India
[2] Department of Physics, National Institute of Technology, Raipur, India

*Corresponding author's e-mail address: smsaini.phy@nitrr.ac.in

**Abstract.** The role of anharmonicity in the electronic heat capacity (EHC) of high temperature superconductors (HTS) has been investigated via electron density of state (EDOS) using quantum dynamical approach. The EDOS may be obtained using correlation function with the help of Zubarev double time temperature dependent electron Green's functions (GF) via a generalized Hamiltonian which consists of (i) unperturbed electron, (ii) unperturbed phonon, (iii) electron-phonon, (iv) anharmonic and (v) defect interactions. The model Hamiltonian includes the force constant changes and mass difference caused by the impurities along with the effects of cubic anharmonicity reveals some striking features of EHC of HTS.

## 1. Introduction

One of the most important equilibrium thermodynamic properties of a crystal is the heat capacity. The measurements of EHC have been widely studied techniques. Experimental observations [1-3] reveal that EHC is highly sensitive at low temperatures. Some of the theoretician [4-7] studied the low temperature impurity induced heat capacity. While the heat capacity of superconducting material at low temperature have been measured by a large number of authors [8-10]. Also a number of reviews of the thermal properties of high temperature superconductors have been given by many authors [11-16]. But none of these workers have considered the anharmonic effects on heat capacity. Only a few of them [17-26] have worked on anharmonic effects, but even that not in a convincing way. They have not used the effect of defects and anharmonicity in an explicit manner. It is mentioned here that defects play an important role in the low temperature region in the analysis of EHC. Also the effect of anharmonic phonon field does not vanish even at absolute zero of temperature and the impurities [23] in a crystal radically change the energy scenario. The theories of EHC proposed till now, however, are more or less based on the free electron theory and could not explain the complete phenomenon [24-28]. There have been several attempts to explain heat capacity of high $T_c$ copper oxides superconductors using a Van Hove singularity in the EDOS [29-36]. In some of these approaches electron-phonon model is used.

In the present theory electronic heat capacity has been developed on the basis of quantum dynamics of electrons [37-38]. It is observed that the electron-phonon coupling plays a crucial role in determining EHC. The change in EHC due to substitutional impurities by the effect of force constant only has been described by many workers [39-41]. But the present theory reveals that this change depends not only on the mass change parameters and the force constant changes, but also on the cross terms of third and fourth order anharmonic parameters with the defect terms. In this scenario we considered an interacting system of electron-phonon fields instead of considering a container of free electron gas or electrons only [42].

## 2. Formulation of the problem

Our approach is based on the quantum dynamics via double-time thermal Green's functions method. The heat capacity may be obtained from crystal energy and DOS [2, 43] without using any quasiharmonic approximation. The shortcoming of the quasiharmonic approximation in the evaluation of thermodynamic properties has been discussed by Barron [44]. Temperature dependence DOS is also one of the important feature of this work.

In the formulation of EHC, the total energy of the system may be obtained in the form [45]

$$E = E_0 + (\pi^2/6)(k_B T)^2 N(\in,T) \tag{1}$$

where $E_0 = \int_0^{\in_f} \in N(\in,T)\, d\in$, is the total energy at $0^0 K$ with $\in_f$ is the Fermi energy. As EDOS cannot be taken independent of temperature, EHC can be evaluated by taking the temperature derivative of equn. (1) in the following form,

$$C_{el}(T) = \frac{\partial E}{\partial T} = C_{el}^{(1)}(T) + C_{el}^{(2)}(T) \tag{2}$$

where

$$C_{el}^{(1)}(T) = (\pi^2/3)k_B^2 T\, N(\in,T) \tag{3}$$

$$C_{el}^{(2)}(T) = (\pi^2/6)(k_B T)^2 \frac{\partial}{\partial T} N(\in,T)\,. \tag{4}$$

The electron density of states ($N(\in,T)$) can be described in the form of Lehman representation as [37, 46]

$$N(\in,T) = -\sum_k \operatorname{Im} G_{q,q}(q,\in) \tag{5}$$

where $G_{q,q}(q,\in) = \langle\langle b^*_{q\sigma}(t); b_{q'\sigma'}(t')\rangle\rangle = -i\theta(t,t')\langle [b^*_{q\sigma}(t); b_{q'\sigma'}(t')]\rangle$ is the electron Green's function with $b_q(b^*_q)$ are electron annihilation (creation) operators with wave vector $q$ .

## 3. Quantum dynamics

The quantum dynamics of electron system is investigated with the help of an almost complete Hamiltonian which includes the effects of electron-phonon, defects, anharmonicities and interference of thereof. The double time temperature dependent Green's functions for electrons and phonons have been obtained in the case of layered compounds. Having obtained the correlation function [47] via Green's function we are able to obtain the electron and phonon self energies, electron and phonon density of states etc, which enable us to study the various physical properties of layered compounds in the frame work of many body theory.

### *3.1. The Hamiltonian and Electron Green's Function*

The complete Hamiltonian (with out BCS) for the said problem may be expressed as

$$H = H_{eo} + H_{po} + H_{ep} + H_A + H_D \tag{6}$$

The details of the contributions to the Hamiltonian are given in references elsewhere [37, 46, 48-49]. The double time temperature dependent Green's function can be evaluated via Hamiltonian (6) using equation of motion technique [47] and Dyson's equation [48, 50] in the final form as

$$G(q,\in) = (2\pi)^{-1}(3\in^N + \in^C)\delta_{q\sigma q'\sigma'}\left[\in^2 - \overline{\in}_q^2 + (3\in^N + \in^C)\Gamma(q,\in)\right]^{-1} \tag{7}$$

where $\in^N$ and $\in^C$ are the normal energy and Cooper pair energy and $\tilde{\in}_q$ is the energy of renormalize mode which takes the form

$$\tilde{\in}_q^2 = (3\in^N + \in^C)^2 - \alpha/(2\pi) - (3\in^N + \in^C)^{-1}\sum_{q'}(g_k + g_{k'}^*)\beta/(2\pi) \tag{8}$$

Also $\bar{\in}_q^2 = \tilde{\in}_q^2 + (3\in^N + \in^C)\Delta(q,\in)$ is the energy of perturbed mode. Here $\Delta^{(e)}(q,\in)$ and $\Gamma^{(e)}(q,\in)$ indicate electron energy shift and line width of electron [48].

### 3.2. Electron Density of States (EDOS)

The evaluation of energy spectrum or density of states for electron is basic building block of dynamical properties of crystalline solids. The electron density of states in the high temperature superconductors can be explored with the help of Green's functions (7) in the form as

$$N^{(e)}(\in,T) = \frac{(3\in^N + \in^C)^2\Gamma(q,\in)}{-\left\{2\pi\left[(\in^2 - \bar{\in}_q^2)^2 + (3\in^N + \in^C)^2\Gamma^2(kq,\in)\right]\right\}} \tag{9}$$

For small value of $\Gamma(q,\in)$ equation (9) may be written as

$$N^{(e)}(\in,T) = \sum_m N_m(\in,T) = \sum_m \frac{(3\in^N + \in^C)^2\Gamma_m(q,\in)}{2\pi(\in^2 - \bar{\in}_q^2)^2} \tag{10}$$

with $m = ep, 3A, 4A, D$. The various contributions to EDOS now can be summarized as [48,50]:

$$N_{ep}^{(e)}(\in,T) = -\sum_{k,q} G_{k,k'}^2(\in^2 - \bar{\in}_q^2)^{-2}\left\{8\xi(\in)N_{q\sigma}\left[4\tilde{\in}_k^2\in_k^{-1}(3\in^N + \in^C)^2 - \in_{k-k}^2\in_k\right]\right.$$
$$\times\delta(\in^2 - \tilde{\in}_k^2) + 4\left[\in_{k-k}^2 n_k + 4\tilde{n}_k(3\in^N + \in^C)^2 + 4(3\in^N + \in^C)\in_{k-k}\tilde{\tilde{n}}_k\right] \tag{11a}$$
$$\left.\times\delta[\in - (3\tilde{\in}^N + \tilde{\in}^C)]\right\}$$

$$N_{3A}^{(e)}(\in,T) = 288\sum_{k,q,k_1,k_2} V_3(k_1,k_2,-k)^2 G_{k,k'}^2(\in^2 - \bar{\in}_q^2)^{-2}\left\{\eta_{12}\xi(\in)N_{q\sigma}\left[S_{+\alpha}\tilde{\in}_{+\alpha}\right.\right.$$
$$\left.\left.\times\delta\left(\in^2 - \tilde{\in}_{+\alpha}^2\right) + S_{-\alpha}\tilde{\in}_{-\alpha}\delta\left(\in^2 - \tilde{\in}_{-\alpha}^2\right)\right] + \delta[\in - (3\tilde{\in}^N + \tilde{\in}^C)]n_{k_1}n_{k_2}\right\} \tag{11b}$$

$$N_{4A}^{(e)}(\in,T) = 2^9\sum_{k,q,k_1,k_2,k_3} V_4(k_1,k_2,k_3,-k)^2 G_{k,k'}^2(\in^2 - \bar{\in}_q^2)^{-2}\left\{\xi(\in)\left[S_{+\beta}\tilde{\in}_{+\beta}\delta\left(\in^2 - \tilde{\in}_{+\beta}^2\right)\right.\right.$$
$$\left.\left. + 3S_{-\beta}\tilde{\in}_{-\beta}\delta\left(\in^2 - \tilde{\in}_{-\beta}^2\right)\right]\eta_{123}N_{q\sigma} + \delta[\in - (3\tilde{\in}^N + \tilde{\in}^C)]3n_{k_1}n_{k_2}n_{k_3}\right\} \tag{11c}$$

$$N_D^{(e)}(\in,T) = 2^6\sum_{k,q,k_1} D(k_1,-k)^2 G_{k,k'}^2(\in^2 - \bar{\in}_q^2)^{-2}\left\{2\xi(\in)N_{q\sigma}\in_{k_1}\delta(\in^2 - \tilde{\in}_{k_1}^2)\right.$$
$$\left. + \delta[\in - (3\tilde{\in}^N + \tilde{\in}^C)]n_{k_1}\right\} \tag{11d}$$

The details of various symbol used in eqns.(9-11) are given in references[50-51].

## 4. Analysis of Electronic Heat Capacity

Using the various contributions of EDOS (eqn. 11) in eqn. 2 and eqn. 3, we can obtain the total EHC with various contributions in the form

$$C_{el}^{EP} = \pi\gamma \sum_{k,k'} G_{k,k'}^{2} \{X(\tilde{\in}_k, \overline{\in}_q) \tilde{\in}_k \in_k^{-1} [4TN_{Q\sigma}(\in_k^4 - 4\tilde{\in}_k^2 \in_{NC}^2) + k_B^{-1}(\in_k^4 - 4\tilde{\in}_k^2 \in_{NC}^2)$$
$$\times (f^n + f^C)] + X(\tilde{\in}_{NC}^2, \overline{\in}_q)\tilde{\in}_{NC}^2 [4T(\in_k^2 n_k + 4\in_k \tilde{n}_k \in_{NC} + \tilde{\tilde{n}}_k \in_{NC}^2) \quad (12a)$$
$$+ k_B^{-1} \tilde{\in}_k \in_k^{-1} (\in_k^3 + 4\in_k^2 \in_{NC} + \tilde{\in}_k \in_{NC}^2)(n_k^2 - 1)]\}$$

$$C_{el}^{3AE} = 36\pi\gamma \sum_{\substack{k,k',\\k_1,k_2}} G_{k,k'}^2 |V_3(k_1,k_2,-k)|^2 \{X(\tilde{\in}_{+\alpha}, \overline{\in}_q)\tilde{\in}_{+\alpha}^2 \eta_{12}$$
$$\times [4TN_{Q\sigma}S_{+\alpha} + k_B^{-1}[N_{Q\sigma}\chi^+ + S_{+\alpha}(f^n + f^C)]] + X(\tilde{\in}_{-\alpha}, \overline{\in}_q)\tilde{\in}_{-\alpha}^2 \eta_{12} \quad (12b)$$
$$\times [4TN_{Q\sigma}S_{-\alpha} + k_B^{-1}[N_{Q\sigma}\chi^- + S_{-\alpha}(f^n + f^C)]] + 2X(\tilde{\in}_{NC}, \tilde{\in}_q)\tilde{\in}_{NC}^2$$
$$\times [4Tn_{k_1}n_{k_2} + k_B^{-1}(n_{k_1}n_{k_2}n_{12} - \alpha_{12})]\}$$

$$C_{el}^{4AE} = 192\pi\gamma \sum_{\substack{k,k',k_1,\\k_2,k_3}} G_{k,k'}^2 |V_4(k_1,k_2,k_3 - k)|^2 \{X(\tilde{\in}_{+\beta}, \overline{\in}_q)\tilde{\in}_{+\beta}^2 \eta_{123}$$
$$\times [4TN_{Q\sigma}S_{+\beta} + k_B^{-1}[N_{Q\sigma}\xi^+ + S_{+\beta}(f^n + f^C)]] + X(\tilde{\in}_{-\beta}, \overline{\in}_q) \quad (12c)$$
$$\times \tilde{\in}_{-\beta}^2 \eta_{123} [4TN_{Q\sigma}S_{-\beta} + k_B^{-1}[N_{Q\sigma}\xi^- + S_{-\beta}(f^n + f^C)]]$$
$$+ 6X(\tilde{\in}_{NC}, \tilde{\in}_q)\tilde{\in}_{NC}^2 [4Tn_{k_1}n_{k_2}n_{k_3} + k_B^{-1}(n_{k_1}n_{k_2}n_{k_3}n_{123} - \alpha_{123})]\}$$

$$C_{el}^{DE} = 16\pi\gamma \sum_{k,k',k_1} G_{k,k'}^2 |D(k_1,-k)|^2 \{\in_{k_1} \tilde{\in}_{k_1} X(\tilde{\in}_{k_1}, \overline{\in}_q)[4TN_{Q\sigma} + k_B^{-1}(f^n + f^c)] \quad (12d)$$
$$+ \tilde{\in}_{NC}^2 X(\tilde{\in}_{NC}, \overline{\in}_q)[8Tn_{k_1} + k_B^{-1}\tilde{\in}_{k_1}(n_{k_1}^2 - 1)]$$

It is observed from eqn. (12) that electron-phonon coupling coefficient ($g_k$) via $G_{k,k'}$ is present in all terms which depicts the importance of electron-phonon interaction in superconductivity phenomenon. The details of the symbol are described in references elsewhere [52-53].

## 5. Results and Discussion

In cubic anharmonicity the dependence of temperature and perturbed mode energy on the dominant term say $F_\alpha$ (Curley bracket terms of eqn. 12b) is shown in figure 1 which reveals that initially there is small effect of temperature on it up to the value of $\overline{\in}_q$ less than 2.0x10$^{-14}$ (erg). As the temperature increases curve start increasing but the value of $F_\alpha$ becomes smaller at higher temperature for smaller value of $\overline{\in}_q$ as shown by figure (2) and figure 3.

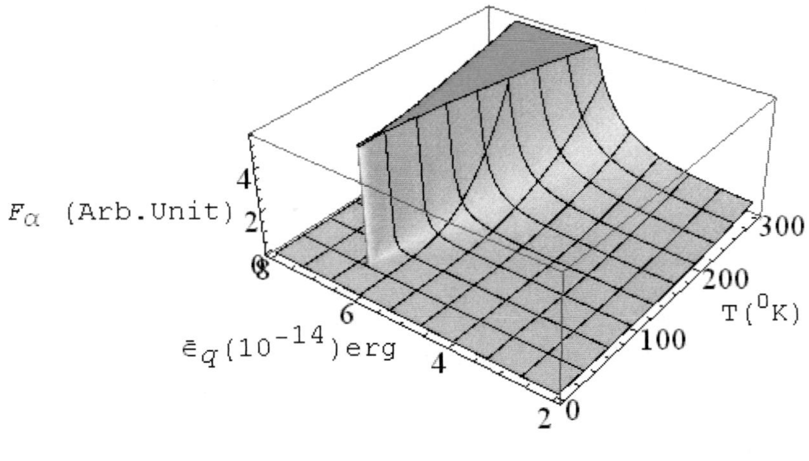


**Figure 1.** Variation of $F_\alpha$ as a function of $\overline{\in}_q$ and $T$.

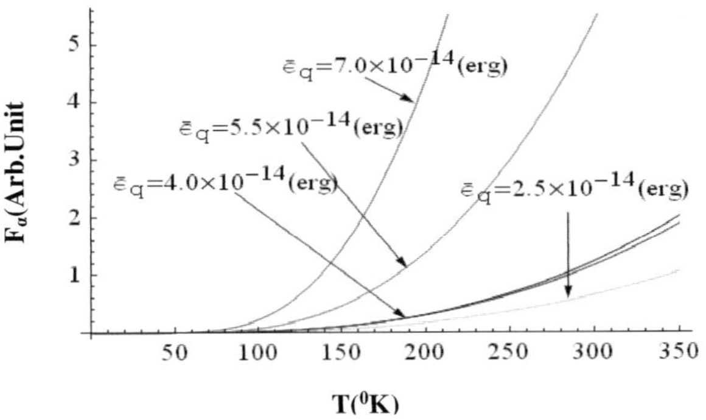


**Figure 2.** Variation of $F_\alpha$ with $T$ at different $\overline{\in}_q$

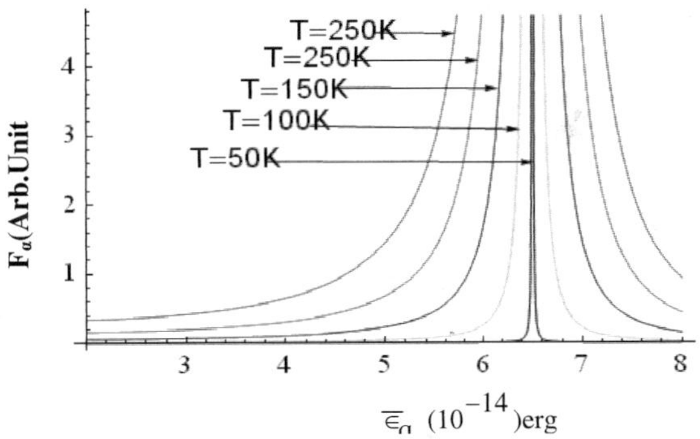


**Figure 3.** Variation of $F_\alpha$ with $\overline{\in}_q$ at different $T$.

It emerges from the present study that the general trends of the thermal and electrical properties of superconductors both in the normal and superconducting states can be successfully explained with the help of present formulation. The various trends of energy and temperature dependences exhibit the new features of the theory in addition to the automatic evolution of Cooper pair energies.

## Acknowledgments

One of the authors Nitin P. Singh is thankful to Prof. B.D. Indu for his help in the theoretical calculation of EHC.

## References

[1] Launay J de 1956 *Solid State Physics* (vol **2**) eds. F Seitz and D Turnbull (New York: Academic Press) p 220
[2] Karlsson A V 1970 *Phys. Rev.* B **2** 3332
[3] Hartmann W M, Culbert H V and Huebener R P 1970 *Phys. Rev.* B **1** 1486
[4] Agrawal B K 1969 *J. Phys.* C **2** 252
[5] Tiwari M D and Agrawal B K 1973 *Phys. Rev.* B **7** 4665
[6] Tiwari M D, Kesharwani K M and Agrawal B K 1973 *Phys. Rev.* B **7** 2378
[7] Indu B D, Tiwari M D and Agrawal B K 1978 *J. Phys.* F **8** 755

[8] Pattalwar S M, Dixit R N, Shete S Y and Basu B K 1988 *Phys. Rev.* B **38** 7067
[9] Urbach J S, Mitzi D B and Kapitulnik A 1989 *Phys. Rev.* B **39** 12391
[10] Chakraborty A, Epstein A J and Cox D L 1989 *Phys. Rev.* B **39** 12267
[11] Junod A 1990 *Physical Properties of High Temperature Superconductors* II, ed. D M Ginsberg (Singapore: World Scientific) p 13
[12] Atake T 1991 *Thermochim. Acta* **174** 291
[13] Phillips N E, Fisher R A and Gordon J E 1992 *Progress in Low Temperature Physics*, (vol **13)** ed D F Brewer (Amsterdam: North Holland) p 267
[14] Fisher R A, Gordon J E, and Phillips N E 1996 *Annu. Rev. Phys. Chem.* **47** 283
[15] Greene LH and Bagley B G 1990 *Physical Properties of High Temperature Superconductors II* ed D M Ginsberg (Singapore: World Scientific) p 509
[16] Malozemoff A P 1989 *Physical Properties of High Temperature Superconductors I* ed D M Ginsberg (Singapore: World Scientific) p 71
[17] Wethamer N R 1970 *Phys. Rev.* B **1** 572
[18] Ida Y 1970 *Phys. Rev.* B **1** 2488
[19] Hui J C and Allen P B 1975 *J. Phys.* C **8** 2923
[20] Zhdanov K R, Belosludov V and Rakhmankulov F S 1985 *Phys. Stat. Sol.* (b) **130** 19
[21] Knapp G S, Bader S D, Culbert H V, Fradin F Y and Klippert T E 1975 *Phys. Rev.* B **11** 4331
[22] Knapp G S, Bader S D and Fisk Z 1976 *Phys. Rev.* B **13** 3783
[23] Indu B D 1992 *Mod. Phys. Lett.* B **6** 1665.
[24] Buckingham M J 1951 *Nature* **168** 281.
[25] Animalu A O E 1977 *Intermediate Quantum Theory of Crystalline solids* (Englewood: Prentice-Hall International).
[26] Eliashberg G M 1963 *Sov. Phys. J. E. T. P.* **16** 780.
[27] Harrison W A 1970 *Solid State Theory* (New York: McGraw-Hill Inc)
[28] Allen P B and Hui J C K 1980 *Z Physik* B **37** 33
[29] Labbe J and Bok J 1987 *Europhys. Lett.* **3** 1225
[30] Labbe J 1989 *Phys. Scr.* **T29** 82
[31] Friedel J 1987 *J. Phys. (Paris)* **48**, 1787; **49** 1435
[32] Combescot R and Labbe J 1988 *Phys. Rev.* B **38** 262
[33] Markiewic R S 1990 *J. Phys. Cond. Matt.* **2** 665
[34] Markiewic R S and Giessen B C 1989 *Physica* C **160** 497
[35] Tsuei C C, Newns D M, Chi C C, and Pattnaik P C 1990 *Phys. Rev. Lett.* **65** 2724
[36] Newns D M, Tsuei C C, Pattnaik P C and Kanes C L 1992 *Comments Cond. Matt. Phys.* **15** 273
[37] Indu B D 1990 *Int. J. Mod. Phys.* B **4** 1379
[38] Painuli C P, Bhatt R M, Raiwani Y P, Srivastava S, Gairola S C, Gairola R P, Indu B D, and Tiwari M D 2004 *Ind. J. Pure & Appl. Phys.* **42** 288
[39] Manheim P D 1968 *Phys. Rev.* **165** 1011
[40] Agrawal B K 1980 *Phys. Rev.* B **22** 6294
[41] Tiwari M D and Agrawal B K 1973 *Phys. Stat. Sol. (b)* **58** 209
[42] Srivastava S, Raiwani Y P, Gairola S C and Indu B D 1996 *Physica* B **223** 538
[43] Ghatak A K and Kothari L S 1971 *An Introduction to Lattice Dynamics* (London: Addison - Wesley Pub. Co.) p 94
[44] Barron THK 1963 *Proc.of Int. Conf. on Lattice Dynamics*( Copenhagen) p 301
[45] Harrison W A 1970 *Solid State Theory* (New York: Mc Graw-Hill Inc)
[46] Bahadur Rita and Sharma P K 1974 *Solid State Commun.* **15** 621
[47] Zubarev D N 1960 *Sov. Phys. Usp.* **3** 320
[48] Ashokan Vinod, Indu B D and Dimri A K 2011 *AIP Adv.* **1** 032101
[49] Sharma P K and Bahadur Rita 1975 *Phys. Rev.* B **12** 1522
[50] Ashokan V, Indu B D 2013 *AIP Adv.* **3** 022108(1)
[51] Gupta A,Verma S K, Kumari A and Indu B D 2018 *Int. J. Mod. Phys.* B **32** 1850237.

[52] Singh Anu, Singh Hempal and Indu B D 2016 *AIP Adv.* **6** 075102-(1)
[53] Singh Anu and Indu B D 2018 *AIP Conference Proceedings* **1953** 120036-(1)

IOP Conf. Series: Materials Science and Engineering 1120 (2021) 012034 doi:10.1088/1757-899X/1120/1/012034

# Moving organ toxicity management in radiation therapy with simple fuzzy logic

**Santosh Kumar Patnaikuni[1,2] and Sapan Mohan Saini[1*]**

[1] Department of Physics, National Institute of Technology Raipur, Chhattisgarh, India.
[2] Department of Radiotherapy, Pt.JNM Medical College, Raipur, Chhattisgarh, India

*Corresponding author's e-mail address: smsaini.phy@nitrr.ac.in

**Abstract.** The aim of present work is to estimate and manage real time toxicity of moving organ bladder in deciding level of damage using a simple fuzzy logic method. The target and bladder inter-treatment motions are considered in radiation therapy of prostate cancer in the present study. Materials and methods: The prostate target and nearby critical structures like bladder varies due to physiological behaviour during the course of radiation therapy treatment. The total displacement errors of prostate target shift ranging from 0-5 mm s.d. was considered in this study while maintain tumor dose conformity. The Mamdani-type Fuzzy Inference System (FIS) was applied for determine biological acceptable margin using TCP and NTCP parameters. The bladder volume variations with weekly analysis of cone beam CT (CBCT) were considered. Using irradiated volume-risk relationship in planning, the overlapped risk volumes were estimated. Fuzzy rules and membership function were used basis on setup errors, asymmetrical nature of organ motion and limitations of normal tissue toxicity in margin formulations. Results and discussion: The prostate target motion management was in a good consistent with conventional margin formulations at 5 mm s.d. displacement error range to maintain conformity index (CI) of target. With application of conventional margins of target volume, the sub-volumes of critical organs such as bladder and corresponding its NTCP values were fuzzified using inter treatment volume variations of critical organs. The risk factor (RF) of critical organs were qualitatively assessed with for each fractional volume of irradiated to total volume and relevant NTCP values. Conclusion: A simple fuzzy logic may be suitable to estimate critical moving organs toxicity like bladder biologically. Using proposed simple and fast method, there is interplay between volume-risk relationship and NTCP of OARs to management of normal-tissue risk level of radiation therapy.

## 1. Introduction

The treatment of cancer using radiation therapy is the process of optimization of maximizing radiation dose to tumour cells whilst sparing healthy tissues and critical organs. In course of treatment, target volumes and critical organs should be outlined clearly. Otherwise inductions of adverse side effects on the normal tissues and critical organs may develop sever complications. There is always trade-off between complications and cure depends on best possible margins to achieve favorable treatment outcomes. The margin formulations were studied based on probabilistic dose distributions by considering a linear relationship between planning target volume (PTV) margin and radiotherapy errors [1]. Currently the PTV margins are symmetric or linear nature for all treatment strategies today. But due to presence of organ motion and other radiotherapy setup errors encountered in actual fractionated radiotherapy treatment phase, delivered dose may differs from planned dose. Further the main disadvantage of currently used margin formulations is that they do not consider the effects of organ motion and surrounding organ at risk when deriving PTV margins. It was found that fuzzy region of PTV to derive exact margin in prostate cancer (PC) radiation therapy from the clinical observations as prostate gland position is likely to change between treatment fractions also critical organs will affect anatomically, adjacent to prostate target. Therefore PTV margins should be precisely defined as well as account for radiobiological effects of tumour and surrounding normal structure at planning level [2] in the treatment of moving organ like prostate. A study was done by Yartsev S et al [3] which illustrate fuzziness of PTV margins of moving targets and may effect on nearby normal overlapped volumes. Also critical organs such as bladder volumes change during the prostate cancer radiotherapy [4]. This has an impact on thus, further radiation induced toxicities [5]. Hence the rigidity of currently using conventional formulation may be limited to adapt for many treatment scenarios such as prostate like dynamic tumour volumes. Introducing fuzzy logic has a distinct advantage as it allows linkage of geometrical and radiobiological parameters of radiotherapy planning through using fuzzy rules and membership functions (MFs). The relationship is difficult between these parameters to quantify mathematically or may have a large degree of variability. Recently, Patnaikuni et al. [6] studied asymmetric margins and practical limitation on the margin size of prostate cancer using VMAT technique and compared with conventional van Herk margin using total displacement standard errors. The present study is based on earlier work by Patnaikuni et al [6].

The objective of present work is to estimate bladder toxicity biologically using a simple novel two-level fuzzy logic method. The VMAT was choice for prostate cancer radiation therapy. To estimate real time bladder toxicities, both prostate target and bladder internal motions are considered in the present study at low displacement error range of organ motion based on earlier work by Patnaikuni et al [6]. The first level fuzzy logic is adopted [6] to apply biological acceptable target margin using TCP and NTCP parameters. In second level fuzzy, bladder volume variation with weekly analysis of cone beam CT (CBCT) was considered here. Fuzzy rules and membership function were used basis on setup errors, asymmetrical nature of organ motion and limitations of normal tissue toxicity in margin formulations. The final risk factor (RF) of critical organs was qualitatively assessed with for each fractional volume of irradiated to total volume and relevant NTCP values.

## 2. Methods and materials

In the present study Mamdani-type fuzzy inference system (FIS) was applied in possible OAR risk management procedure. Because Mamdani-type FIS is expected output suited to human thinking and so widely accepted for capturing expert knowledge which is very significant particularly in real time organ toxicity assessment of radiation therapy treatment. The localized prostate radiotherapy patients (n=08) were selected for study with dose prescription of 73.5 Gy. The preferred planning method was volumetric modulated arc therapy (VMAT). Planning Objectives and Dose Constraints were mentioned using Eclipse 15.6 Treatment planning system (TPS), Varian Medical Systems (VMS) with gEUD based parameters setting. The radiobiological parameters TCP and NTCP were calculated in Matlab R2018a based simulation tool [7] using the equivalent uniform dose (EUD) modelling with required statistical values from all plans. The framework of our study involves two levels along with standard setup errors were incorporated in each level. In the first level, possible PTV margin was estimated based on target motion displacement error s.d $<$ 5 mm. In the second level, the possible OAR toxicity was estimated in the same error range on employing fuzzy PTV margin from first level.

### *2.1. Fuzzy logic framework based on organ motion*

In the first level fuzzy framework, target margin derivation procedure was started with statistical analysis of target motion using in-house using tracking methods such as image guidance analysis [8-9]. All asymmetric margins were generated for target PTV from minimal to maximum motion limits based on nature of displacement of target which was observed statistically. Here 1mm stepped size margin was manually added using TPS to subsequent asymmetric PTV margin up to maximum limit. All VMAT plans were performed from minimum acceptable value of PTV margin to maximum. Biological parameters TCP and NTCP were calculated for each plan. After each stepped increment margin of PTV, recalculation of new TCP and NTCP values were done and compared with base value which is corresponding plan with minimal PTV margin. Subsequent changes in TCP and NTCP due to target volume displacements were used these initial TCP and NTCP values to estimate subsequent loss in TCP (i.e., ΔTCP) and increase in NTCP (i.e, ΔNTCP). The FIS consisted of two inputs as ΔTCP and ΔNTCP while one output as PTV margin. The fuzzification of inputs (TCP and NTCP) and defuzzification of output (PTV margin) were done in Mamdani-type FIS with adopted rules and membership functions. Fuzzy rules were devised based mainly on limitation that increase in NTCP is compensated for by reducing PTV margin while loss in TCP is compensated for by increasing PTV margin size. The 3D output surface [6] represents combination of ΔTCP, ΔNTCP and PTV margin values that indicates uneven changes in PTV margin with required TCP/ NTCP relation while a standard uncertainty 0.5±0.2 mm was considered as error in PTV margin. All results were studied below 5 mm s.d. of target displacement errors which was considering as lower error region. Finally the margin obtained from fuzzy model was applied in actual VMAT treatment planning for bladder toxicity estimation and management.

The second level fuzzy framework was started with VMAT plans generation using fuzzy PTV margin. The overlapped bladder volumes and corresponding NTCP values were fuzzified in second level fuzzy using bladder volume variations. The final risk factor (RF) [10] of OARs was qualitatively assessed with for each fractional volume of irradiated to total volume and relevant NTCP values. The risk or toxicity estimation of any critical organ around the tumour depends upon its volume gets irradiated during irradiation. Sub-volume of volume of interest ($V_{VOI}$) based NTCP was calculated for each CBCT. It is directly proportional to the fragment or sub-volumes of OAR which gets irradiated by its tolerance dose. Also toxicity or risk depends on its entire volume ($V_{TVO}$) of OAR. If the total volume of OAR is more, then its risk of damage or risk factor (RF) is less during irradiation and vice versa. RF is inversely proportional to the total volume of the organ. Bladder volume varies significantly on a daily basis, so the actual radiation prescribed dose received by the bladder may be expected different in comparison to the planned radiation dose. To assess the volumetric variation of bladder, cone beam computed tomography (CBCT) was used for each patient on weekly basis. In each patient, five CBCT-scans obtained in 5 day interval each through the treatment course. The bladder volume was manually outlined on CBCTs along with actual planning CT (CTp) so resulting in 6 bladder contours for each patient. All bladder contours obtained from each patient were imported and merged with CTp as reference. Recalculation of bladder volume and dose was done for each CBCT using CTp with same set up beams. Fuzzy rules were framed according to OAR bladder toxicity limits so that each CBCT vs. CTp was assessed to estimate real-time sub-volume related NTCP. For assessing the risk of damage rationally, MFs were selected for inputs (i.e; fractional volume and its NTCP) and output (i.e; RF). MFs of input NTCP were selected as Medium, High, vary high risk regions while MFs of input $V_{VOI}$/ $V_{TVO}$ were selected 0-25%, 25-50%, 50-75%, 75-100% respectively. The MFs of output RF were defined as medium, high and very high.

### *2.2. OAR bladder toxicity estimation and clinical assessment*

The risk or toxicity estimation of any critical organ around the tumor depends upon its volume gets irradiated during irradiation. The MFs of output RF were defined as medium, high and very high. Near to 0% then the OAR is completely safe which was impractical for moving organs. So MFs for RF were practically considered on clinical situations. Medium RF: If the numerical value of RF is 50% or less, then the OAR will get partially damaged during the course of irradiation. High RF: If the numerical value of RF is 75% or higher which means OAR will get completely damaged during the course of irradiation. In our study MFs for bladder risk (RF) were defined by risk factor scale by Ansari et al. For MFs of NTCP as medium (0-2%), high (2-4%) and very high (above 4%). The acceptance of bladder toxicity limits may vary patient to patient and existing co-morbidities. If toxicity limits exceeds based on intended NTCP, target volume or normal structure contours will be altered for re-optimization and planning. The selection of MFs and rules were mainly based on high displacement error s.d., organ type such as serial organ or parallel organ or serial–parallel organ. The NTCP values were based on quantitative analysis of normal tissue effects in the clinic (QUANTEC) dose constraints.

## 3. Results and discussion

The VMAT plans were performed with 6 mm fuzzy PTV margins corresponding to total displacement standard errors < 5 mm s.d. PTV dose distribution and dose-volume histogram (DVH) of fuzzy PTV-plan was compared to currently using Van-Herk margin models. All plans were clinically acceptable in the present study in view of target volume (PTV) dose conformity so there were no significant changes found in PTV objectives when compared with currently using margin models. This may be due to fuzzy margin was found to be more or less similar compared to currently using van-Herk margin methods in our radiotherapy centre. However taking the modelling uncertainty into account, results showed good match between fuzzy PTV margin calculated and currently using margin formulation in first-level fuzzy PTV margin derivation of moving target of up to displacement error 5 mm s.d. Figure 1 represents combination of ΔTCP, ΔNTCP and PTV margin values that indicates uneven changes in PTV margin with required TCP/ NTCP relation. With application of fuzzy margin obtained from first level fuzzy, overlapped bladder volumes and corresponding NTCP values were fuzzified in second level fuzzy using bladder volume variations. The final risk factor (RF) of OARs was qualitatively assessed with for each fractional volume of irradiated to total volume and relevant NTCP values. Figure 2 represents risk factor estimation of with required NTCP and sub-volume relationship of OAR bladder using second level fuzzy logic. The bladder risk or toxicity (%) corresponding to NTCP values for one patient under low displacement errors s.d, was demonstrated for one patient as in Table-1. From the table it was observed all NTCP values were well within toxicity limits so that RF of bladder is acceptable. Hence no action was recommended for re-planning of treatment. Though all NTCP values and its associated RF values were well within limit but at CBCT3, RF is relatively higher than CTp. The reason may be depends on different bladder volumes due to organ motion related physiological behavior of OAR bladder at the time of acquiring CBCTs. The effect of real-time organ motion related RF was may not be feasible until there might be manual assessment approach which takes usually longer time in routine radiotherapy treatment. Because in the routine radiation therapy, the treatment plan is normally performed for one time at the beginning of treatment course. So with help of FIS tool in our study, it may be feasible to manage radiation induced risk of moving organ like bladder biologically.

The currently using standards of deriving margins in radiotherapy have their limitations, they have been used for many patients over the years and the survival rates are generally satisfactory. The early and or late effects on critical organs and normal tissues is however variable. These methods work but they can be clearly improved to get better outcomes. No 'gold standard' method currently exists in this field which addresses all the previously mentioned limitations to be used for comparing the models from this study. The novelty of the method proposed in this study lies more in that they allow for the calculation of individualised patient margins, which is currently very difficult to accomplish due to manual setup of current techniques. These margins are expected to lead to better clinical outcomes. With help of simple fuzzy logic, it may be expected to assess the possible OAR toxicity on plan of the day or week basis. However clinical trials are required to fully valid ate this observation. It is concluded that the RF is a comprehensive evaluation tool for conformal radiotherapy plans which encompasses a wider range of clinically relevant parameters, isodose volumes and tolerance dose of OARs. It is expected as advance tool to check the qualitative nature of a conformal plan including the assessment of degree of damage of the OARs. The major advantage of current study is that it gives information about the conformity of dose to tumour and severity level of damage of OARs at a glance in clinical application. The major limitation of our work was the estimation of toxicity level under small standard errors s.d only with small number in the sample over course of treatment.

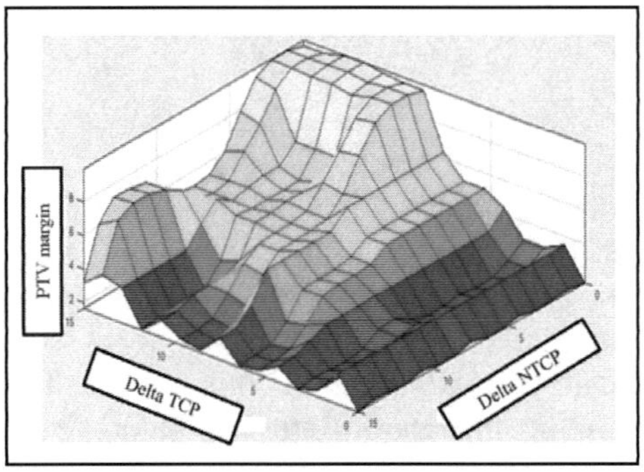


**Figure 1.** Fuzzy output as 3D-surfaces in first-level fuzzy PTV margin derivation of moving target [6].

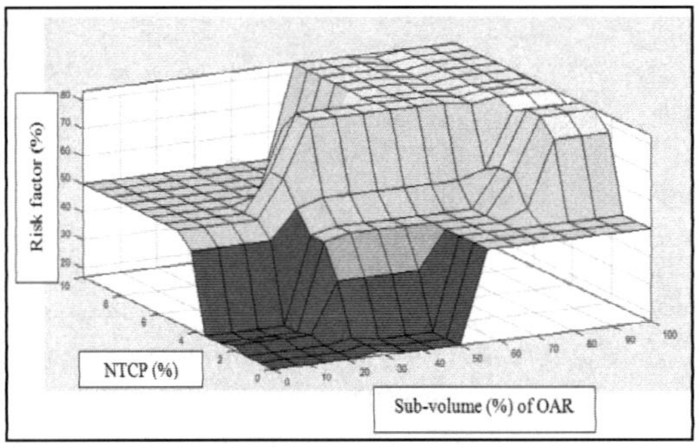


**Figure 2.** Fuzzy output as 3D-surfaces in second-level fuzzy for risk factor estimation of OAR bladder.

**Table 1.** The bladder risk or toxicity (%) corresponding to NTCP values for one patient under low displacement errors s.d, was demonstrated quantitatively for one patient case.

| Parameter | CTp | CBCT1 | CBCT2 | CBCT3 | CBCT4 | CBCT5 |
|---|---|---|---|---|---|---|
| **Sub-volume(%)** | 20 | 15 | 14 | 30 | 18 | 10 |
| **NTCP(%)** | 3.1 | 2.6 | 2.4 | 4.1 | 2.9 | 1.98 |
| **Bladder RF** | 0.41 | 0.344 | 0.342 | 0.49 | 0.491 | 0.2 |

## 4. Conclusions

The various limitations of current using margin recommendations can be addressed by the use of computational techniques to give margin models which include organ motion and other setup errors. The feasibility of the new models have been tested using clinical studies and the fuzzy models have been found to quality improvement method for better critical organ sparing and toxicity management while maintaining tumor control probability the in the presence of errors.

**Acknowledgements**

Authors are thankful to Department of Physics, National institute of technology, Raipur and Department of radiotherapy, Pt JNM Medical College for providing support at each level to complete this work.

## References

[1] Byrne T E 2005 *Med Dosim.* **30** 155
[2] Deasy J O, Charles S M and Colin G O 2015 *Med Phys.* **42** 2753
[3] Yartsev S and Bauman G 2016 *Br J Radiol.* **89** 20160312
[4] Chen Z, Yang Z, Wang J and Weigang H 2016 *Radiat Oncol.* **11** 103
[5] Varadhan R, Hui S K, Way S and Nisi K 2009 *J. Appl. Clin. Med. Phys.* **10** 2883
[6] Patnaikuni SK, Saini SM, Chandola RM, Chandrakar P, and Chaudhary V 2020 *J Med Phys.* **45** 88
[7] Mzenda B, Hosseini-Ashrafi M E, Palmer A, Hodgson D F, Liu H and Brown D J 2009 *J. Radiother. Pract.* **8** 41
[8] Yartsev S and Bauman G 2016 *Br J Radiol.* **89** 20160312
[9] Zahra N, Monnet C et al. 2015 *Radiother. Oncol.* **115** S908
[10] Ansari S, Zope M K and Yadav N 2020 *J. Radiother. Pract.* 1 DOI: https://doi.org/10.1017/S1460396920000205

IOP Conf. Series: Materials Science and Engineering 1120 (2021) 012035 doi:10.1088/1757-899X/1120/1/012035

# DFT study of adsorption and diffusion of Mg, K, Ca over monolayer antimonene

**S Upadhyay and P Srivastava**

Nanomaterials Research Group- CNTL (NRG- CNTL),
ABV-Indian Institute of Information Technology, Gwalior, M.P., India-474015

*Corresponding author's e-mail address: sneha@iiitm.ac.in

**Abstract.** Density Functional theory (DFT) based calculations are carried out to investigate adsorption and diffusion properties of Magnesium (Mg), Potassium (K) and Calcium (Ca) over single layer antimony (Sb) i.e. antimonene. Ca has highest value of absolute adsorption energy followed by K and Mg respectively. Charge transfer is greater in case of K adsorbed antimonene as compared to both of the other metals. This affects the electronic properties such as bandstructure and density of states. Antimonene is semiconducting in its pristine form. The adsorption of K leads to a shift to metallic nature, whereas Mg and Ca adsorption still maintain the semiconducting nature after reducing the bandgap. Diffusion of these metals shows that K can easily diffuse through the antimonene surface as compared to the Mg and Ca. The value of diffusion barrier for Mg, K and Ca is 0.30, 0.12 and 0.43 eV respectively. The effective adsorption and charge transfer along with low values of diffusion barrier suggests that phosphorene can be explored further as potential candidate for metal-ion batteries.

## 1. Introduction

The Two-dimensional (2D) materials are being constantly researched due to their manifestation of novel properties at nano level. These properties emerge due to the quantum confinement imposed over them in two-dimensions [1,2]. Phosphorene, derived from black phosphorus [3,4] was the first 2D material and it led to synthesis and analysis of many others too, such as silicene [5], germanene [6], borophene [7], arsenene, antimonene, bismuthene [8,9], Transition Metal Dichalcogenides (TMDs) [10,11] 2D tellurium [12] and nitrides [13,14]. Antimonene, the monolayer of antimony (Sb) [15,16] is the material of interest in this work. Among the four allotropes of antimonene [17] β-antimonene is the most stable and thus, it has been used in this research work. Various methods have been developed for its synthesis such as epitaxial growth on Ag (1 1 1) substrate [18], germanene substrate [19] and on various other substrates [20], liquid phase exfoliation [21,22] and mechanical exfoliation [23]. The 2D pnictogens due to their semiconducting nature provide an advantage over graphene which is semimetal with a zero bandgap. The bandgap of these materials can be tamed easily via doping, defects, multi layers or by presence of any foreign element in their vicinity due to which they find place in numerous applications [2,8,17]. Here, we investigated adsorption and diffusion of Magnesium (Mg), Potassium (K) and Calcium (Ca) atoms (metal (M)-atoms) over antimonene monolayer surface. Due to the bigger atomic radius of Sb and buckled structure of antimonene, these large metal atoms can conveniently react over its surface without much disturbing the lattice which may not be the case with graphene [24]. The obtained results related to adsorption energy ($E_{ads}$), activation energy ($E_a$), charge transfer and band structure analysis indicate that adsorption of M-atoms shift the semiconducting antimonene towards metallic properties providing better conductivity and ease of diffusion. This opens up its prospects in applications such as electronic/optical sensors, batteries and smart switching.

## 2. Computational Details

The visualization and simulation of adsorption and diffusion of metal atoms on antimonene have been done using Density Functional Theory (DFT) [25,26] based Atomistic -Virtual Nanolab (ATK-VNL)[27] simulator. All of the structural relaxations and the calculations incorporate the use of Generalized Gradient Approximation (GGA)[28] functional with Perdew–Burke–Ernzerhof (PBE) exchange correlation and a k-point sampling of 10×10×1 points [29]. During optimization the maximum force of convergence was 0.05 eV/Å. To calculate van der Waals (vdW) effect, Grimme DFT-D2[30] correction has been applied and the density mesh cut-off energy used was 150 Ryd. A vacuum padding of 24 Å in z-axis has been used to neglect the effect of interaction with other layers.

## 3. Results and Calculations

### *3.1 Adsorption of Mg, K and Ca over Antimonene and the Associated Properties.*

Mg, K and Ca are adsorbed over antimonene individually at different sites as shown in the Figure 1. These sites are named H-site, HEX-site, B1-site, B2-site and T-site for reference. Single atom of these metals was placed above the antimonene surface and the structure was allowed to relax. It results in rearrangements of these sites which are site and metal specific. The lattice parameters of antimonene are a = b = 4.3Å and Δc = 1.59 Å [31]. The lattice structure remains undisturbed and these parameters remain same after the successful adsorption of M-atoms. However, the adjacent Sb-Sb bonds in supercell were affected a little in some cases which can be traced back successfully after desorption. The $E_{ads}$ (eV) is calculated using equation,[32]

$$E_{ads} = E_{(Sb+M)} - E_{Sb} - E_M \qquad (1)$$

Where, $E_{Sb+M}$ is the total energy of M-atom adsorbed antimonene, $E_{Sb}$ is total energy of pristine antimonene and $E_M$ is total energy of M-atom in molecular configuration. Values of $E_{ads}$ in each and every case come out to be a negative value as listed in table 1. This infers that all the adsorptions done are exothermic and energetically favourable.

When K is adsorbed, H-site comes out to be the most stable adsorption site since it has highest absolute value of adsorption energy. For Mg and Ca adsorption, Hex-site is the most energetically stable site. In case of Mg, B1 and B2 site converted to Hex-site while all the other sites were retained. For Ca adsorption, only H- and Hex-sites were retained and rest were converted to Hex-site after optimization. However, in case of K adsorption, all the sites were retained after optimization. The height after optimization ($\Delta h_{opt}$) (Å) and charge transferred ($\Delta Q_M|e|$) between M-atoms and antimonene is mentioned in table 1 for each case. These results confirm that, for each particular atom lowest $h_{opt}$ and maximum $\Delta Q_M|e|$ obtained corresponds to the most stable site reported.

IOP Conf. Series: Materials Science and Engineering 1120 (2021) 012035 doi:10.1088/1757-899X/1120/1/012035

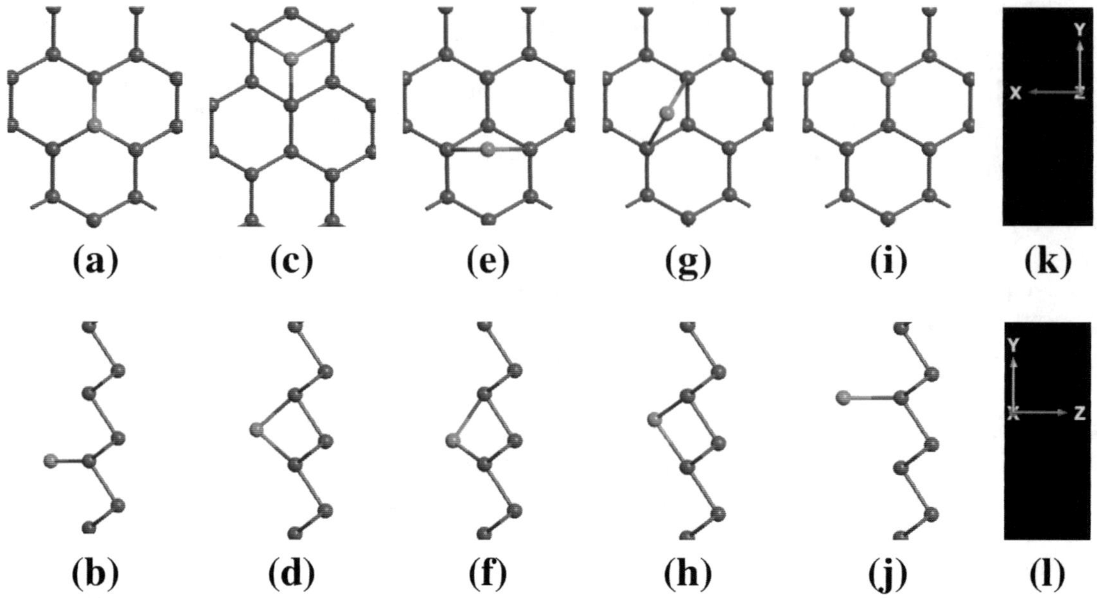


**Figure 1.** (colour): Front and cross sectional view of adsorption sites for M-atoms; (a, b) H-site; (c, d) Hex-site; (e, f) B1-site; (g, h) B2-site; (i, j) T-site, (k, l) axis system. Where blue colour represents Sb atoms and pink colour represents M-atoms.

**Table 1.** Adsorption energy ($E_{ads}$), vertical height of M from antimonene surface after optimization ($\Delta h_{opt}$), M-antimonene bond distance ($\Delta d$) (the numbers in bracket represent number of bonds with same values) and total absolute charge transferred between M and antimonene ($\Delta Q_M|e|$) for all the identified adsorption sites.

| Site | | H | HEX | B1 | B2 | T |
|---|---|---|---|---|---|---|
| | Mg | -0.92 | -1.15 | -1.11 | -1.12 | -0.36 |
| **$E_{ads}$(eV)** | K | -1.78 | -1.77 | -1.70 | -1.70 | -1.30 |
| | Ca | -2.17 | -2.78 | -2.73 | -2.73 | -2.62 |
| | Mg | 2.10 | 1.59 | 1.43 | 1.50 | 3.40 |
| **$\Delta h_{opt}$ (Å)** | K | 2.20 | 2.36 | 2.53 | 2.53 | 2.28 |
| | Ca | 1.40 | 1.01 | 1.20 | 1.50 | 1.20 |
| **$\Delta d$(Å) M-Sb (Å)** | Mg | 3.32(3) | 3.03(3) | 2.98(2), 2.99 | 2.98(3) | 3.40 |
| | K | 3.45(3) | 3.44(3) | 3.37(2) | 3.37(2) | 3.29 |
| | Ca | 3.08(3) | 3.01(3) | 3.01(2), 3.02 | 3.01(2), 3.02 | 3.02(2), 3.00 |
| | Mg | 0.03 | 0.09 | 0.09 | 0.09 | 0.02 |
| **$\Delta Q_M|e|$** | K | 0.12 | 0.14 | 0.14 | 0.14 | 0.23 |
| | Ca | 0.02 | 0.04 | 0.01 | 0.01 | 0.01 |

IOP Conf. Series: Materials Science and Engineering 1120 (2021) 012035 doi:10.1088/1757-899X/1120/1/012035

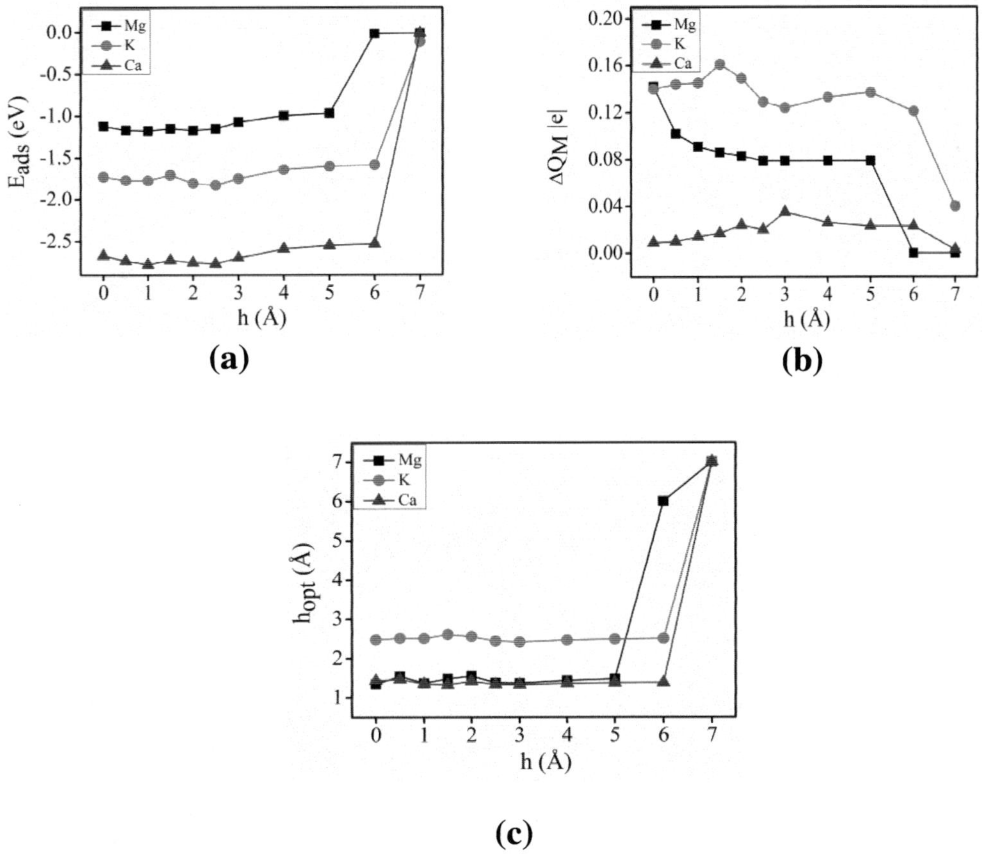


**Figure 2.** (colour): (a) adsorption energy ($E_{ads}$), (b) Charge transferred between M-atom and antimonene ($\Delta Q_M|e|$) and (c) vertical height of M from antimonene surface after optimization ($\Delta h_{opt}$) when the M-atom has been placed at varying heights over antimonene for adsorption and the structure is then relaxed. Values shown in black refer to Mg, red to K and blue to Ca related properties.

For better understanding of adsorption phenomena, we tried analysing adsorption when the M-atom was placed at different heights. The resulting graphs are shown in figure 2. At lower heights, the adsorption took place successfully but after a specific height (6 Å for Mg and 7 Å for K and Ca) the atom stops interacting with the antimonene. That is why $E_{ads}$ and $\Delta Q_M|e|$ tends to become zero with increasing distance in figure 2(a) and figure 2(b) respectively. The corresponding $h_{opt}$ eventually becomes equivalent to the distance at which atom was placed initially in figure 2(c). Mg stops interacting at smaller height due to the smaller size of Mg atom compared to the other two atoms. For every atom, values of $h_{opt}$ remain almost constant at whatever distance it would have been placed. The average $h_{opt}$ is 1.3 Å for Mg and Ca and 2.5 Å for K which can be attributed to the larger atomic radii of K. Similarly, the average bond length for Mg and Ca is 3.0 Å and for K, it is 3.3 Å. The number of bonds in each case of adsorption for each atom varies after optimization and it has been given in table 1 for perusal.

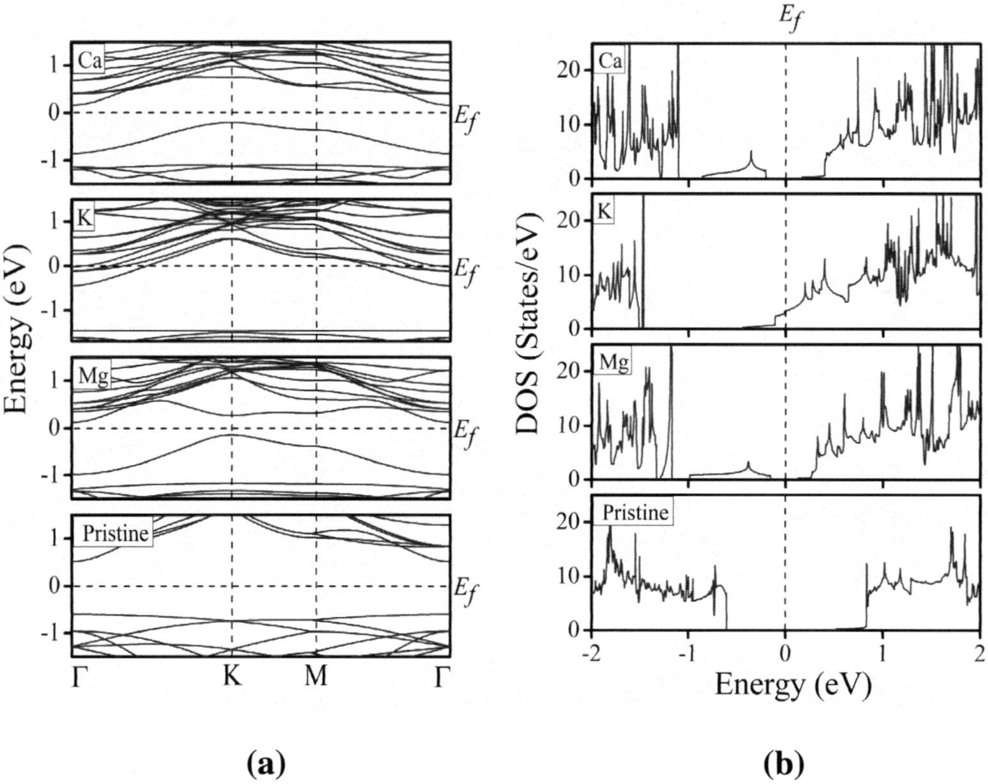


**Figure 3.** The (a) bandstructure and (b) Density of states (DoS) plot for pristine antimonene and antimonene after the adsorbtion of single M-atom over it.

Antimonene, in pristine condition is a semiconductor with 1.1 eV energy bandgap [31]. When a single M-atom is adsorbed over it, the bandstructure experiences some shifts as can be seen from the plots given in figure 3(a). The bandgap of the structure experiences a reduction after the adsorption of Mg and Ca atoms. The value of the new bandgap observed is approximately 0.4 eV. K adsorption, however, changes the electronic property of antimonene altogether. We can see from the graph that K adsorption induces metallic properties and the antimonene is no longer a semiconductor. As seen from the bandstructure graph, fermi energy level is moving towards the conduction band in all cases. This, alongwith the reduced bandgap (or conversion to metallicity) indicates that M-atom adsorption leads to higher conductivity in antimonene. The DOS analysis shown in figure 3(b) is also in coordination to what we found out in bandstructure diagram. The reduced bandgap is clearly visible in case of Mg and Ca adsorption. In case of K adsorption, some bands occupy and cross the fermi level. This eletronic behavioural shift can be backed up with the charge transfer analysis as given in table 1. Charge transferred between K and antimonene is higher in comparison to other two adsorbents. This gives us a clear picture that why K adsorption leads to metallic antimonene while Mg and Ca adsorbed antimonene still remains semiconducting in nature.

*3.2 Diffusion of Mg, K and Ca over Antimonene.*

Movement of M-atoms over antimonene is analyzed using Climbing Image Nudged Elastic Band (CI-NEB) method [33,34]. Three diffusion paths (path I, II, III) were identified over the antimonene surface as shown in figure 4. Since, Hex-site was most suitable for Mg and Ca, so the initial and final images taken in these cases were of Hex-sites (set B in figure 4) while for K atom, H-sites were used for the same (set A in figure 4). When an atom moves over the substrate surface, it faces energy barrier throughout that path which resists its diffusion. This energy barrier is known as activation energy ($E_a$). For each identified path, graph can be plotted between the diffusion coordinates and the $E_a$ as shown in figure 5. The highest value of $E_a$ at peak in graph decides whether the path is preferred by atom or not. While diffusing, the adsorbent prefers the path of lowest $E_a$ i.e. minimum energy paths.

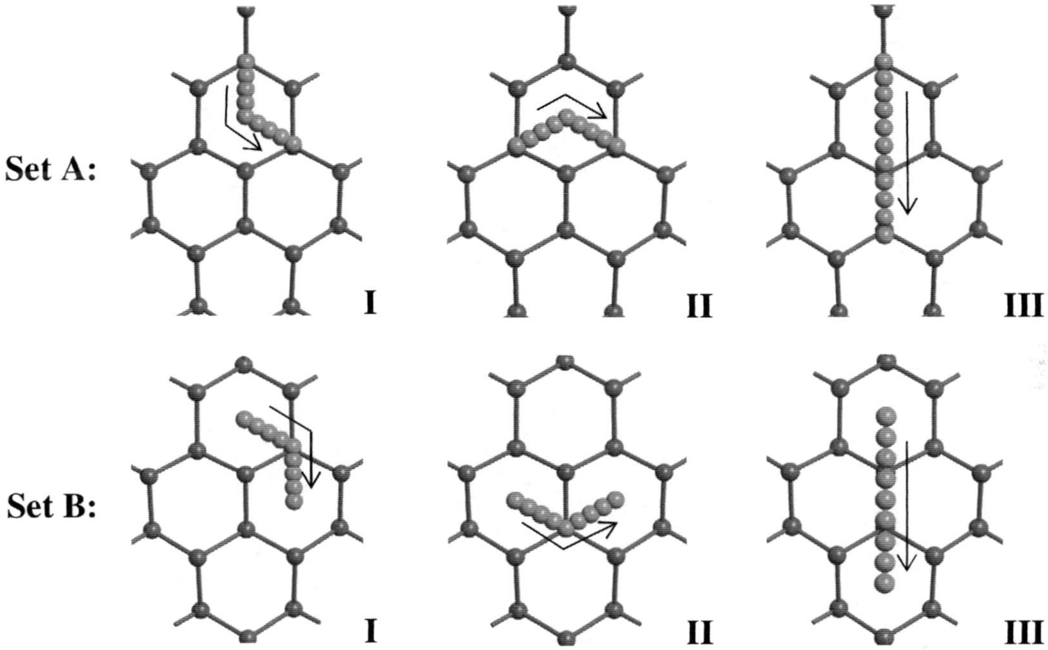


**Figure 4.** (colour): Front view of various diffusion paths (path I, II, III) for each M-atom over antimonene layer. (set A) H-site to H-site paths for K-diffusion. (set B) Hex-site to Hex-site paths for Mg- and Ca- Diffusion. Where blue colour represents Sb atoms and pink colour represents M-atoms.

Among all of the three paths, path III has highest $E_a$ for all three atoms because in path III, the M-atom has to cross one Sb atom which requires a lot of energy. For path I and II, the graph looks identical due to the similar placement of saddle points and Sb-Sb bonds in both of these paths. Saddle point is an intermediate stable adsorption site which lies near the defined path. Since the site is stable, relaxation of nearby image of atom during CI-NEB relaxation tends to bring it to that site. This introduces a trough between the energy peaks as can be seen in figure 5(a,b). For Hex sites, H site is the intermediate site and vice-versa. The only exception is when Ca moves along path I or II. Saddle point is absent in both of these plots because Ca does not move to the stable intermediate site, instead it goes rather straight affecting some Sb-Sb bonds. The lowest $E_a$ is 0.3 eV for Mg, 0.12 eV for K and 0.43 eV for Ca. The exact values of energies at saddle points and secondary peaks along with the lowest $E_a$ values are given in table 2 for each atom on each path. We have discussed in earlier section that K adsorption increases conductivity of antimonene more compared to the other two atoms. As a consequence of this, here we see that K diffuses rather easily than the other two atoms. K is followed by Mg and Ca respectively in ease of diffusion. The peak and saddle point in case of K-diffusion for path I and II is very small because the adsorption energies of Hex- and H-site was approximately same as mentioned in table 1.

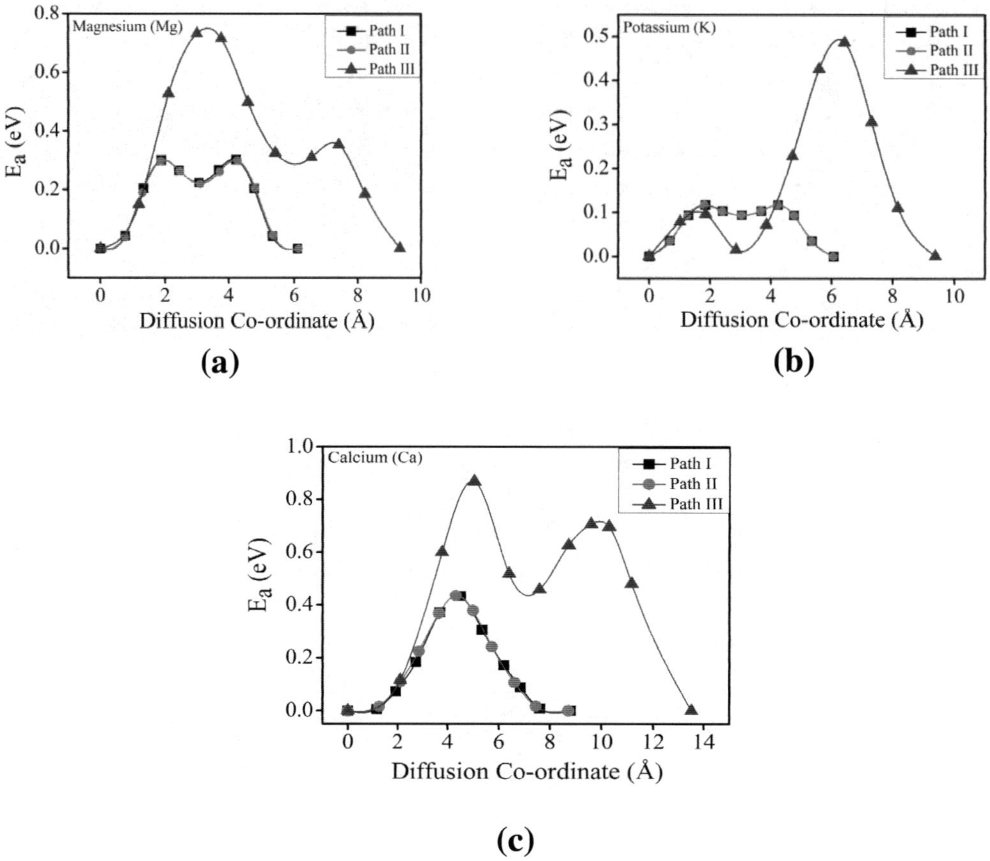


**Figure 5.** (colour): Diffusion path vs $E_a$ plot at all the three paths, path I (black), path II (red) and path III (blue) for metal atoms diffusion over antimonene.

**Table 2.** Diffusion parameters i.e. Activation Energy ($E_a$) (eV), Energy at saddle point (eV) and Energy barrier at Secondary Peak (eV) of antimonene across each path for each M-atom.

| Parameter | Path | Mg | K | Ca |
|---|---|---|---|---|
| | I | 0.30 | 0.12 | 0.43 |
| **Activation Energy ($E_a$) (eV)** | II | 0.30 | 0.12 | 0.43 |
| | III | 0.75 | 0.49 | 0.86 |
| | I | 0.22 | 0.09 | -- |
| **Energy at saddle point (eV)** | II | 0.22 | 0.09 | -- |
| | III | 0.29 | 0.01 | 0.43 |
| | I | -- | -- | -- |
| **Energy barrier at Secondary Peak (eV)** | II | -- | -- | -- |
| | III | 0.36 | 0.10 | 0.72 |

## 4. Conclusion

The adsorption and diffusion of M-atoms (Mg, K and Ca) over antimonene monolayer has been studied thoroughly and their properties have been analyzed. The M- atoms can interact with antimonene from a distance as large as 6 Å. The adsorption process is exothermic in nature and the bond formation depends upon at which site the atom resides over surface. The charge transfer is greater between K and antimonene and thus, antimonene starts behaving metallic. In case of Mg and Ca, charge transfer is insignificant but they incorporate in reducing the bandgap of antimonene. As a result of these phenomena, K diffuses easily over the antimonene surface compared to that of Mg or Ca, making the structure more conducting in nature. Hence, it can be emphasized that M-atoms can be successfully adsorbed over antimonene surface and K-adsorption provides better diffusivity and conductivity to antimonene than Mg and Ca.

## References

[1] Geim A K and Grigorieva I V 2013 *Nature* **499** 419
[2] Ersan F, Keçik D, Özçelik V O, Kadioglu Y, Aktürk O Ü, Durgun E, Aktürk E and Ciraci S 2019 *Applied Physics Reviews* **6** 021308
[3] Carvalho A, Wang M, Zhu X, Rodin A S, Su H and Castro Neto A H 2016 *Nat. Rev. Mater.* **1**
[4] Bagheri S, Mansouri N and Aghaie E 2016 *Int. J. Hydrogen Energy* **41** 4085
[5] Vogt P, De Padova P, Quaresima C, Avila J, Frantzeskakis E, Asensio M C, Resta A, Ealet B and Le Lay G 2012 *Phys. Rev. Lett.* **108** 155501
[6] Bianco E, Butler S, Jiang S, Restrepo O D, Windl W and Goldberger J E 2013 *ACS Nano* **7** 4414
[7] Alducin D, Myers B D, Liu X and Fisher B L 2013 Re s ear ch | r e p o r t s **16** 151
[8] Pumera M and Sofer Z 2017 2D *Adv. Mater.* **29** 1605299
[9] Zhang S, Guo S, Chen Z, Wang Y, Gao H, Gómez-Herrero J, Ares P, Zamora F, Zhu Z and Zeng H 2018 *Chem. Soc. Rev.* **47** 982
[10] Hu Z, Wu Z, Han C, He J, Ni Z and Chen W 2018 *Chem. Soc. Rev.* **47** 3100
[11] Manzeli S, Ovchinnikov D, Pasquier D, Yazyev O V and Kis A 2017 *Nat. Rev. Mater.* **2** 17033
[12] Xie Z, Xing C, Huang W, Fan T, Li Z, Zhao J, Xiang Y, Guo Z, Li J, Yang Z, Dong B, Qu J, Fan D and Zhang H 2018 *Adv. Funct. Mater.* **28** 1705833
[13] Song L, Ci L, Lu H, Sorokin P B, Jin C, Ni J, Kvashnin A G, Kvashnin D G, Lou J, Yakobson B I and Ajayan P M 2010 *Nano Lett.* **10** 3209
[14] Mahmood J, Lee E K, Jung M, Shin D, Choi H, Seo J, Jung S M, Kim D, Li F, Lah M S, Park N, Shin H-J, Oh J H and Baek J-B 2016 *Proc. Natl. Acad. Sci.* **113** 7414
[15] Ares P, Palacios J J, Abellán G, Gómez-Herrero J and Zamora F 2018 *Adv. Mater.* **30** 1
[16] Wang X, Song J and Qu J 2019 *Angew. Chemie - Int. Ed.* **58** 1574
[17] Wang G, Pandey R and Karna S P 2015 *ACS Appl. Mater. Interfaces* **7** 11490
[18] Shao Y, Liu Z L, Cheng C, Wu X, Liu H, Liu C, Wang J O, Zhu S Y, Wang Y Q, Shi D X, Ibrahim K, Sun J T, Wang Y L and Gao H J 2018 *Nano Lett.* **18** 2133
[19] Fortin-Deschênes M, Waller O, Menteş T O, Locatelli A, Mukherjee S, Genuzio F, Levesque P L, Hébert A, Martel R and Moutanabbir O 2017 *Nano Lett.* **17** 4970
[20] Ji J, Song X, Liu J, Yan Z, Huo C, Zhang S, Su M, Liao L, Wang W, Ni Z, Hao Y and Zeng H 2016 *Nat. Commun.* **7** 1
[21] Gibaja C, Rodriguez-San-Miguel D, Ares P, Gómez-Herrero J, Varela M, Gillen R, Maultzsch J, Hauke F, Hirsch A, Abellán G and Zamora F 2016 *Angew. Chemie - Int. Ed.* **55** 14345
[22] Huo C, Yan Z, Song X and Zeng H 2015 *Sci. Bull.* **60** 1994
[23] Ares P, Aguilar-Galindo F, Rodríguez-San-Miguel D, Aldave D A, Díaz-Tendero S, Alcamí M, Martín F, Gómez-Herrero J and Zamora F 2016 *Adv. Mater.* **28** 6332

[24] Wang Y X, Chou S L, Liu H K and Dou S X 2013 *Carbon N. Y.* **57** 202
[25] Kohn W and Sham L J 1965 *Phys. Rev.* **140** A1133
[26] Hohenberg, P., Kohn W 1964 *Phys. Rev. B* **136** 864
[27] Anon Atomistix ToolKit Virtual NanoLab (ATK-VNL) QuantumWise Simulator.
[28] Perdew J P, Burke K and Ernzerhof M 1996 *Phys. Rev. Lett.* **77** 3865
[29] Monkhorst H J and Pack J D 1976 *Phys. Rev. B.* **13** 5188
[30] Grimme S 2006 *J. Comput. Chem.* **27** 1787
[31] Upadhyay S and Srivastava P 2020 *Mater. Chem. Phys.* **241** 122381
[32] Srivastava P, Abhishek, Sharma V and Jaiswal N K 2020 *Appl. Phys. A Mater. Sci. Process.* **126** 1
[33] Henkelman G and Jónsson H 2000 *J. Chem. Phys.* **113** 9978
[34] Henkelman, G., Uberuaga, B. P., Jónsson H 2000 *J. Chem. Phys* **113** 9901

# AUTHOR INDEX

**Institute of Physics**
Dirac House, Temple Back
Bristol BS1 6BE UK

ISSN: 1757-8981
ISBN 978-1-7138-3406-9

CURRAN ASSOCIATES INC.
proceedings.com